Consumer Electronics Component Handbook

Consumer Electronics Component Handbook

How to Identify, Locate, and Test Consumer Electronic Components

Homer L. Davidson

McGraw-Hill

New York San Francisco Washington, D.C. Auckland Bogotá
Caracas Lisbon London Madrid Mexico City Milan
Montreal New Delhi San Juan Singapore
Sydney Tokyo Toronto

Library of Congress Cataloging-in-Publication Data

Consumer electronics components handbook : how to identify, locate, and test consumer electronic components / Homer L. Davidson.
p. cm.
Includes index.
ISBN 0-07-015807-X (hardcover)
1. Electronic apparatus and appliances —Handbooks, manuals, etc.
TK7870.D35 1998
621.381—dc21 98-3964
CIP

McGraw-Hill

A Division of The ***McGraw-Hill*** *Companies*

1 2 3 4 5 6 7 8 9 0 DOC/DOC 9 0 3 2 1 0 9 8

ISBN 0-07-015807-X

The sponsoring editor for this book was Scott Grillo, the editing supervisor was Sally Glover, and the production supervisor was Sherri Souffrance. It was set in ITC Century Light per the EL1 Specs by Michele Bettermann of McGraw-Hill's Professional Book Group compositions unit, Hightstown, N.J.

Printed and bound by R. R. Donnelley & Sons Company.

This book is printed on recycled, acid-free paper containing a minimum of 50 percent recycled, de-inked fiber.

Dedication

I dedicate this book to my children, who have been equally successful within their own chosen professions: Larry, Colleen, Bruce, Julie, Cathy, Darin, and Nancy.

Love you.

Dad

Contents

Introduction

Consumer Electronics Component Handbook provides an alphabetical listing of all electronic components found in the consumer electronics field. This book contains more practical data than theory. The book shows you what the electronic component is, how it operates, where the part is found in a particular circuit or electronic product, what to do when the component breaks down, and how to test the part with the different test instruments. The most important and popular electronic components in the consumer electronics field are found within the pages of this book.

This book is broken down into 13 different chapters. How the component operates and what the part is made of are listed under the "What" heading. Where the component is located in the circuit and in what electronic product it is used are explained in the next section. With smaller, more common components, sometimes the "What" and "Where" features are listed as one.

Appearing next is the "When" heading, which lists the various symptoms, what to look for, breakdown problems, and what happens when the component fails in the circuit. How the component becomes dead, leaky, shorted, or intermittent within the electronic circuit are also explained. When the component breaks down, the various voltage and resistance changes in the circuit indicate a defective part.

How to test the suspected component is the most important and the last step in locating the defective part. Taking the various voltage, resistance, and current measurements with the VOM, DMM, and FET-VOM are practical methods of locating the suspected component. How to make the various tests with different test instruments is discussed under the "Test" heading. A lot of simple and practical tests learned in more than forty years of working in the electronics field are found here. How to test and locate the defective component turns out to be the most prolific, jubilant, and fun experience in the consumer electronics field.

The book has 13 chapters, starting with antennas and ending with zener diodes. Chapter 1 shows how to test electronic components from A to B, including adapters through buzzers. How to test the various amplifier circuits is listed from RF–IF (audio amplifiers, IC audio, and TV audio amp circuits). How to test the different batteries and buzzers concludes the first chapter.

How to test electronic components C, which includes the most popular part—the capacitor—is found in Chapter 2. Bypass, coupling, decoupling, filter, electrolytic, high-frequency, high-voltage, memory, trimmer, surface-mounted, and variable capacitors are found in the various circuits of the many electronic products. When the various capacitors break down, you'll find how to test and replace them in this important chapter.

Testing electronic components R, which includes the many types of resistors, is covered in Chapter 10. The carbon, chip, fixed, flameproof, metal film, metal oxide, power, thick film, and wire-wound resistors are given throughout this chapter. How to locate, test, and replace the various resistors is found here.

Chapter 11 is the most important chapter on testing electronic components that begin with S, from saw filters to semiconductors. All semiconductors (which includes diodes, ICs, CMOS, TTLs, Hall-effect ICs, LEDs, transistors, signal transistors, FETs, MOS-FETs, unijunctions (UJT) and TV transistor circuits) are discussed in detail, covering when and what electronic products they are located in, what happens when they break down, and how to locate and test suspected semiconductors. How to find the defective semiconductor with signal, waveform, leakage, voltage, and resistance tests is explained. Locating and replacing the defective semiconductor components might solve 85 percent of the troubles in the electronic chassis.

This book can be a supplement or backup to any electronics book covering how-to tests. The book contains more than 450 pages, with illustrations, photos, and partial schematics to help in testing electronic components. This book should be stocked on the service bench for a quick reference test of electronic parts. Just look up the defective part, find out how to test it, and above all, have some fun in locating and testing the electronic component.

Consumer electronics technician's suggested reading list

Magazines

Electronic Servicing and Technology
76 N. Broadway
Hicksville, NY 11801
Telephone: 1-516-681-2922

Electronics Now
500 B1-County Boulevard
Farmingdale, NY 11735
Telephone: 1-516-293-3000

Electronics books

Troubleshooting and Repairing Audio Equipment (3rd Ed.)
Homer L. Davidson
McGraw-Hill
11 West 19th Street
New York, NY 10011

Troubleshooting and Repairing Camcorders (2nd Ed.)
Homer L. Davidson
McGraw-Hill
11 West 19th Street
New York, NY 10011

Troubleshooting and Repairing Compact Disc Players (3rd Ed.)
Homer L. Davidson
McGraw-Hill
11 West 19th Street
New York, NY 10011

Troubleshooting and Repairing Computer Printers (2nd Ed.)
Stephen Bigelow
McGraw-Hill
11 West 19th Street
New York, NY 10011

Troubleshooting and Repairing Computer Monitors (2nd Ed.)
Stephen Bigelow
McGraw-Hill
11 West 19th Street
New York, NY 10011

Troubleshooting and Repairing Microwave Ovens (4th Ed.)
Homer L. Davidson
McGraw-Hill
11 West 19th Street
New York, NY 10011

Troubleshooting and Repairing PCs (3rd Ed.)
Michael Hordeski
McGraw-Hill
11 West 19th Street
New York, NY 10011

Troubleshooting and Repairing Solid-State TVs (3rd Ed.)
Homer L. Davidson
McGraw-Hill
11 West 19th Street
New York, NY 10011

Troubleshooting and Repairing Consumer Electronics Without a Schematic
Homer L. Davidson
McGraw-Hill
11 West 19th Street
New York, NY 10011

Lenk's Video Handbook (2nd Ed.)
John D. Lenk
McGraw-Hill
11 West 19th Street
New York, NY 10011

Troubleshooting and Repairing VCRs (4th Ed.)
Robert L. Goodman
McGraw-Hill
11 West 19th Street
New York, NY 10011

Electronic Troubleshooting and Repair Handbook
Homer L. Davidson
McGraw-Hill
11 West 19th Street
New York, NY 10011

1

CHAPTER

Testing electronic components from A to B

(Adapters to buzzers)

Electronic tests covered in this chapter include voltage, resistance, continuity, signal injection, and signal tracing. The voltohm-milliammeter (VOM) and digital multimeter (DMM) are used in taking voltage, resistance, and continuity measurements. Audio waveforms are taken with the oscilloscope to locate the correct signal, loss of signal, amplification, intermittent signal, and distortion. Besides the above tests, you can check for burn marks and overheated components, feel overheated parts, smell unusual odors, and hear different noises and distortion in the speaker.

Adapters

Ac and dc

What

Most adapters found in ac and dc units provide dc voltage to operate various consumer electronic products. The ac adapter provides a dc output voltage from 3 to 24 volts dc (Vdc). A multioutput adapter can provide 3, 4.5, 6, 7.5, 9, and 12 volts dc. The output current from the adapter may provide 100 mA to 3 amps of power (Fig. 1-1).

Where

The ac-dc adapter provides power to operate small radios, cassette players, portable CD players, telephone answering machines, wireless telephone systems, cordless power tools, computers, paper shredders, and video games. The desktop adapter may contain a fuse with capacitor filtering, regulation, or a switching power supply.

1-1 The portable RCA CD player is operated in the home with an ac adapter.

The ac-dc circuit may consist of a small power transformer with fullwave rectification, using two or four silicon rectifiers (Fig. 1-2). A bridge rectifier circuit consists of four silicon diodes. You may find no capacitor filtering in the dc charging circuits of the cordless power tool. Usually, one or two small electrolytic capacitors provide filtering in consumer electronic products' power adapters.

The small power transformer provides ac voltage to the fullwave rectifiers, with no on/off switch. The ac-dc power supply is enclosed in a plastic container, except power cord and plug. Most adapters plug directly into the power line with ac spade tongs projecting from the plastic container. Several different types of plugs of diverse output voltages are available to power different electronic products.

When

Check the output voltage of the ac-dc adapter when the electronic product appears dead or will not operate. Place the VOM or DMM terminals at the center hole and outside the metal shield of the dc output plug. The center terminal is usually the positive terminal. Remember that the dc voltage output will always be higher when no load or product is connected. Check the measured dc voltage against the voltage marked on the plastic container.

Testing

If there is no dc voltage, place the ohmmeter terminals of the meter across the ac plug, with the adapter removed from the ac receptacle. Rotate the DMM to the 200-ohm scale. A quick continuity measurement indicates that the primary winding of the power transformer is probably good. Most primary windings will go open with a heavy load or shorted component inside the electronic product. The measured primary resistance should be between 10 and 100 ohms. Discard the power adapter if the power transformer has an open winding, since the transformer replacement costs more than the original adapter.

Simple repairs can be made on the power adapter when the transformer primary winding tests normal. The secondary winding of the transformer is wound with larger-diameter wire and very seldom opens up. Next, check the ac cord as it seems to break where it enters the dc plug or plastic container. Place a needle or safety pin in each wire close to where it enters the plaster body to check for a broken wire. Rotate the meter to the 200-ohm scale and place a one-meter clip on the needle or pin and at either terminal of the dc power plug. Check both wires inside the rubber cord for continuity. Shake the cord for intermittent breaks. Cut off the cord when found broken at the plug, and replace the cord with one found at your local Radio Shack store.

Remove the plastic cover if the transformer cord and plug are normal. Place the point of the soldering iron on the area where the plastic is joined together, and hold the flat blade sideways to go down the seam. Remove excess plastic until the plastic sides will come apart.

Test each silicon diode for leakage with the diode test of the DMM (Fig. 1-3). Check the continuity of the secondary winding of the step-down transformer. Measure

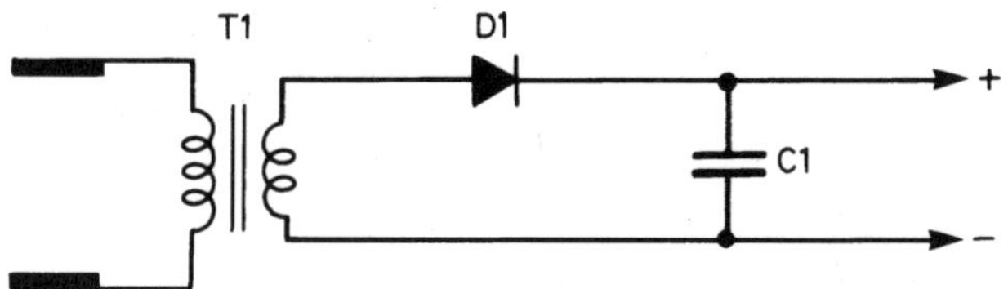

1-2 The circuit of a typical ac-dc adapter used to power small electronic products.

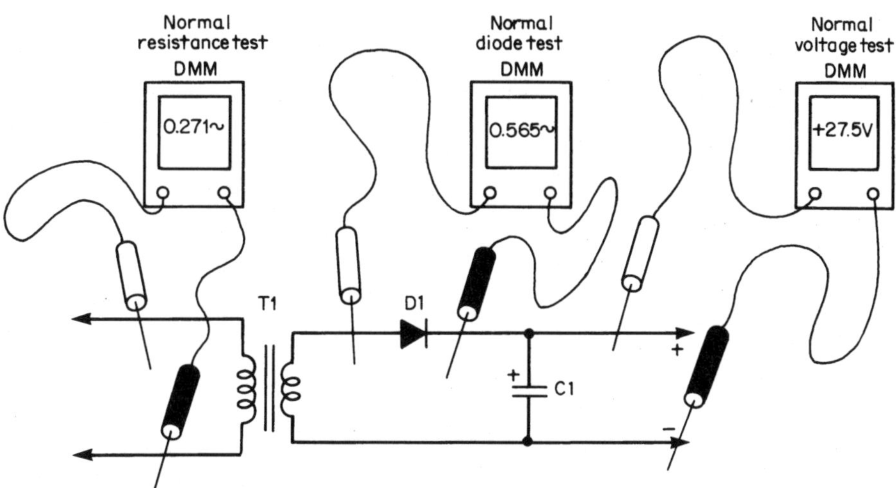

1-3 Check each silicon diode with the diode test of the DMM.

the output voltage across the small electrolytic capacitor. If voltage is low at the capacitor, replace the electrolytic capacitor. A normal voltage at the filter capacitor indicates a break in the cord where it enters the plastic case. Cut off the end of the defective cord and resolder the terminals at the plug.

Make sure the output voltage polarity is correct at the dc voltage plug. Usually, the center terminals have a positive voltage. Remember, this voltage without a load will be a few volts higher than the voltage found on the side of plastic case. For instance, a 15-Vdc output adapter will measure 19.2 volts without a load.

Amplifiers

An amplifier is a device that increases an applied signal. An amplifier takes a low-input and provides a greater output signal. In the consumer electronics field, the amplifier may amplify RF (radio frequency), audio, current, and voltage. A current amplifier amplifies the input current to a larger output current. The power amplifier may be tube, transistor, and IC operated in the output stages of an audio amplifier. Voltage gain occurs with a low-input signal voltage to provide a higher-output signal voltage. Within the small table radio, you will find RF (radio frequency), IF (intermediate frequency), AF (audio frequency), and audio output amplifier components. The RF amplifier may be a RF or FET (field-effecttransistor) or IC (integrated circuit).

RF and IF amplifiers

The first stage in a superheterodyne circuit is the radio frequency (RF) stage. The RF stage consists of components that amplify radio frequencies, while the intermediate frequency (IF) is amplified by the IF stages. The IF frequency is the difference between the incoming signal and that of the oscillator stage resulting in IF amplification. One or more IF stages may be found in the radio receiver. Today, most AM, FM, and shortwave receivers contain a superheterodyne circuit with RF and several IF stages.

In the early solid-state radio receivers, the RF stage may be included in the AM converter stage. The RF broadcast signal is picked up by a ferrite rod antenna, tuned, and tied into the base circuit of Q1 (Fig. 1-4). C/A tunes in the proper radio station with coil L1 and L2 of the RF amplifier circuit.

The oscillator circuit is found in the collector and base circuits. The oscillator coil (L3) is tuned with the same variable-gain capacitor C1B. The difference between the oscillator and RF frequencies is called the intermediate frequency. The IF frequency of most broadcast receivers is 455 kHz, while in the car radio the IF frequency is 262 kHz, and in the FM receiver the IF is 10.7 MHz. Besides the IF transformer, you may find 455-kHz ceramic filters in IF amplifier stages. Today, one IC component may contain the RF, oscillator, and two IF circuits.

Audio amplifiers

What

The audio amplifier amplifies the audio signal from the AM-FM radio circuits, cassette deck, and CD players. The audio frequency (AF) amplifier operates in the

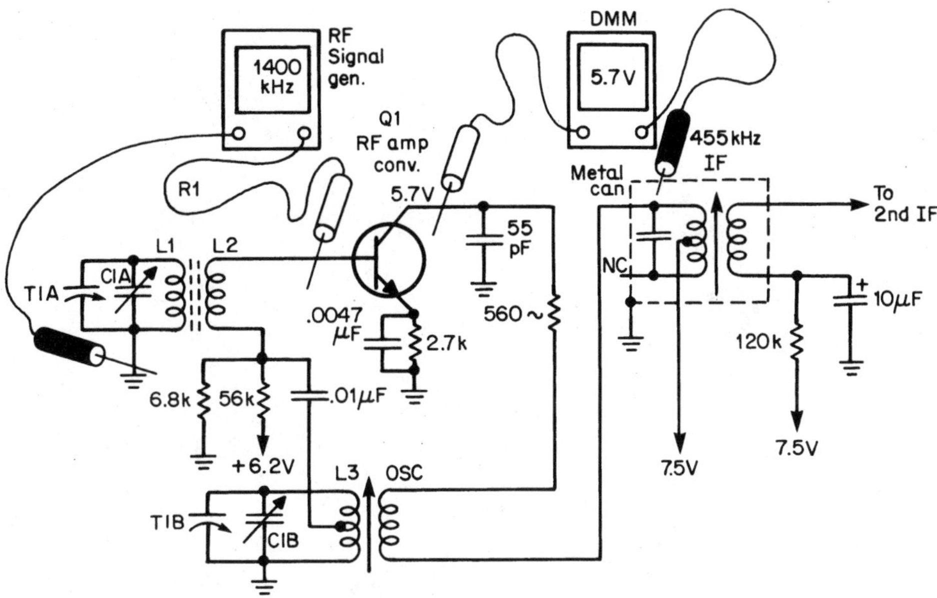

1-4 How to take voltage tests to determine if the RF stage is operating.

20 to 20-kHz audio range. You may find more than one AF amplifier in the audio circuits. The AF driver amplifies and drives several transistor or IC components in the audio output stages. The stereo amplifier has two separate channels of audio signals.

Where

Audio amplifiers can be found in table and clock radios, high-wattage audio-visual (A/V) receivers, shortwave receivers, cassette and CD players, phonographs, auto radio and CD players, public address (PA) systems, and high- and low-powered amplifiers (Fig. 1-5). Audio amps are also found in TV chassis, camcorders, telephone answering machines, CBs and other communications equipment, video tape recorders (VCRs), and test equipment.

When

The RF, IF, and audio circuits can become dead, weak, intermittent, or distorted. These circuits must be tested with quick troubleshooting methods. Most solid-state circuits are tested with voltage, resistance, and signal-testing methods. The VOM, DMM, oscilloscope, RF-IF, and audio signal generators can quickly spot a defective stage.

A weak RF stage with an open or leaky transistor and IC can produce weak radio reception. Check for open or leaky transistors with a voltage test between the base and emitter terminals (NPN-0.6V). Often, only a local broadcast station can be heard with a weak RF stage. When a test prod or screwdriver blade is touched to the base of an RF transistor or IC and the volume of the broadcast station increases, check the RF circuits.

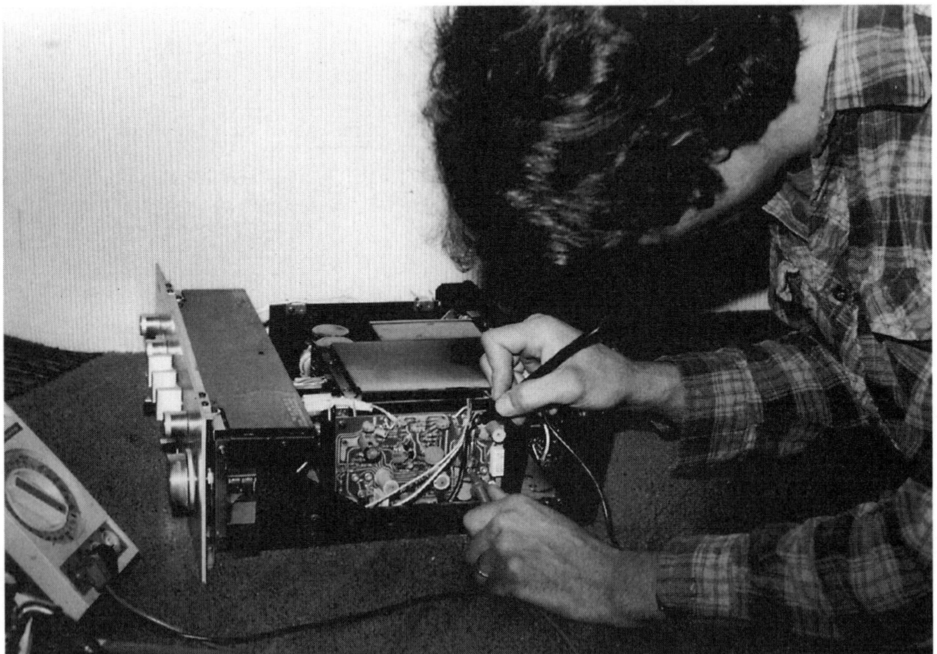

1-5 The electronics technician repairing a high-powered audio amplifier circuit.

A weak AM or RF stage may be caused by a leaky or open RF transistor, broken ferrite core, broken coil leads, or improper supply voltage from the low-voltage power supply. Intermittent radio reception can result from a poor coil connection or intermittent transistor or IC.

Check for correct voltage on the RF transistor or IC. Test the transistor within the circuit with the diode-transistor test of the DMM or transistor tester. Remove the suspected transistor and test it out of the circuit. Test the coil for continuity and resistance of the emitter resistor.

Test the oscillator circuit of the receiver by placing another radio nearby and rotating the tuning dial through the various broadcast stations. You should hear the local oscillator beating against the other radio circuit with a squeal or whistle in the radio, indicating that the oscillator is operating. Another method is to place the probes of a vacuum-tube voltmeter (VTVM) or field-effect transistor voltohm-milliammeter (FET-VOM) tester across the base and emitter terminals of a converter transistor. When the different stations are tuned in, the meter hand will vary, indicating that the oscillator is performing.

You can test both the oscillator and IF circuits by connecting the AM signal generator to the base of the RF transistor and common ground (see Fig. 1-4). Adjust the signal generator to 1400 kHz. Rotate the tuning capacitor to 1400 kHz on the dial, and you should hear the 1400-kHz modulated tone, indicating that the RF transistor, oscillator, IF, and audio sections are normal. You can also check the RF, oscillator, mixer, and IF stage of complete receiver alignment with the AM or FM signal generator.

Audio amp tests

The audio amplifier circuits can be tested with critical voltage and resistance measurements. Make sure the supply voltage is quite close to the low-voltage source. The weak audio signal can be caused by a defective transistor or IC, leaky bypass and electrolytic capacitors, leaky or open coupling capacitors, and improper voltages. Check all transistor emitter and bias resistors for correct value. When a leaky or open transistor is found, check the base and emitter bias resistors before replacing with a new part. Remove one end of the suspected diode or resistor from the PC board for accurate leakage and resistance measurements.

The audio amplifier can be tested with an audio signal generator and tracer. Inject a 1-kHz audio signal at the input terminal and notice if a tone is heard in the speaker. Signal trace each stage and locate the loss of signal with the oscilloscope or audio signal tracer (Fig. 1-6). The audio signal tracer is actually another external audio amplifier.

Extreme distortion or very weak audio stage can be located with the audio signal generator and scope. Check for distortion by inserting a 3-kHz square wave audio signal generator at the base terminal and checking the output at the collector terminal. Insert a 10-kHz sine wave signal at the input and go from stage to stage with the scope and notice if the sine wave is clipped, flattened, or rounded off, indicating poor frequency response and distortion.

A simple audio test without any meter is the finger screwdriver test. Take a small screwdriver and start at the middle terminal of the volume control with volume wide open. Touch the forefinger on the small screwdriver blade and start at the volume control. Proceed towards the base terminal of the transistor or IC component. You should hear a loud hum. The hum should get louder as you work towards the base of

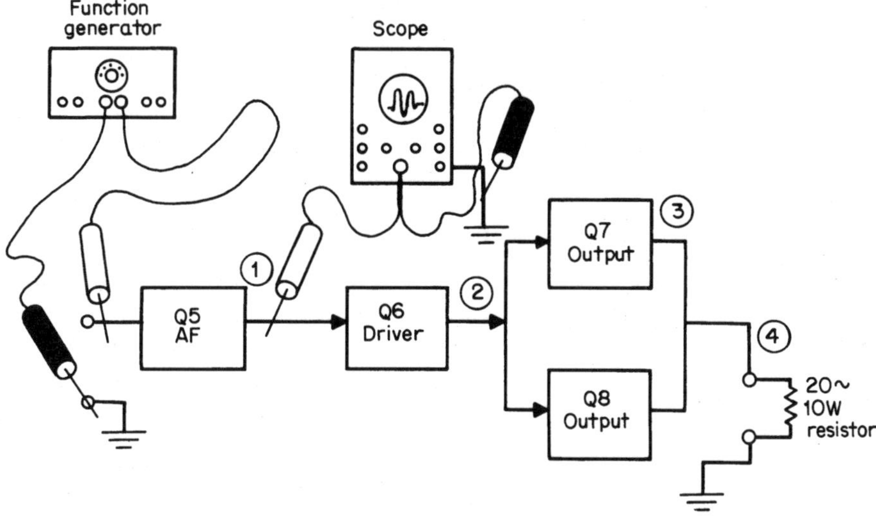

1-6 Checking the high-powered amplifier amp stage by stage with a function generator, scope, and speaker load resistor.

an AF transistor or input terminal of an IC component. When no hum is heard, suspect that stage.

Check the audio stages with an audio or function sweep generator and voltmeter. Provide a 1-kHz sine or square wave from the audio generator at the input of the audio amplifier. Replace the speaker with an 8-, 10-, or 20-ohm, 10-watt load resistor. The resistor provides accurate matching load. Connect a VTVM, FET VOM, or VOM across the resistor terminals. Switch the voltmeter to a low ac voltage scale. If no signal is found at the ac meter, proceed to the base of each audio transistor or IC until a measurement is located. Another method is to place the scope terminals across the load resistor as indicator. The ac meter and scope can be used as indicator in AM-FM-MPX receiver alignment.

IC audio tests

The audio IC circuit may be contained in one IC for monaural or stereo operation. Signal trace the audio circuits up to the audio power output circuit with the scope or external audio amplifier. If the signal is present at the volume control, the dead, weak, or distorted sound is found in the power output circuit. Determine if one or both stereo output channels are defective. A defective power output IC may have only one normal stereo channel, or both channels are dead.

Check the supply source voltage at pin 12 (Fig. 1-7). Test all voltage at pins 1, 4, 8, 12, and 13 and compare with the schematic. Low supply voltage at pin 12 may be caused by a leaky IC1100 or improper voltage source. Remove pin 12 from the PC board and take another voltage measurement. Suspect the supply voltage source if the voltage is still low. When the supply source increases above normal, suspect a leaky IC. Check all pin terminals for low resistance to common ground, indicating a defective component tied to the IC terminal.

TV audio amplifier

The audio amplifier within the TV chassis may consist of one large or two separate IC components. In the latest low-priced TV chassis, the audio output circuits may contain three audio transistors. Several ICs and transistors are found in the stereo TV systems. The audio output IC may operate one speaker, while the stereo channels are fed from one large IC.

Testing

Take critical voltage measurements on each transistor terminal. Check each transistor in the circuit for open or leaky condition (Fig. 1-8). Double-check the transistor out of the circuit when one is found leaky or open. Sometimes the intermittent transistor may test normal when removed and test normal out of the circuit. Measure the forward bias voltage between the base and emitter terminals of each transistor. The forward bias voltage should be around 0.6 volts on a NPN silicon transistor. Suspect a leaky or open push-pull transistor when the supply voltage is lower than normal.

Check each diode with the diode test of the DMM. Remove one end if the diode has a low resistance in both directions. Always remove one end of the bias resistors

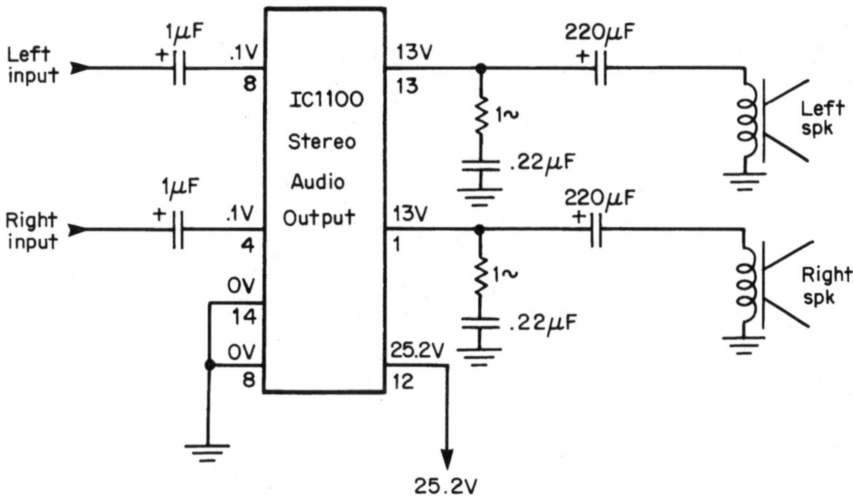

1-7 Take critical voltages on each output IC terminal to determine if IC1100 is defective.

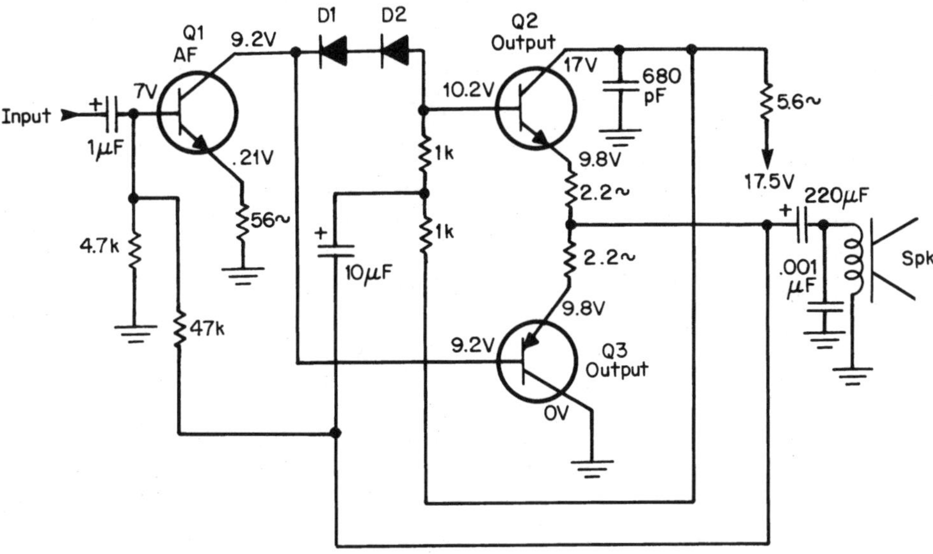

1-8 In the audio output circuits of a TV set, test each audio output transistor after taking critical voltage measurements.

for correct resistance tests. The audio signal can be checked in and out of the small TV audio amplifiers with the external audio amp. You may find one output transistor leaky and the other open. Replace both output transistors when one is found leaky or shorted. Often, the 5.6-ohm resistor will open up when Q2 and Q3 appear leaky or shorted.

Antennas

What

The antenna picks up the received signal, or the signal is radiated from the antenna, like a CB antenna. A rod antenna is found with the car or CB receiver, while a folded dipole picks up the broadcast signal for the radio, FM receiver, and TV set (Fig. 1-9). The antenna impedance should match the transmission line and receiver input. A folded TV dipole antenna feeds into a flat wire of 300 impedance or transformer-coupled 75-ohm coaxial cable.

Dipole antenna

The dipole antenna is nothing more than a rod with shielded lead-ins, bringing broadcast stations into the auto radio. Single dipole antennas are found with portable radio, auto, CB, shortwave, telephone, and scanner receivers. The damaged auto dipole can be replaced with the car manufacturer's or universal replacement. These dipole antennas may be constructed of fiberglass and stainless-steel whip units. Replace the portable radio, telephone, and scanner dipole antenna with the manufacturer's part number or a universal replacement found at most electronic parts stores.

TV antennas

The TV indoor antenna may be rabbit ears or an antenna that sits on top of the TV set and tunes in strong local TV stations. These antennas must be rotated and adjusted to eliminate ghosts in the picture. Better TV reception can be obtained with an outside TV antenna mounted as high as possible. The most common outdoor antenna is the yaggi type for directional TV reception. The yaggi antenna may have up to 56 elements for 100 miles of reception. A low-priced yaggi antenna may have three or four directors, a folded dipole, and a reflector at the back of the antenna. Today, a UHF antenna can be mounted ahead of the VHF antenna on the same antenna mast (Fig. 1-10). Point the small end of the antenna (directors) toward the broadcast station with the longest rods (reflectors) at the rear of the antenna mast.

The yaggi antenna has impedance of 300 ohms, and regular flat antenna wire is a perfect match at 300 ohms. The flat antenna lead-in cable has a tendency to pick up noise from autos, neon lights, power transformers, and noisy ac power lines. A 75-ohm coaxial foam cable is shielded and will not pick up noise or another TV station. The 75-ohm shielded cable must be matched at the antenna and TV receiver with 300- to 75-ohm transformers for maximum reception. You can mount a TV booster, with one section at the antenna mast and the other behind the TV set, for greater long-distance reception (Fig. 1-11).

The UHF TV antenna covers the 14 to 83 channel UHF bands. This smaller antenna may consist of a corner reflector, yaggi, or combination UHF dipole mounted in front of the VHF antenna. The fixed or cut UHF dipole antenna is designed for long-distance TV reception. Today, most UHF antennas have a corner reflector yaggi antenna for greater reception. The corner reflector is at the back of the antenna mast with the small dipoles pointing toward the UHF station to be received.

1-9 Replace the broken or bent FM antenna with the exact or universal replacement.

1-10 Here the two top yaggi antennas are for the UHF band, and the large bottom yaggi antenna picks up stations from 2 to 13.

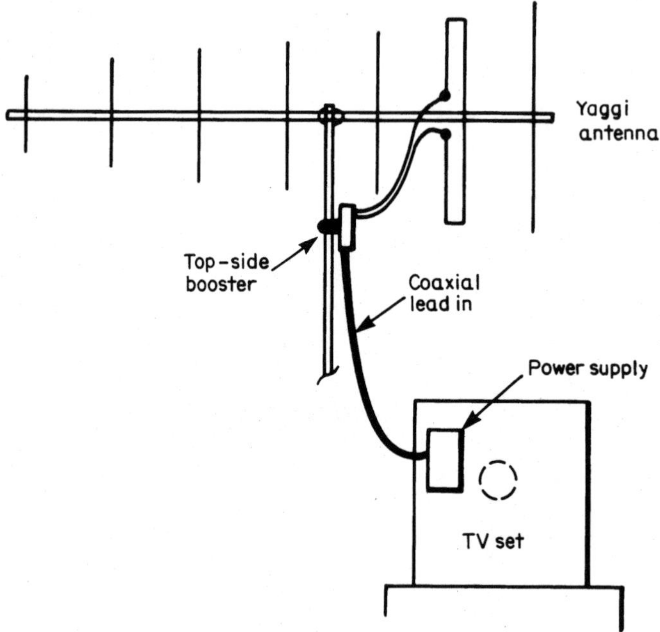

1-11 A top-side booster amplifier is mounted just below the antenna mast and the power supply on the rear of the TV set to boost the picked-up TV signal.

DSS antenna dish

The RCA Digital Satellite System (DSS) uses direct broadcast from a satellite. It enables millions of people to receive over 100 channels of high-quality digital programs from anywhere in the United States. The system provides digital data, video, and audio to the various homes via high-powered KV-band satellites. The broadcast signal is transmitted to the DSS satellite (orbiting 22,000 miles above the earth) from a site in Colorado Springs, Colorado. The signal is relayed back to earth and decoded with a receiver unit that rests on top of the TV set. Today, Primestar, RCA, Sony, and others provide DSS antennas.

The 18-inch-diameter dish with a slight oval shape picks up the satellite KV-band antenna (Fig. 1-12). The LNR is offset so it is out of the way and does not block any surface area of the dish. The low-noise converter converts the 12.3-GHz to 12.7-GHz satellite signal to the 950-MHz to 1450-MHz signal picked up by the receiver. A dual-output LNR is found with a deluxe model so two different TVs can operate from the same antenna.

All satellite antenna systems can be checked with an expensive satellite signal meter on the ground or in the air. The RCA DSS receiver contains a signal and audio tone indicator. The signal-strength meter is represented by lines and will light up, showing how strong the reception is. The signal should read above 72 for good reception. When the dish is pointed at the satellite, a continuous tone is heard. When not pointed at the signal, short bursts of tone can be heard. Excessive tree growth in the summer or heavy rainstorms can interrupt the TV reception. Make sure no buildings or trees are in front of the dish and satellite.

Testing

TV antennas with a folded dipole can be checked with a low ohmmeter continuity measurement. When open dipoles are found on the yaggi antenna, check the antenna reception with a TV signal-strength meter. This test instrument can also be used to adjust the antenna towards a long-distance TV station. A high resistance measurement of the coaxial cable may spot a poor cable connector or water inside the 75-ohm line. Replace the TV lead-in when the cable is cracked, checked, or broken. A broken or poor lead-in connection can produce intermittent TV reception.

Batteries

A battery provides electrochemical action that generates dc voltage. Batteries produce direct-current electricity. The battery may contain several cells connected together or in series to supply a certain voltage. Batteries may be made up of lead acid (wet), alkaline, nicad, and lithium cells. The most common batteries in the commercial electronic field are the 1.5-, 6-, and 9-volt types. Batteries can be placed in series for greater voltage or in parallel for higher current-drawing products. Batteries have two terminals; one is positive (+), the other negative (−).

1-12 The RCA 18-inch DSS dish picks up direct satellite TV stations from a satellite.

Alkaline battery

What

The early flashlight cells consisted of a carbon-zinc cell of 1.5 volts. The heavy-duty zinc-chloride battery may provide longer service with leakproof protection. Today, the alkaline battery may deliver up to seven times the service of a standard battery. A battery can be rated in ampere (A), the unit of measuring electrical current. Usually, the current measured in battery operations are rated in milliamps (mA), thousands of an ampere. The ampere-hour (Ah) is the measure of battery capacity. Ampere-hours equal current in amperes multiplied by time. The higher the number, the longer the battery will deliver useful power.

Where

The alkaline battery is a primary cell with a negative electrode mixed with a potassium hydroxide (alkaline) electrolyte. The positive (+) electrode is in the center of the cell and is a polarizer in electrical contact with the outer metal can cell (−). This alkaline battery is ideal for smoke alarms, remotes, testers, battery backup systems, and high-current devices such as CBs, walkie-talkies, RC toys, and portable TV sets.

Nicad batteries

What

The nickel-cadmium cell is made up of nickel-cadmium with the anode cadmium, the cathode nickel hydroxide, and the electrolyte potassium hydroxide. The nicad cell is rechargeable and can be used over and over. These cells have only 1.2 volts dc compared to the 1.5-volt regular battery. Although nicad cells cost more than flashlight cells, nicads are a far more economical power source for portable devices that are constantly operated. The nicad battery can be placed in a battery charger and charged overnight. A lead-acid cell is also a chargeable battery.

Where

The nicad battery is found in portable consumer electronic products such as portable radios, camcorders, cellular phones, telephones, RC vehicles, walkie-talkies, scanners, CD and tape players, and button-type cells for watches and calculators. The nicad AA, AAA, C, D, and N cells have a 1.2-volt output. The 9-volt cell is only 7.2 volts with a Nichido battery. A battery pack of 4.8, 6.0, 7.2, and 9.6 volts operates cellular, camcorder, and notebook computers. The new 7.2-volt nicad battery pack is designed to operate handheld scanners.

You may find a 2.4-volt nicad battery for a special Sony Discman portable and a 6-volt nicad for various CD players. A 9V lithium battery is designed for reliability and longer life and is used in smoke detectors, wireless alarm systems, and garage-door openers. Renewal batteries combine the long life and high performance of alkaline batteries with the reusability of nickel cadmium. They can be charged 25 times or more in a charger.

Camcorder/VCR batteries

What

Batteries used in the camcorder can be nickel-cadmium and lead acid types. Both are rechargeable. These batteries can be charged from thirty minutes to two hours. The larger the playing time (Ah), the longer the battery needs to be charged. The nicad battery can produce 6, 7.2, 9.6, 10, and 12 volts with 1.3-Ah to 2.6-Ah current (Fig. 1-13). These nicad batteries plug right into the camcorder for quick removal. A universal 12-volt eliminator for camcorders or VCRs can plug into the auto cigarette lighter for charging the batteries. Most camcorders come with an ac-dc charger to charge up the nicad batteries. Most camcorders have a battery indicator that tells when the battery should be recharged. Other than the battery pack supplied by the manufacturer, you can purchase universal battery packs that will operate the VCR or camcorder.

Cellular phone batteries

Most common cellular telephone batteries provide 1.5, 6, or 7.2 volts from 600 mA to 2.6 amps of current power. A nicad battery pack or AA lithium batteries can be found in the cellular telephone. These batteries can be recharged over and over again in the ac-dc battery charger.

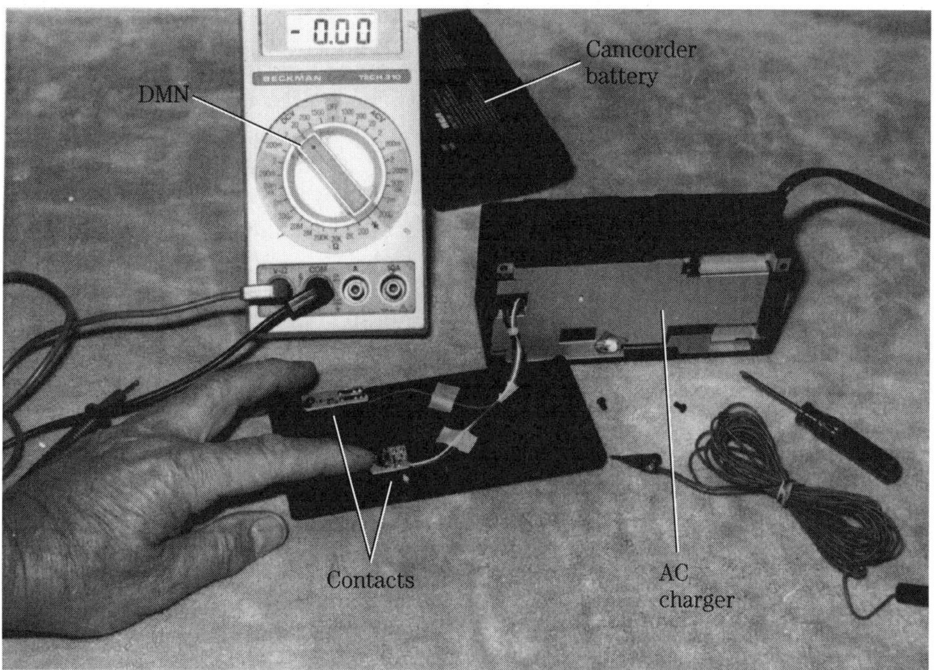

1-13 The camcorder battery is made up of several Ni-Cd cells in series to power the camcorder.

Cordless telephone batteries

The cordless telephone contains chargeable nicad battery packs with voltage ranging from 3.6, 3.75, 4, and 4.80 volts at 0.250 to 0.650 Ahs. The battery packs might be soldered in or have a small cord with a plug on the end (Fig. 1-14). Double-check the plug for correct replacement of battery polarity. Most portable telephone batteries have three 1.2-volt cells wired in series, inside a plastic covering. If the new replacement has the wrong plug, cut off the plug of the old battery that cannot take a charge anymore, solder, and tape to the new battery wires. Always check the battery polarity. These batteries can be charged on an outside nicad battery charger with alligator clips to charge output connected to the battery wires. Exact battery telephone replacements can be obtained from most electronic, drug, or outlet stores.

Button batteries

The special-purpose batteries are button types found in hearing aids, watches, and calculators. You can also find the button-type batteries in cameras, clocks, alarm remotes, games, and measuring instruments. The high-energy lithium batteries deliver a greater and longer output life. The voltage varies between 1.2, 1.25, 1.55, and 3 volts in button types.

Replace the batteries when one begins a howling noise or weak reception. When the computer loses its memory, suspect a weak or dead lithium battery. Replace the button-type battery in your watch when it begins to lose time or quits.

Radio batteries

Most portable batteries are made from extra-life zinc-chloride material with longer storage life, low cost, and good performance. The miniature radio may have three or more AA 1.5-volt batteries. Large portables might operate from four to six 1.5-volt "C" or "D" cells (Fig. 1-15). The 9-volt rectangular battery is found in some small receivers. Heavy-duty zinc-chloride batteries provide longer service life, while the nicad battery can be subbed and charged over and over again for a greater life.

Testing

Batteries can be tested with a small battery tester or with the voltmeter. The DMM is more accurate in checking voltage of a suspected battery. All batteries should be checked under load. When tested out of the electronic product, the battery may test normal. The TV or VCR remote control might not operate or might become intermittent when the voltage of each cell falls to only 1.25 volts with three or four AAA cells wired in series. A cordless telephone might operate at a shorter distance with a weak nicad battery. A weak camera battery might indicate only one line on an indicator, and the shutter or motor will not operate to take a picture or advance the film.

When batteries are tested and no battery tester is on hand, check the battery voltage right in the electronic product. Turn the switch on and take a voltage measurement across each battery (Fig. 1-16). If a 1.5-volt battery has decreased to 1 volt, discard the battery. Recharge the renewable nicad or lithium batteries in a battery charger. Remember, batteries work best at 70 degrees Fahrenheit. A cold

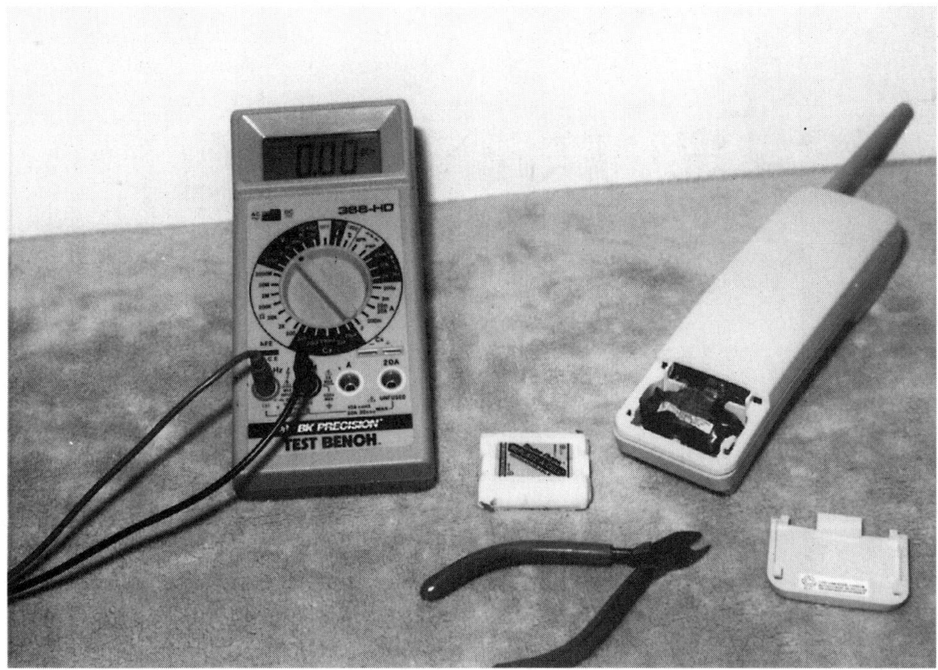

1-14 A telephone battery may be made up of three 1.2-volt Ni-Cd batteries wired in series with or without a plug.

1-15 Large multiband portable radios are powered from four to six 1.5 C or D cells.

1-16 Testing the batteries under load in a shortwave receiver with the DMM.

battery will have less energy and will not last as long. Nicad batteries left in very cold rooms or outdoors may lose the charge in two or more months without being used or charged. Observe the polarity when charging batteries.

Battery charger

What

A battery charger recharges wet, nicad, and renewable batteries from the power line. The circuit may consist of a step-down transformer, rectifier, and filter capacitor. In some power tool chargers, the filter is eliminated. These battery chargers produce a higher dc voltage to charge up the weak batteries (Fig. 1-17). You will find solar cells charging up an array of batteries in home or cabin power systems.

Actually, the battery charger is a simple low-voltage rectified circuit containing only a few electronic components. The power line voltage is stepped down to a very low voltage, rectified by silicon diodes, and filtered with a small electrolytic capacitor (Fig. 1-18). Some battery chargers are more complicated than others. The battery charger for the automobile battery produces more power and higher current to bring up the charge in a weak battery. The RC turbo 7.2-volt nicad battery pack can be charged from a fast dc charger connected to the auto battery.

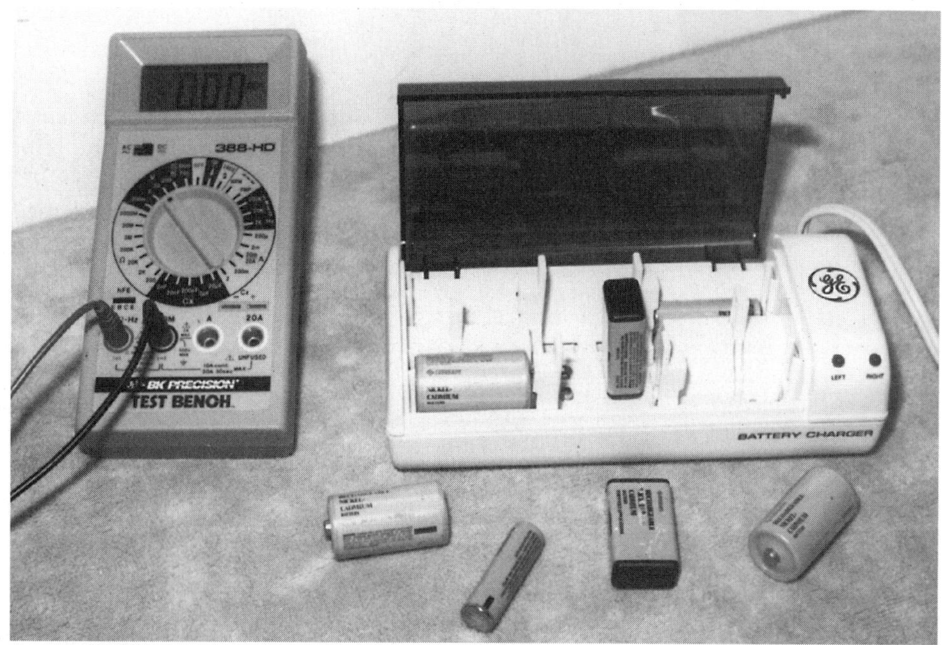

1-17 A Ni-Cd battery charger can charge several nickel-cadmium batteries overnight.

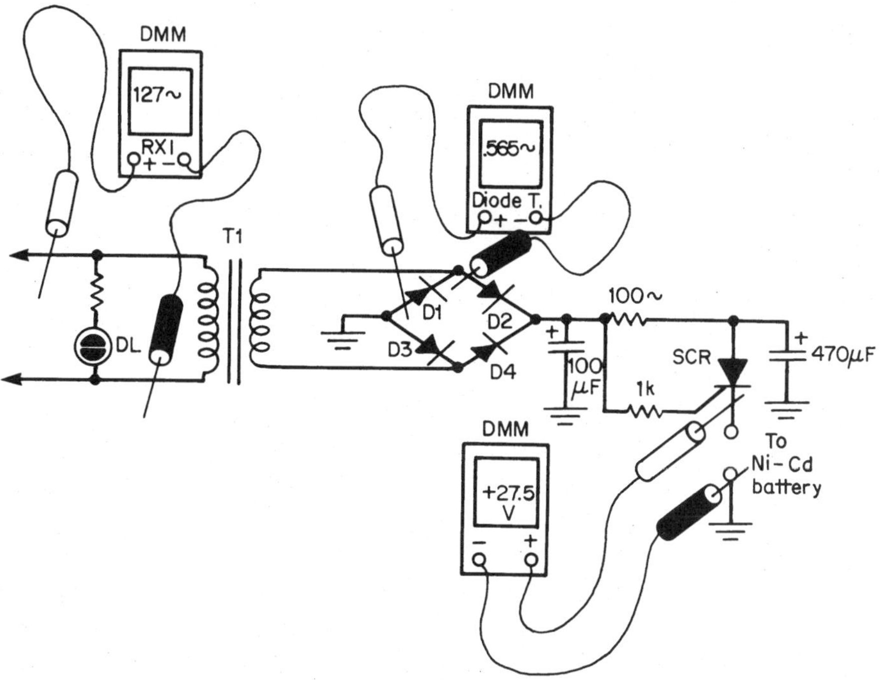

1-18 A simple Ni-Cd battery charger circuit.

Where

The battery charger can be found to charge up small radios, portable tape and CD players, power tools, camcorders, RC turbo racing cars, automobile scanners, CBs, electric razors, and telephone batteries. The telephone charging circuits are found within the base unit that the telephone rests in, and they charge the telephone battery when not in use. Usually, a camcorder battery charger is included when a new camcorder is purchased. Plug the power tool charger into the body of the power tool and charge overnight. Fast battery-charging features can charge the battery in one hour.

Camcorder battery charger

The ac camcorder battery charger may include a plug to the battery, or the battery is inserted on top of the charger (Fig. 1-19). Most camcorders have a battery indicator circuit that tells when the battery is weak or low in voltage. Some camcorders will still operate with the very weak battery, so the unloading and loading functions operate until the battery is replaced.

Before tearing into the ac adapter, check all cords and plugs. Clean all slip-pressure-type contacts with cleaning fluid. Inspect the dc power output cord for breaks or poor connections. On some camcorder chargers, place a load (20-ohm, 10-watt resistor) across the charging output terminals and measure the output voltage with a DMM or VOM.

Determine if the charger is dead in both voltage and charge procedures. If the unit supplies voltage but no charging of batteries, suspect a dirty voltage/charge switch. If the adapter is entirely dead, remove the top cover and inspect the fuse. Replace the fuse with exact amperage. Take a peek; sometimes there is more than one fuse in these adapters.

To check the output voltage of the ac adapter, place a 10- or 20-watt low-ohm resistor or a 100-watt bulb across the battery terminals. It's best to solder alligator clips and leads to the resistor or bulb so it can be easily clipped across the battery terminals. Make sure there are no sense or interlock switches that must be jumped. Do not forget to remove jumpers or clips after the adapter has been repaired.

Testing

Check the simple battery charger the same way as with adapter circuits. Measure the output voltage or ac adapter at the small plug (Fig. 1-20). If there is no voltage, remove the plastic body (as in ac adapter removal) and check each silicon diode with the diode test of a DMM. Next check the ac voltage on the secondary winding of the power transformer. Take a continuity resistance measurement of the primary winding of the transformer. If open, purchase a new charger. Charger transformers are too costly to replace. When every component tests normal and low voltage is measured at the output plug, sub another electrolytic capacitor. Remember, the output voltage measured at the plug should be a few volts higher with no load or battery attached for charging.

Battery eliminator

What and where

Most battery eliminators are classified as ac adapters, and the ac adapter is used in place of a battery. An ac adapter provides a dc voltage to operate an electronic device or circuit. The small portable radio or CB unit may operate from a solar panel

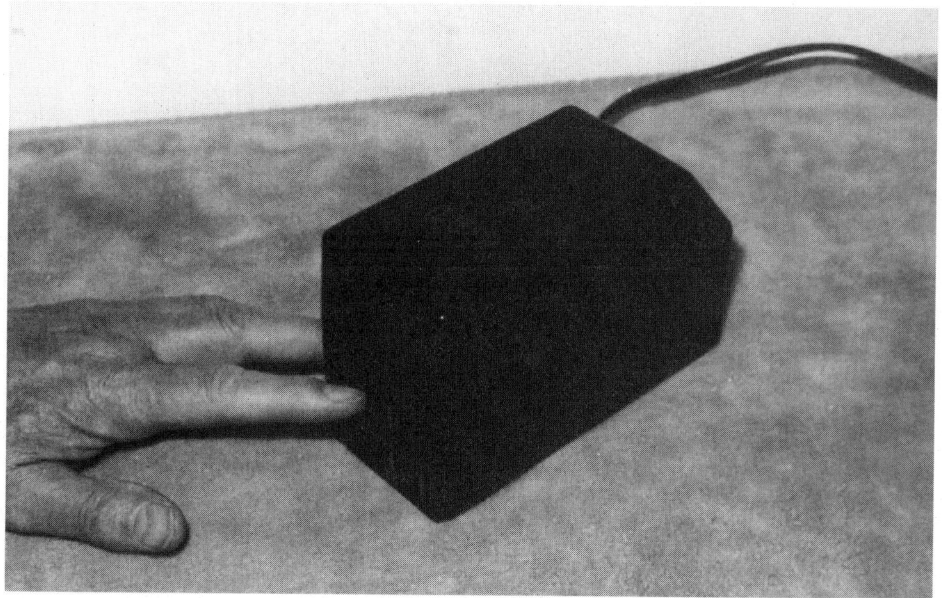

1-19 Here the camcorder battery is plugged on top of the Ni-Cd battery charger.

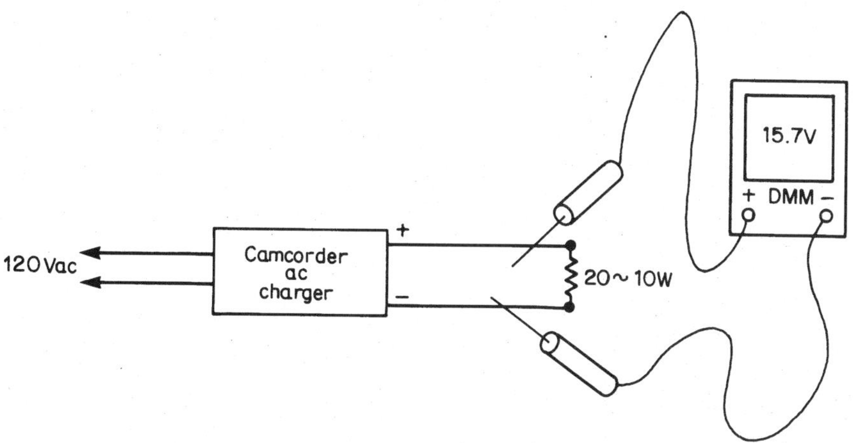

1-20 Checking the fuse and output voltage in a camcorder battery charger with a 20-ohm load resistor.

producing voltage from the sun. Test and service the battery eliminator the same as the ac adapter unit.

A battery eliminator can operate a camcorder, cellular telephone, and VCR from the auto cigarette lighter (Fig. 1-21). A cellular telephone battery eliminator completely replaces the cellular telephone battery while traveling in your car. Simply plug the unit into the cigarette lighter and allow use of the phone while simultaneously charging the battery. A camcorder or VCR can be operated from the cigarette receptacle with a battery eliminator voltage/charger adapter (Fig. 1-22).

Belts

A belt is generally a rubber device that transfers mechanical movement from a motor pulley to an idler, wheel, push roller, or rotating mechanisms (Fig. 1-23). Several different belt sizes can be found in the tape player or VCR. Belts may have a round, flat, or square surface.

What and where

The rubber belt within the tape and CD player and VCR may stretch and become loose and worn, producing slow or sometimes no operation. Rubber belts in the VCR are used in loading the tape from the loading motor pulley to the loading mechanism, and from motor pulley to the capstan drive assembly and drive wheels. Likewise, in the automatic CD player, a rubber belt may be used in loading, moving the pickup head to the outside and inside, and rotating the turntable in a CD changer. A flat rubber belt rotates the turntable in the automatic phonograph. You might find round, flat, and square belts rotating the capstan and take-up reels in the cassette player.

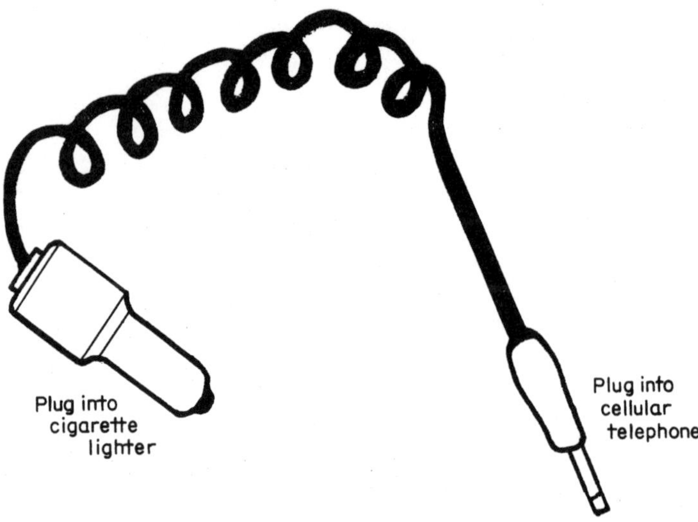

1-21 You can operate the camcorder, CD player, or cellular telephone from the auto cigarette lighter receptacle.

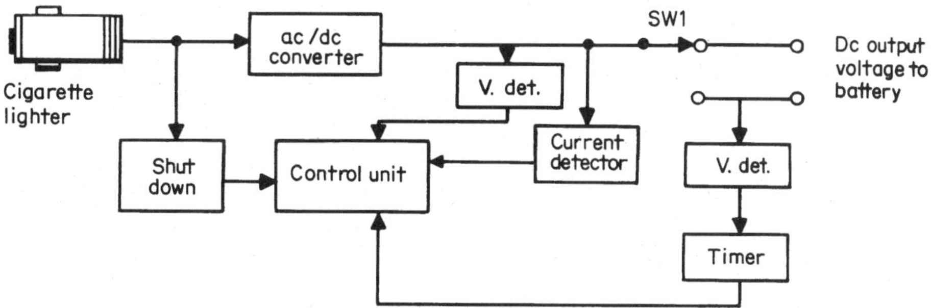

1-22 Block diagram of a car battery voltage/charger adapter.

1-23 Two different belts are found in this auto cassette player.

When

If the tape or phonograph player slows down, becomes erratic, and will not play, suspect a worn, cracked, or broken drive belt. A slipping drive belt will show gray or shiny marks on the inside of the belt. Pieces of rubber or dark areas on the motor pulley indicate that the belt is slipping as it rotates. The loading belt within this VCR will not load when external gum or candy wrappers are pushed in the loading area by children. Sometimes the belt is found broken or has slipped off of a motor pulley, causing no loading or playing function.

Worn, cracked, or broken belts can be replaced by the part number or with universal belts. The circumference of a VCR belt can be measured using two pencils and drawing a circle. Do not stretch the belt in measuring. Select a replacement that is 3% to 5% smaller than the stretched original. A phono drive belt can be measured with a flat ruler (Fig. 1-24). Choose a belt that is $\frac{1}{4}$ inch less than the original. Belts can also be measured from a belt gauge ordered from most electronics stores. Do not interchange a flat belt for a round one. Replace the new belt with a smaller-circumference belt size and width. There are several universal belt kits on the market for belt replacement. Sometimes cleaning up all belts and pulleys with cleaning fluid can solve a lot of speed and operational problems.

Breadboards

What and where

The breadboard can be a perforated board or chassis where wires and parts can be inserted to form a circuit. The first breadboards were wooden ones. Today the breadboard is made up of holes and contacts on a piece of plastic chassis. Breadboards are ideal for experimental, prototypical, and educational purposes (Fig. 1-25). These boards allow connection of DIPs of all sizes as well as most discrete components with leads up to .032 in diameter. The various connections can be made with number 22 gauge solid wire. You can purchase a jumper kit with various lengths of wire for making quick connections. Some breadboards have red and black voltage terminal posts.

When

When the plug-in tie terminal will not hold a wire, proceed to the next set of holes. IC components are plugged into holes that straddle the larger lines of holes.

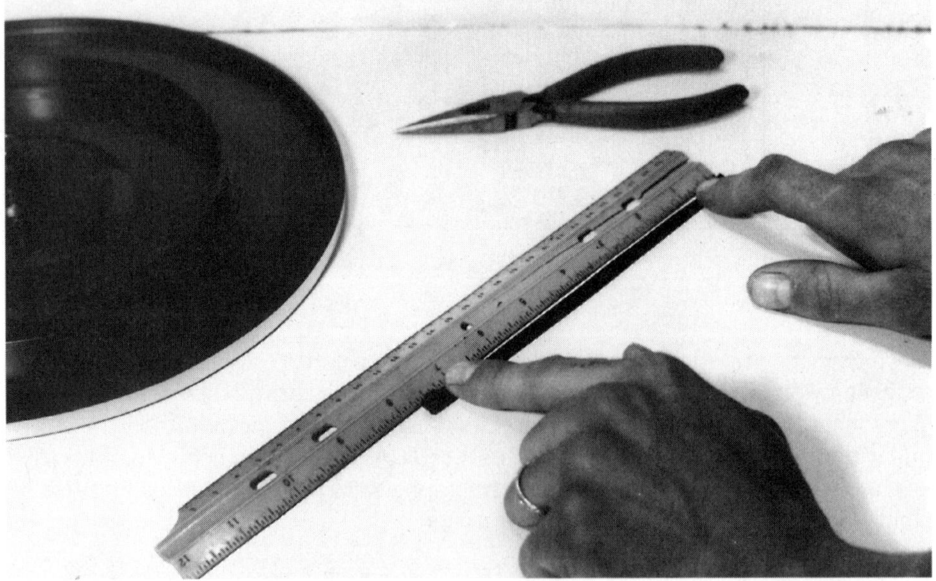

1-24 Place the phonograph motor drive belt alongside a flat ruler and measure the length of the broken or cracked belt.

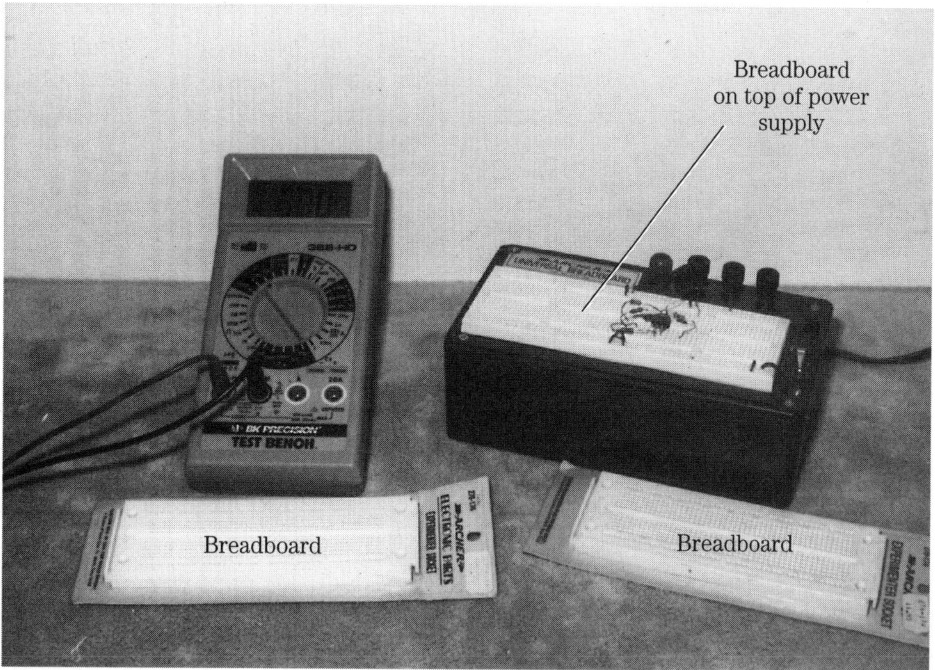

1-25 The different size breadboards for building and experimenting with quick and easy electronic circuits.

Connecting wires or components can be plugged into each hole at right angles with each IC terminal. A $2\frac{1}{4} \times 6\frac{1}{2}$-inch breadboard may have up to 630 plug-in tie points with four distribution strips, each with 50 plug-in tie points. You can quickly and loosely make trial hookups when building or designing your own projects. Breadboards can be used over and over again without any problems.

Bridge rectifiers

(See rectifiers.)

Bulbs

A bulb is a globe-type component of different shapes and lengths that encloses elements such as a lamp bulb or LED. The incandescent lamp is found in a glass bulb, while the LED is inside a small plastic container. The bulb or LED may be found in flashlights, panel lights, and as indicators (Fig. 1-26). The incandescent bulb operates only on lower voltages within electronic circuits.

What and where

The flashlight bulb may contain screw, bayonet, plug-in, fuse-type, surface-mounted, or solder terminal connections. The premium-grade incandescent bulb may operate from 1.2 to 28 volts and pull from 40 to 500 milliamperes. The superbright

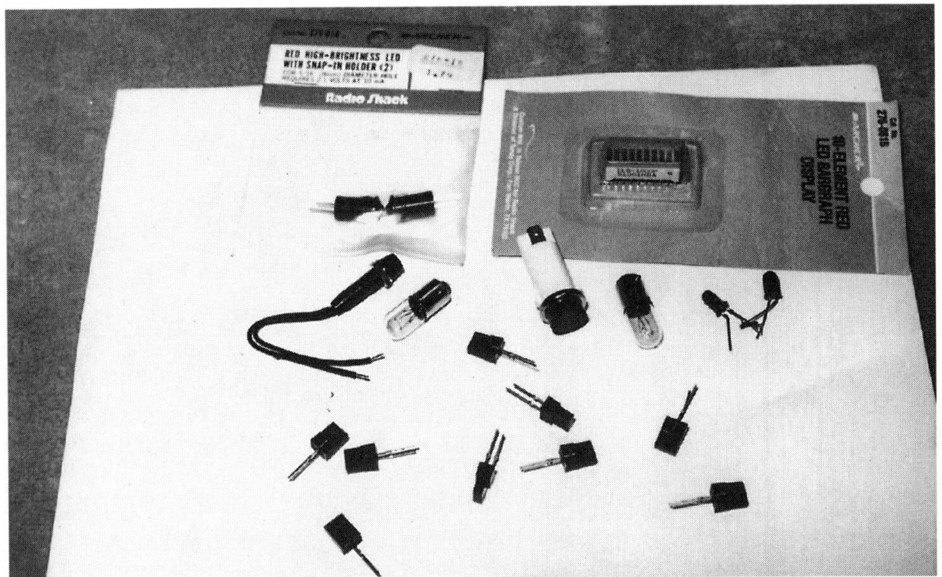

1-26 Bulbs come in many different sizes and shapes.

krypton lamps for flashlights with flange bases can replace PR-types of standard flashlight lamps. These lamps may operate from 2.33 to 7.2 volts and from 480 to 800 milliamperes of current. Halogen flashlights and lanterns provide greater light from a 3-volt to 12-volt source. Minilamps draw around 25 mA and operate on 1.5, 6, and 12 volts, with small hookup wire leads.

A neon bulb is small, is usually filled with neon gas, has a pink glow, and operates from a much higher voltage than incandescent bulbs. Neon bulbs are used as panel indicators, signal circuits, testers, and ac power indicators. Neon bulbs operate from LEDs at a higher voltage than the incandescent or LED bulbs. The NE-2H bulb must add a resistor in series with one lead to operate from the ac power line. The NE-2 bulb has a built-in resistor (22K ohms) (Fig. 1-27). New bulbs operate from a higher voltage than the incandescent or LED bulbs.

The LED comes in jumbo or standard sizes. A light-emitting diode (LED) is a semiconductor device that emits visible light when forward voltage is applied. The laser diode found in the CD pickup belongs to the LED family. LEDs come in many different colors.

What and where

The standard LED may appear in red, yellow, green, orange-red, yellow-green, orange, and orange-green. A jumbo LED has more brightness and may come in red, green, yellow, and orange. The blinking LED has a red and green on/off light. A new LED may emit a separate red, green, and yellow light in one bulb.

The LED must be installed with correct polarity. The positive terminal lead is positive, while in others the flat side of the bulb at the bottom indicates the cathode.

The anode terminal of an LED is positive (+) and the cathode is negative (−). The maximum voltage applied to most LEDs is from 1.75, 1.85, 2, 2.1, 2.5, 2.6, 2.8, and 3 volts. Special LEDs have a maximum voltage of 5 and 12 volts. Most LEDs operate at a current from 10, 20, 30, 40, and 50 milliamperes. LEDs are also found in optocouplers, infrared detector modules, segment digital displays, arrays, and infrared phototransistor applications.

Testing

LEDs can be tested by the diode tests of a DMM. A normal LED will test like a silicon diode. In one direction, you will measure resistance, and reversing test leads will not register with a normal LED (Fig. 1-28). A shorted or leaky LED will show a measurement in both directions.

Incandescent light bulbs can be tested with continuity or resistance measurements. Rotate the DMM to the 200 scale, and a normal bulb will show a low resistance reading. The open filament will have no measurement. Flashlight light bulbs can be tested with two 1.5-volt batteries wired in series and clipped to the bulb terminals. A weak, low, or bright light indicates the bulb is normal. Although the bulb lights up, replace the bulb when the bulb area has burned black and when you replace the batteries.

1-27 Connect a 22K-ohm resistor in series with one lead of a NE-2 neon bulb to operate from the power line.

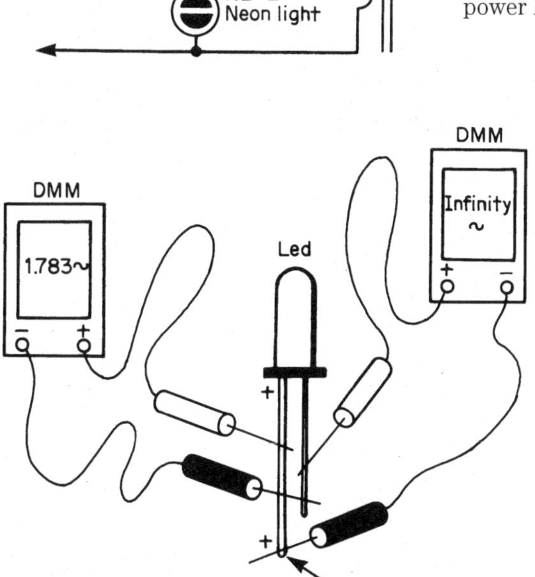

1-28 How to test a normal LED with the diode test of the DMM.

Buzzers

What and where

A buzzer produces audible sound when energized. The buzzer may be a mechanical vibrator type operating from a battery or ac voltage, like the front or rear doorbell. The electronic piezo or piezoelectric (transducers) can be signal devices like those found in the microwave oven indicating that the cooking cycle is finished (Fig. 1-29). A dc buzzer has a vibrating-reed electromagnetic coil, and when voltage is applied the reed is attracted to the core and moves away from contact, while the magnetic core ceases and the reed springs back to the contact once again. This action keeps repeating itself with a buzzing sound until the voltage is removed from the circuit.

Piezo electronic buzzers

The piezo or solid-state buzzers are ideal for smoke detectors, burglar alarms, fire alarms, home appliances, radio beepers, electronic games, cameras, car and industrial backup alarms, microwave oven indicators, and alarm detectors. The piezoelectric device is made of a ceramic material which delivers a voltage when stretched or compressed and changes its shape when voltage is applied. A piezoelectric sensor is a crystal transducer that employs a piezoelectric crystal as the sensitive element. The crystal earphone, loudspeaker, microphone, pickup, and vibration are crystal transducers.

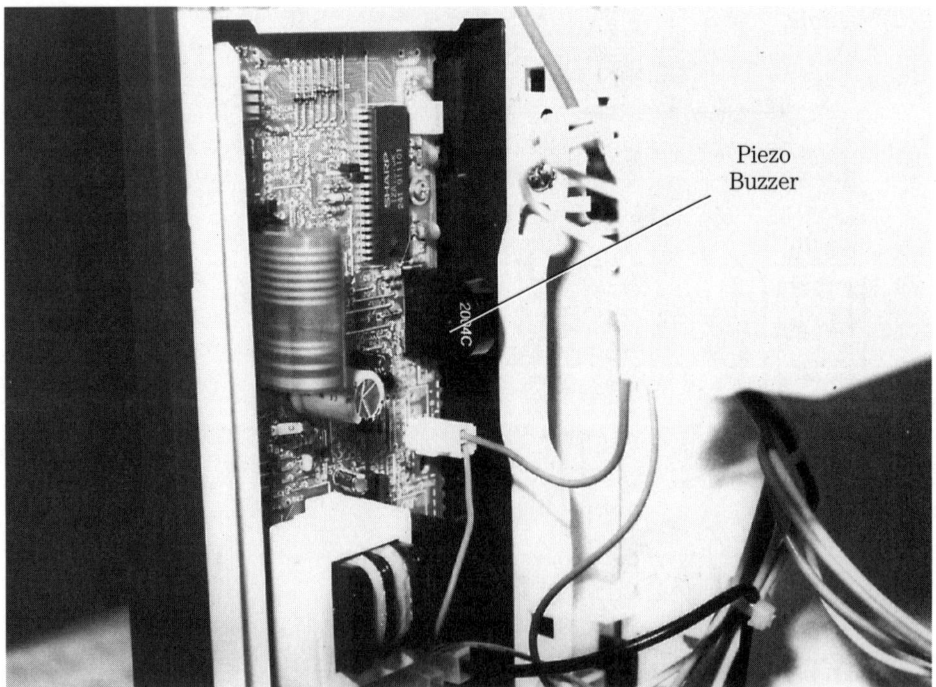

1-29 A piezo buzzer found in the microwave oven indicating that cooking time is up.

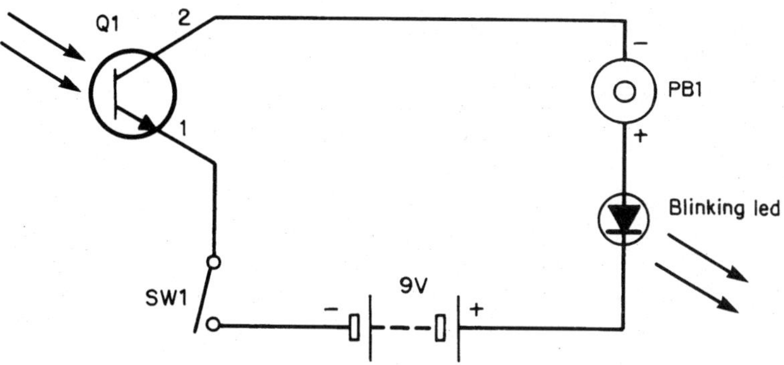

1-30 PB1 is a solid-state piezo buzzer used in a simple infrared remote control tester.

The piezo electronic buzzer may operate from 3–28 volts dc, while the solid-state buzzer operates from 1.5 to 16 volts (Fig. 1-30). Magnetic and piezo buzzers add sound to various projects. The piezo buzzer may operate with a tone from 1500 to 3700 Hz. The piezo or sold-state buzzer has transistor or IC components built right inside the unit.

Testing

Magnetic buzzers can be tested by taking a continuity or resistance test at the voltage terminals. Piezo or solid-state buzzers can only be checked by applying a low dc voltage to the terminals. Check the voltage applied across the piezo buzzer terminals (Fig. 1-31). Be careful, as most piezo or solid-state buzzers have a + and − terminal. The piezo buzzer will not operate with reversed polarity. Buzzers are so small and inexpensive that it's best to replace them rather than try to repair.

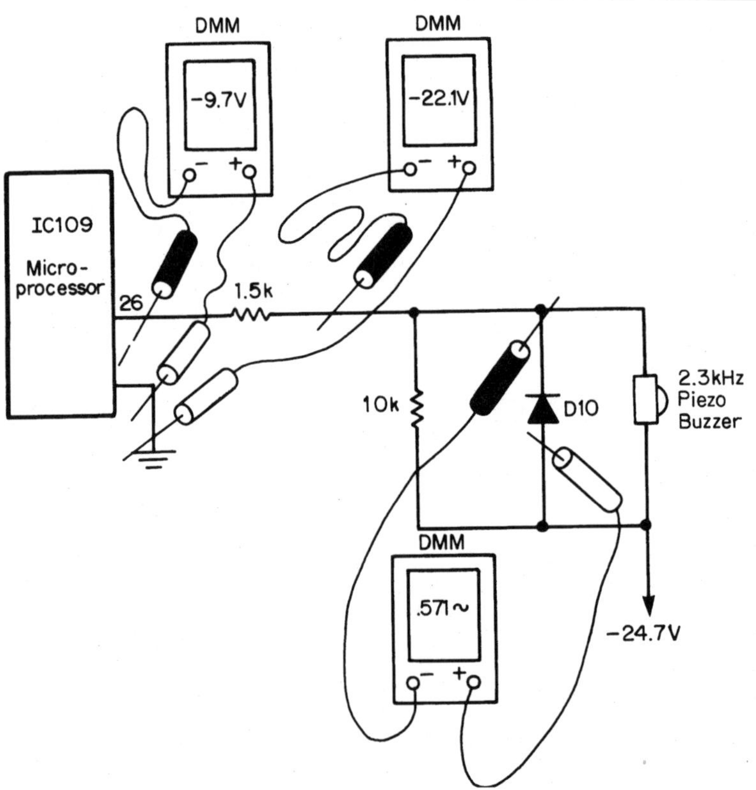

1-31 Check the voltage on the piezo buzzer terminals.

2
CHAPTER

Testing electronic components C

(Capacitors)

The capacitor symbol is C, and the unit is measured in farads. A capacitor has two conductors separated by a dielectric, where an electrical charge is stored between the two elements. The capacitor stores electrons. The capacitor and resistor are two well-known components found in electronic circuits (Fig. 2-1). Capacitors are found in just about every electronic circuit. The larger the size of the capacitor, the greater the charge it will hold, and the longer it will hold the charge. Decoupling capacitors are used to filter coupling, bypass, tuning, timing, and phase-shift circuits. The common filter capacitor is found in low-voltage power supply circuits. A coupling capacitor is found between the AF and driver stage of an audio amplifier. In early receivers, the tuning capacitor contained variable rotating plates that tuned in broadcast stations with table radio or auto radio. Bypass capacitors shunt certain frequencies to ground potential, while decoupling filter capacitors are found in voltage-resistance circuits of voltage sources.

Capacitor conductors can be made up of two separate plates of aluminum or aluminum foil, while the dielectric material can be air, vacuum, plastic, film, paper, ceramic glass, and mica. In the electrolytic capacitor, the electrode may be aluminum foil separated by a paper-paste dielectric insulator. The paper is saturated with an electrolyte liquid or paste. Bypass capacitors might be made up of ceramic dielectric material with silver paint or metal on each side as conductors. Both conductors have a pigtail or wire attached so it can be soldered into a circuit.

The capacitor can be charged by simply applying voltage across the two terminals. Current flows while the capacitor charges. The capacitor is fully charged when

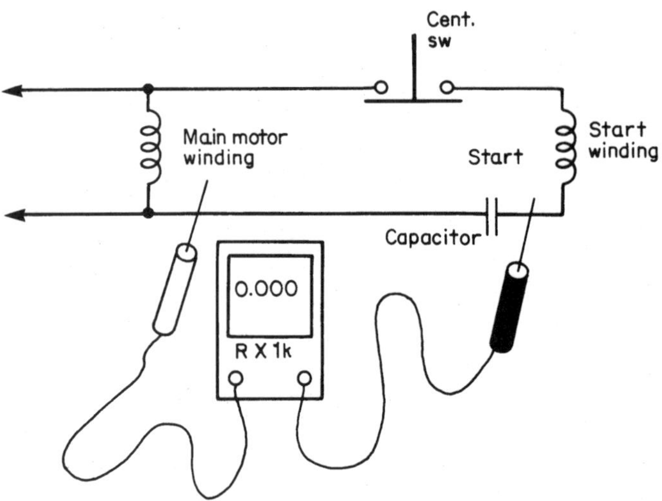

2-1 The electronics technician taking voltage measurements to locate a defective capacitor.

the voltage across the capacitor equals the voltage source. When the voltage source is disconnected from the two conductors, current flows in the opposite direction while it discharges. Current in a capacitor ceases to flow when it is fully charged or discharged.

Most capacitors are rated in microfarads (µF) (one millionth of a farad). A picofarad (pF, one-millionth of a microfarad) capacitor is the smallest rating of capacitors found in RF and very high frequency circuits. Capacitors are rated in tolerances of + or − 1%, + or − 5%, and + or − 10%. The most common tolerance is 10%. Always use a 5% capacitor in high-frequency or critical circuits. Capacitors connected in parallel add up the capacity while wired in series decreases the capacity of the capacitors. The voltage rating of an electrolytic capacitor is the working or operating voltage. Always, observe the correct capacity, voltage rating, and polarity when replacing a defective capacitor.

Ac capacitors

What and where

The ac capacitor is found in alternating current circuits such as the motor-start capacitor. Bypass capacitors found in the power line circuits of a TV or radio receiver may shunt extreme noise to ground. The fixer capacitor found in the ac motor circuit is used with an auxiliary winding to bring the motor up to a normal running speed (Fig. 2-2). The ac capacitor is switched out of the circuit when motor speed is normal. The nonpolarized capacitor can also be found in speaker circuits. The ac capacitor has a larger capacity and may be a paper capacitor.

Testing

The ac capacitor can be tested on a commercial capacitor tester or with the resistance test of a VOM or DMM. A defective or leaky capacitor will show some resistance leakage. The normal capacitor will charge up and discharge when the ohmmeter leads are reversed on the capacitor terminals. The VOM or FET volt-ohm meter will show the hand going up when charging and down when discharging. The VOM, FET VOM, and DMM can detect a leaky or shorted capacitor, while the VOM and FET VOM indicate the charging and discharging of the capacitor. If in doubt, remove one end of the capacitor for correct tests. Check the capacitor for leakage with the 1K range of the ohmmeter (Fig. 2-3).

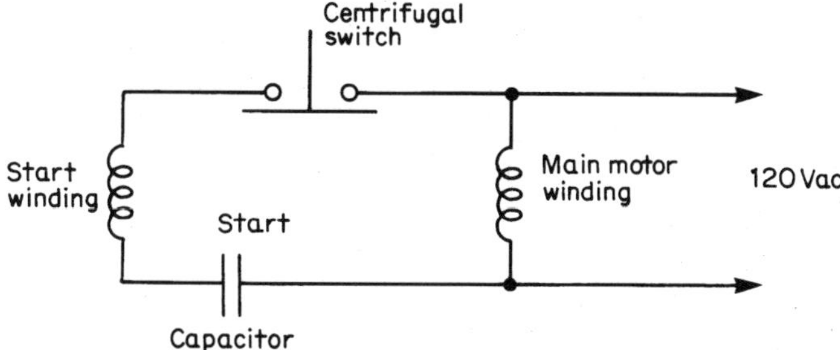

2-2 The starting capacitor and start winding are switched out of the circuit once the ac motor is up to correct speed.

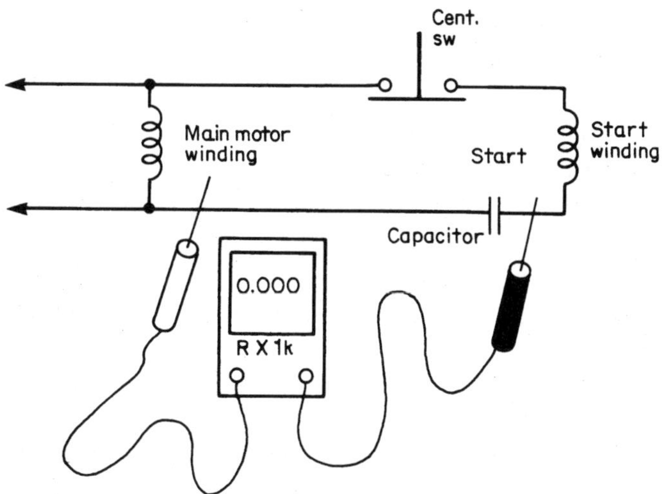

2-3 Check the capacitor for leakage with the RX1K range of the ohmmeter.

Audio frequency capacitors

What and where

Capacitors found in the audio amplifier circuits are classified as audio frequency capacitors. The audio frequency is usually from 20 Hz to 20 kHz. The fixed audio capacitor may be classified as paper, ceramic disc, plastic-film, tantalum, and electrolytic. The fixed capacitor can be nonpolarized or polarized. A two-lead configuration can be axial and radial mounting.

The paper capacitors were found in the early radios and TV chassis (Fig. 2-4). A ceramic disc capacitor has a very low loss, forms a round disc, and values range in a few picofarads to 0.5 µF. The plastic-film capacitors are made up of polyester, polyethylene, and polystyrene material, such as a mylar capacitor. Plastic capacitors are found in audio and radio frequencies, with high efficiency, and may range from 0.001 µF to 10 µF. The tantalum and electrolytic capacitors are found in bypass, coupling and decoupling circuits of the audio amplifier. A tantalum capacitor uses tantalum rather than aluminum, while the electrolytic employs aluminum as electrodes.

Testing

Small ceramic disc, paper, and plastic film capacitors can be tested in a small capacitor tester or combination digital multimeter. These small capacitors can be checked with the ohmmeter by rotating to the 10K-ohm range. You may see only a quick charge of the capacitor. Just reverse the leads and see the capacitor charge up again. The charging and discharging of a fixed disc capacitor is short and quick. The

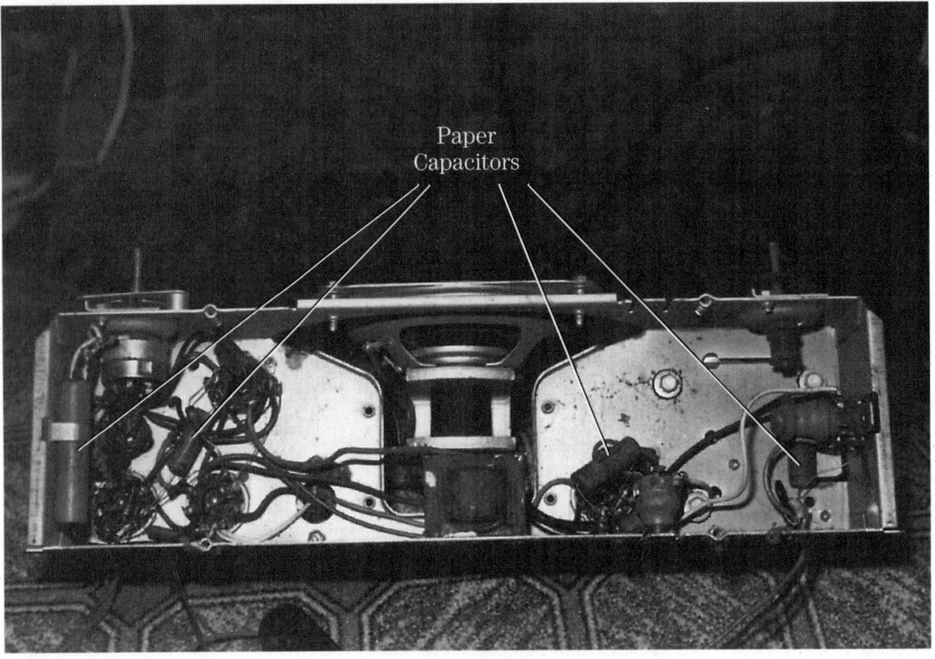

2-4 Paper capacitors were found in the early radios and amplifiers.

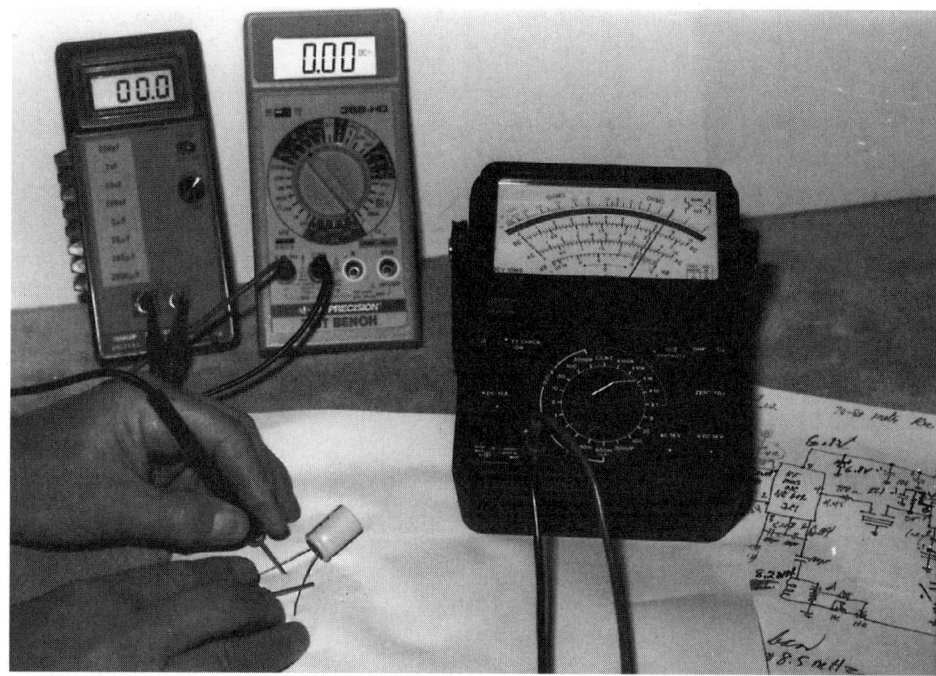

2-5 Notice the charging of the capacitor on the meter of the VOM or FET-VOM.

ohmmeter range of the VOM or FET-VOM provides movement of the meter while the capacitor is charging up (Fig. 2-5). A leaky or shorted capacitor will give a low resistance measurement in both directions. The normal capacitor will charge up and discharge without any signs of leakage. The open capacitor will not charge or have infinite reading. When the capacitor will not charge up in the circuit, remove one end and test again.

Bypass capacitors

What and where

The bypass capacitor in the early radio was found in the RF, IF, and audio stages. These early bypass capacitors were called paper or ceramic-end type bypass capacitors (Fig. 2-6). A bypass capacitor allows ac to bypass around a component or several parts. RF frequency may be bypassed to ground within the RF stage to separate it from the audio signal. Usually, ac is separated from the dc to allow ac to pass through a component.

The bypass capacitor might be paper, ceramic, mica, plastic-film, or electrolytic. In the radio receiver, a bypass capacitor can be found in the RF, IF, audio and power supplies. Bypass capacitors found in the TV may be located in just about any circuit of the chassis. Likewise, bypass capacitors are found in many circuits of the CD player, including motor circuits. Bypass capacitors are found in the emitter and collector voltage circuits of a solid-state chassis to common ground (Fig. 2-7).

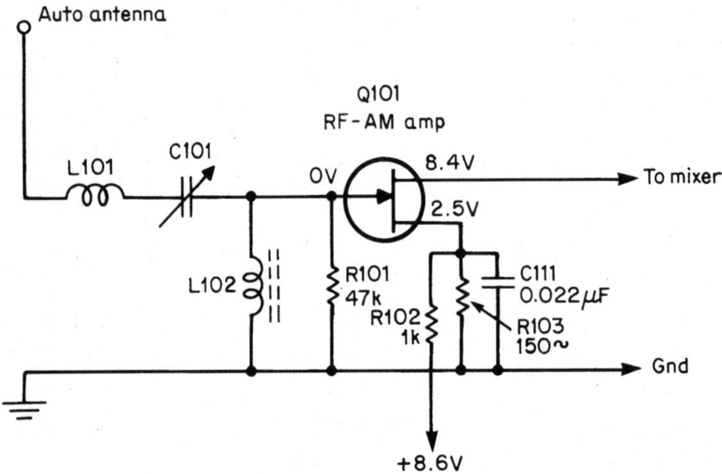

2-6 A paper bypass capacitor (C111) is found in the early auto radios.

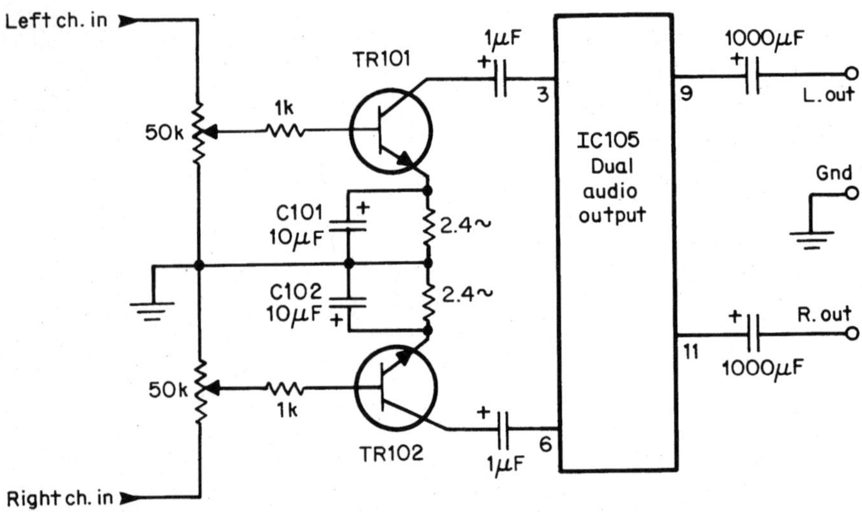

2-7 The bypass capacitors (C101 and C102) are connected across the emitter terminals to ground in the AF circuit of the audio amplifier.

When

The bypass capacitor might open up, become intermittent, or leaky. The leaky capacitor upsets the voltage measurement in a given circuit. You might find a weak or dead stage with a leaky bypass capacitor. The open bypass capacitor may cause oscillations within a RF or AF circuit. The signal might be weak with an open bypass capacitor. The radio signal may cut up and down with an intermittent bypass capacitor. Sometimes it is difficult to locate an intermittent bypass capacitor. Sometimes it is difficult to locate an intermittent or open bypass capacitor in the RF, IF, audio, or low-voltage circuits.

Testing

The bypass capacitor can be located with critical voltage and oscilloscope tests. Often the voltage will change with a leaky bypass capacitor across the emitter resistor in the solid-state circuits. If no voltage is found across the emitter resistor, suspect a leaky bypass capacitor to ground. Critical resistance tests across the bypass capacitor can locate a leaky capacitor. Extreme noise and RF trash can be detected with the scope in the RF, IF, and low-voltage circuits. Critical bypass capacitors can be soldered or clipped across to determine if one is found defective.

Remove one end of the capacitor while testing a suspected bypass capacitor in the circuit (Fig. 2-8). Now check for correct capacity with the capacity tester. Although the charging and discharging test of a bypass capacitor on the VOM is not as accurate as the capacity tester, you can locate an open or leaky capacitor. Remember, capacitors found in many electronic circuits can be 10 percent off of the value stamped upon the capacitor. Extreme and critical circuits might contain 5% or 1% capacity tolerance.

Coupling capacitors

When and where

Coupling is linking two different circuits together. *Capacity coupling* is transferring ac energy between the two circuits. The coupling capacitor in the audio stage may be between driver and audio output transistors, volume control to the base of a transistor or input terminal of an output IC, and from the sound output stage coupled to the speaker (Fig. 2-9). A coupling capacitor may be paper, ceramic, or electrolytic.

Within the TV chassis, the horizontal oscillator stages can be coupled to the horizontal driver transistor with a coupling capacitor. The vertical output waveform can be coupled to the yoke or yoke-return circuits. The IF stage may have capacity coupling between the IF and saw filter circuits. You may find capacity, resistance, or

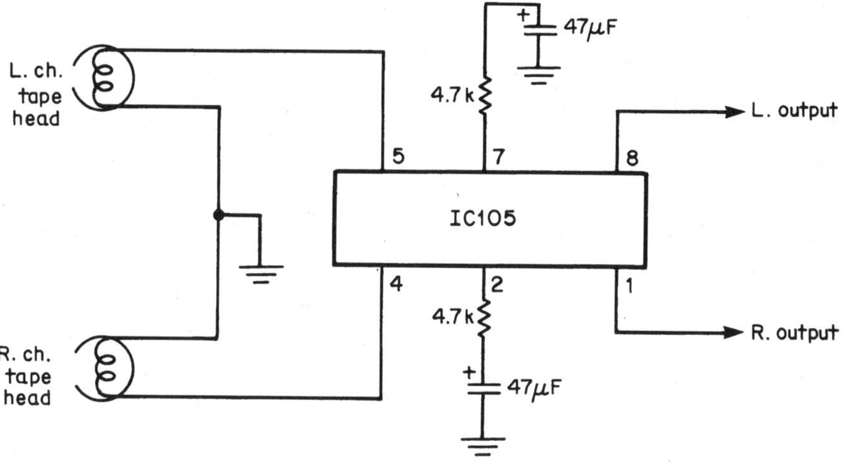

2-8 Remove one end of the suspected capacitor when making in-circuit capacitor tests.

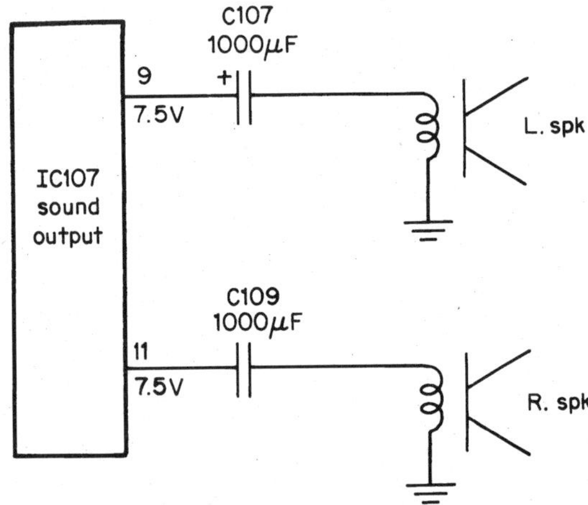

2-9 The electrolytic coupling capacitors (C107 and C109) are coupled to the speakers and prevent dc voltage from damaging the voice coils.

capacity-resistance coupling within the video circuit of the TV chassis. Coupling capacitors can be found in many electronic circuits within the consumer electronics field.

When

The sound may appear distorted, intermittent, or dead if the coupling capacitor becomes defective from the sound output IC to the speaker voice coil. When the audio output coupling capacitor goes open, the sound is dead at the speakers. If the capacitor becomes intermittent, the audio will cut up and down. A leaky or shorted coupling capacitor can cause distorted sound in the speaker.

The defective coupling capacitor within the AF circuit of the audio amplifier can cause extreme distortion, weak or low audio, or a combination of both. If the capacitor goes open, very little sound can be heard in the speaker. When the coupling capacitor becomes leaky, dc voltage increases on the base circuit of the driver or audio transistor (Fig. 2-10). A leaky coupling capacitor from volume control to the input IC terminal can change the voltage on the IC. In directly driven transistor circuits, a coupling capacitor to these transistors might upset the voltages of each subsequent transistor.

A defective coupling capacitor from the IC audio output to the speaker can produce weak and distorted sound. The voltage at the output terminal will become lower in value applied to the voice coil of speaker to ground. When the coupling capacitor becomes intermittent, the audio at the speaker produces intermittent sound. Often, a broken lead inside the coupling capacitor occurs with intermittent sound. If the capacitor goes open, very little sound will be heard at the speaker. Just clip another electrolytic capacitor of the same value across the suspected capacitor to locate either an intermittent or open coupling capacitor.

Testing

The coupling capacitor can be checked with voltage, resistance, capacity tester, signal, and substitution methods. Usually, coupling capacitors with semiconductor audio output circuits are electrolytic types. Test the capacitor with the capacity tester or sub another across the suspected one. A quick voltage test on both sides of the capacitor to common ground can easily spot a leaky or shorted capacitor (Fig. 2-11). A higher voltage measured on the base or input terminal of a transistor or IC might indicate a leaky capacitor.

Remove one end of the suspected capacitor and take a resistance measurement across the terminals. A normal electrolytic capacitor will charge up and discharge on the FET VOM or VOM meter. Set the FET VOM on the ohmmeter 10K or 1K range.

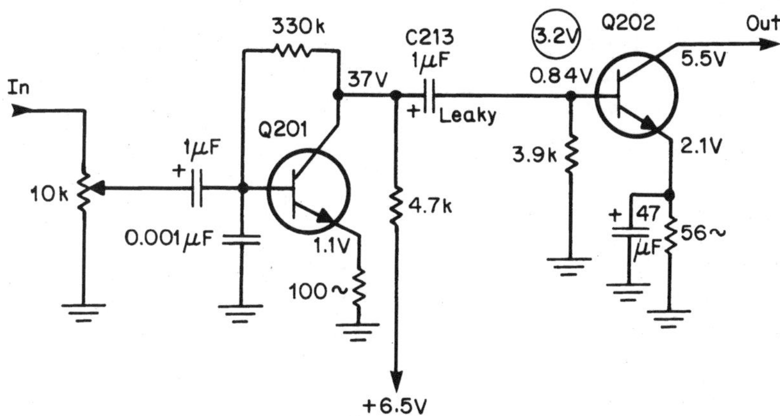

2-10 The measured circled voltage indicates that C213 or Q202 are leaky. Remove the negative end of C213 for a leakage test.

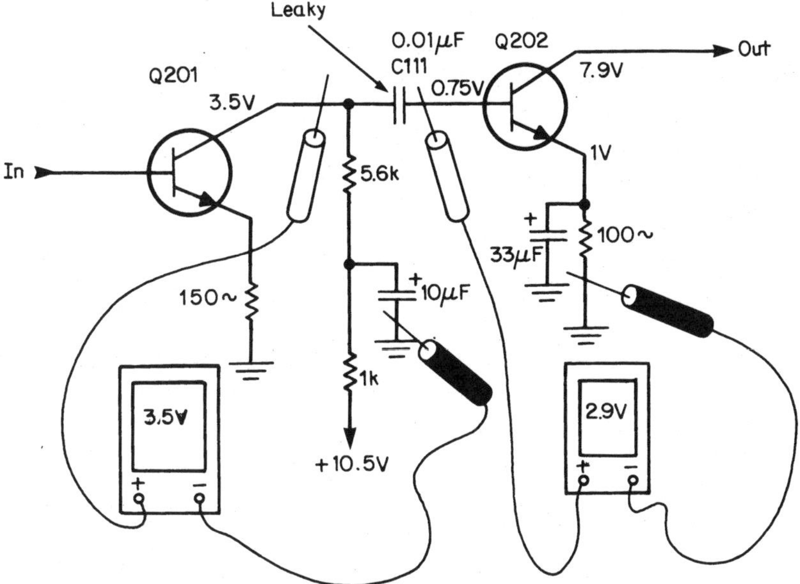

2-11 A quick voltage test indicates that C111 is leaky with voltage found on both sides of the coupling capacitors.

If the capacitor will not show a charge, suspect an open electrolytic capacitor. A low ohmmeter resistance measurement in both directions indicates a leaky or shorted capacitor. While one end is out of the circuit, test the capacitor on a capacitor checker for correct value.

The electrolytic capacitor within the audio circuits can be checked with audio applied or with a signal generator. Check the signal on both sides of the capacitor for correct audio with the scope or external amplifier. When no signal is found on one side of the capacitor, suspect an open capacitor. Suspect a leaky capacitor if the signal is weaker on one side of the capacitor. The scope can easily spot an intermittent or open electrolytic capacitor with audio signal from the radio receiver circuits or cassette tape, or with the audio signal from a function generator (Fig. 2-12). Compare the signal from the audio generator with the scope as indicator to locate a weaker channel, distorted signal, and intermittent reception. Check each stage in the same manner with an external audio amp, if a scope is not handy.

The coupling capacitor can be tested without any test instrument. Simply sub another capacitor of the same value, or one close in capacity and voltage, and clip it across the suspected one. Observe correct polarity with the electrolytic capacitor. If the signal is weak or intermittent, the sound will return to normal. At the same spot in the audio circuit, check the same amount of audio with the normal channel in a stereo audio output circuit.

Decoupling capacitors

What and where

A decoupling capacitor provides a low impedance to ground to prevent regular coupling between stages of an electronic circuit. Often, the decoupling capacitor is found between a couple of resistors to ground. Decoupling capacitors are usually electrolytic types. The decoupling capacitor may prevent interaction in amplifier stages through a common power source. Sometimes the decoupling capacitor acts like a regular filter or bypass capacitor.

The decoupling capacitor in a table-model radio can be found in the low-voltage circuits applied to the IF and AF circuits. Capacitor C40 is found between R37 and R31 serving as a decoupling filter capacitor. R37 feeds the AM and FM IF circuits, while R31 provides voltage to audio circuits (Fig. 2-13). The IF stages might oscillate with an open C40 decoupling capacitor. This voltage at the junction of R31 and R37 might become lower, producing a weak or loss of volume to the audio circuits.

A decoupling capacitor can be found in the VCR at many different circuits. C7010 provides decoupling action between resistor R7030 (100K) and R7028 (33K) to the base of transistor Q7004. The count up/down signal is fed at pin 4 to the resistor network controlling Q7004 and other operation and timing circuits. C7010 (4.7-µF) capacitor provides a constant voltage and signal applied to the control transistor (Fig. 2-14).

Besides the radio and VCR circuits, decoupling electrolytic capacitors can be found in the horizontal, pincushion, IF, and audio TV circuits. The decoupling capacitor may be a paper or ceramic capacitor in the TV decoupling circuits. Often,

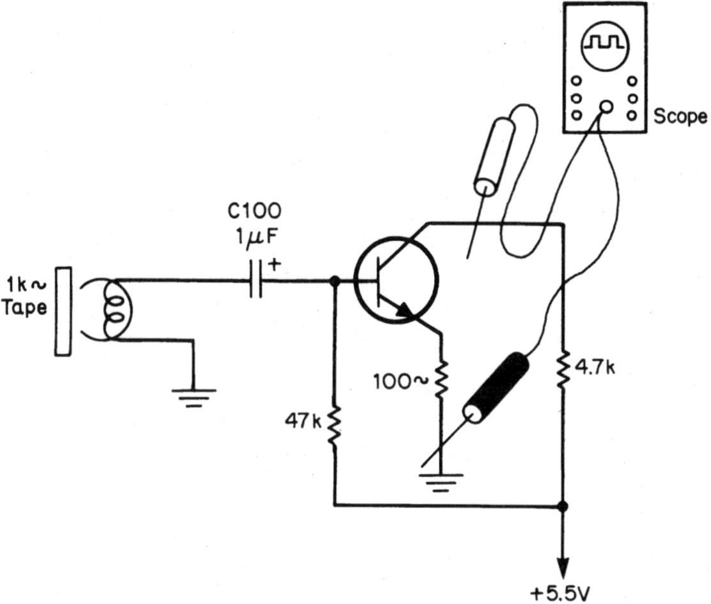

2-12 When C100 is open, no tape signal is found on the collector terminals of Q109 in the tape player.

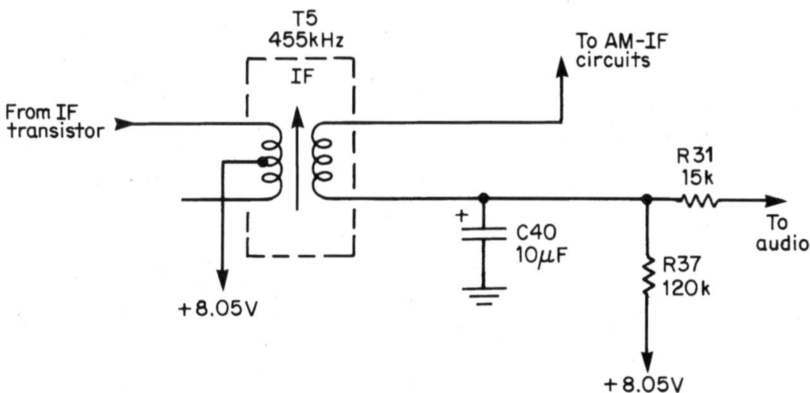

2-13 C40 provides decoupling action in the AM-FM IF and AF audio circuits of a radio receiver.

the small electrolytic capacitor is found in the various voltage circuits. Electrolytic capacitors in the voltage sources feeding other circuits are called *filter capacitors*.

When

Usually, unstable audio, video, or oscillations occur when a decoupling capacitor opens up. The picture may want to tear or bounce around within the TV chassis. The audio may appear weak with some distortion when a decoupling capacitor appears

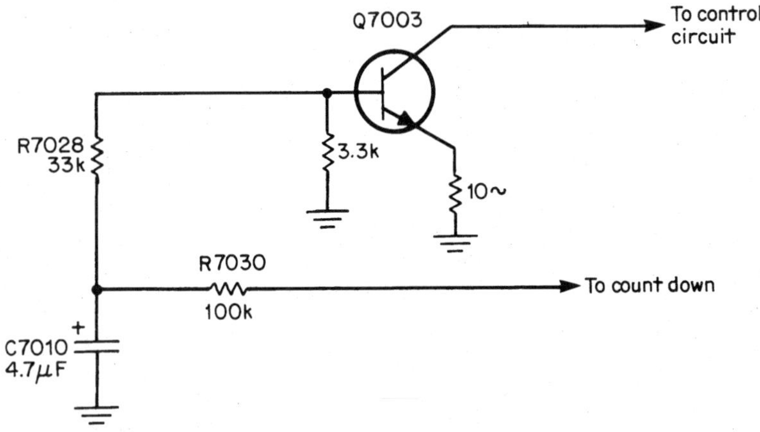

2-14 A decoupling capacitor (C7010) is in the control and tuning circuits of a VCR.

leaky. If the decoupling capacitor shorts out or has high leakage, very little voltage is applied to the connecting stages. Intermittent decoupling capacitors are very difficult to locate without the oscilloscope. A low hum may be noticed within the speaker with an open or leaky decoupling capacitor.

Testing

Monitor the affected stage with the oscilloscope and voltage measurement. Then shunt each decoupling capacitor with a known electrolytic capacitor. If the voltage returns to normal with stable sound and video, remove one end of the capacitor for testing. Notice if the audio is clean and sharp after the decoupling capacitor is shunted. Shut the chassis down and clip the capacitor into the circuit, so as not to damage sold-state components. Check the electrolytic with a capacitor tester. Most electrolytic decoupling capacitors are under 50 microfarads, with 4.5- and 10-µF capacitors found in most circuits. These small electrolytic capacitors can be tested in a small capacitor tester or one included with the DMM.

Filter capacitors

What and where

A *filter* is a circuit or device that passes one frequency and signal while blocking out others. The filter capacitor provides capacity resistance within the low-voltage power supply and also blocks dc from ground potential. The filter capacitor is found in all low-voltage power supply circuits (Fig. 2-15). The electrolytic type capacitor provides considerably greater capacity in the low voltage circuits. A filter capacitor levels off the ac ripple within the dc power voltage sources. The polarized electrolytic capacitor has a positive and negative terminal.

The filter capacitor has a paste-type electrolyte between the two aluminum foils or electrodes. A thin oxide film is formed on the surface of the positive plate. The film serves as a dielectric insulator between plates. Filter capacitors may vary from 0.47 µF to above 50,000 microfarads. The electrolytic capacitor can be operated at a higher working voltage (1000 V) than most capacitors. When electrolytic capacitors become old, they dry out and lose capacity. Electrolytic capacitors have radial or axial terminal leads. Radial capacitors stand up, and axial capacitors lie down.

Always observe correct polarity when replacing the electrolytic capacitor. If not, the capacitor becomes warm, overheats, and may blow up in your face. The minus (−) sign and black line should be at ground potential (Fig. 2-16). Discharge the capacitor by shorting both terminals before removing from the chassis. Always discharge the electrolytic before taking resistance measurement within the same circuit. Do not leave a charged capacitor on the service bench, as one might pick it up, become shocked, and knock test equipment off of the service bench.

When

The defective electrolytic capacitor can dry out and lose capacitance, become leaky or shorted, go open, and become intermittent. When the filter capacitor opens or dries out, you can hear a loud hum in the speaker. The open capacitor may show one or two hum bars on the TV picture tube (Fig. 2-17). An open or dried up capacitor can produce a lower voltage source. The leaky or shorted filter capacitor in the low-voltage power supply can damage silicon rectifiers and sometimes the primary winding of the small power transformer. The intermittent electrolytic capacitor can produce intermittent voltage and sound in the audio output stages.

2-15 There are many different sizes and shapes of filter capacitors found in the AM-FM-MPX stereo receiver.

2-16 A black line with a negative sign indicates the ground terminal of a filter capacitor.

2-17 Two hum bars found on the TV screen indicate poor filtering in the low-voltage power supply.

Electrolytic capacitors can also be found in the secondary voltage source of the TV chassis. There are many filter and decoupling electrolytic capacitors found in the scan-derived (secondary) winding of the flyback or horizontal output transformer. You may find up to 10 or more voltage sources taken from the secondary winding.

Often, zener or transistor regulators are found with resistor and electrolytic capacitors in each voltage source (Fig. 2-18). If one or more capacitors become leaky, a separate regulator transistor or silicon half-wave rectifier can be overheated and damaged. When one of these electrolytics opens or dries out in a certain voltage source, the voltage will always be lower than that found on the schematic. The defective electrolytic within the secondary sources can cause hum in the sound and picture, cause the horizontal and vertical circuits to become unstable, and produce no or little voltage from the voltage source. Check the electrolytic with a waveform taken before, and then after, to notice how the waveform flattens out when taken across the suspected electrolytic.

Testing

The best method to locate a defective filter capacitor in the power supply is to shunt it with a known electrolytic capacitor. Most electronics technicians test the power supply voltage source first. Look for the largest filter capacitor found on the electronic chassis (Fig. 2-19). Measure the dc voltage across the capacitor terminals. Check for correct voltage and compare it to the schematic. If a diagram is not

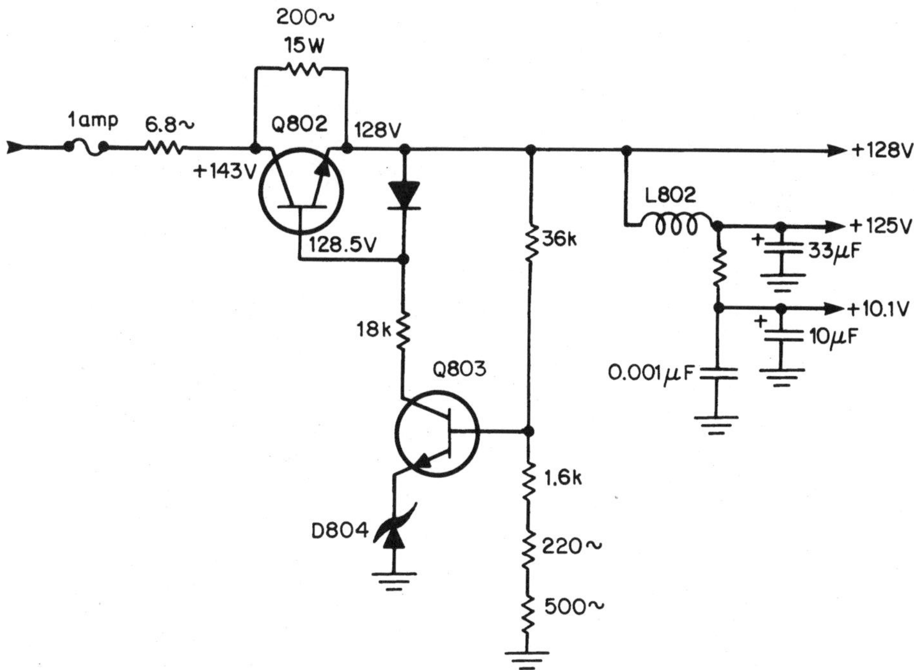

2-18 The transistor and zener diode regulators may have several electrolytic capacitors within the low-voltage source.

2-19 A shorted filter capacitor knocked out two silicon diodes in an AM-FM-MPX receiver.

handy, the working voltage marked on the capacitor will indicate that the voltage should be a little less within the circuit.

Shunt the electrolytic capacitor with another capacitor of the same or higher capacity and working voltage. Never shunt the electrolytic with a lower voltage, as the capacitor will become very warm. When hum is heard in the speaker or hum bars are found on the TV screen, shunt all electrolytic capacitors found in the low-voltage power supply. Shut the chassis down. Clip the new capacitor across the suspected one, so as not to damage other solid-state devices.

Check the electrolytic for leakage with a shorted or leaky silicon rectifier (Fig. 2-20). The capacitor should charge up and then down with an ohmmeter measurement. Replace the leaky capacitor with a low measurement. If a low measurement of 1K ohms is found across the capacitor terminals, remove the positive lead on the capacitor and take another reading. Sometimes the circuits connected to the same voltage sources can cause a low ohmmeter measurement. An overloaded circuit connected to the same capacitor can cause a low reading. Then check where the voltage source ties to the various circuits to determine which component is lowering the voltage and resistance.

The electrolytic filter capacitor can be tested within a commercial capacitor tester that can handle large-capacity electrolytics. Select a capacity test range that will test up to 2000 µFs. Most DMMs that have a built-in capacity tester will only go up to 10 or 20-µF ranges.

You can check the electrolytic capacitor for charging and resistance measurements, if only a VOM or FET-VOM is handy (Fig. 2-21). Remove the positive terminal from the capacitor or remove the entire filter capacitor and take a resistance measurement. A normal electrolytic will charge up the meter scale and slowly discharge. A leaky capacitor will have a low measurement under 1K ohms with reversed test leads. If the capacitor will hardly charge up, discard and replace.

Replace electrolytic capacitors with the same capacity and working voltage rating as the original or replace with a universal replacement. For instance, a 5-µF 10-volt capacitor can be replaced with a 10-µF 15-volt electrolytic. Universal capacitors can be paralleled for greater capacity. When two or more electrolytic capacitors are found in one container, replace the entire unit when one is found defective.

Sub universal capacitors when the exact one is not available. If you have a dual 470 µF and 870 µF at 50 volts in one can, and the original is not available, replace it with a 500 µF and 1000 µF at 100 volts that is on hand (Fig. 2-22). Do not replace a defective capacitor with a lower capacity or working voltage. Remember, the capacity can be increased by paralleling capacitors in the circuit.

High-frequency capacitors

What and where

High-frequency capacitors are designed to work within the 3–30-MHz band or above. The high-frequency capacitor might consist of silver mica, ceramic disc, and

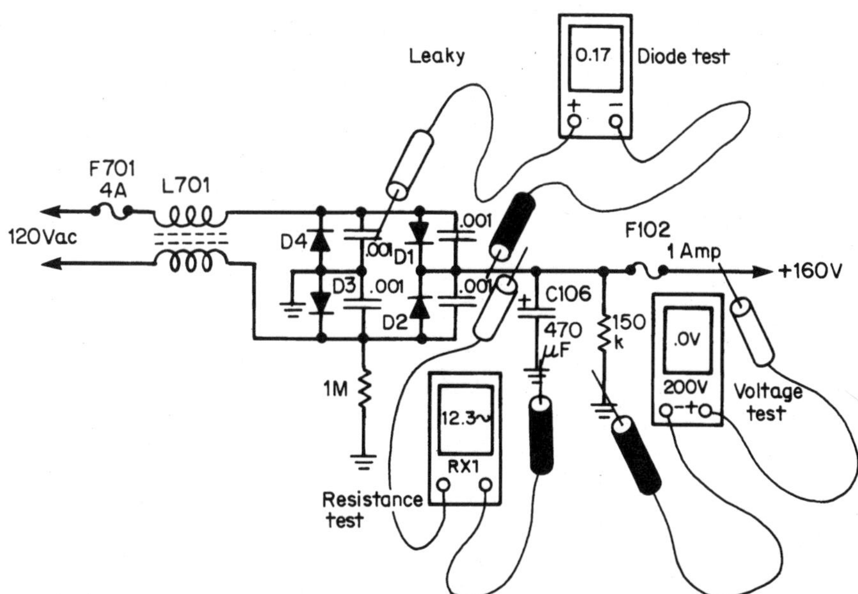

2-20 Leaky C106 (470 µF) destroyed D1, D2, and a (4-amp) fuse in the TV low-voltage power supply.

2-21 Check the large filter capacitor for a charge up and discharge on the FET-VOM.

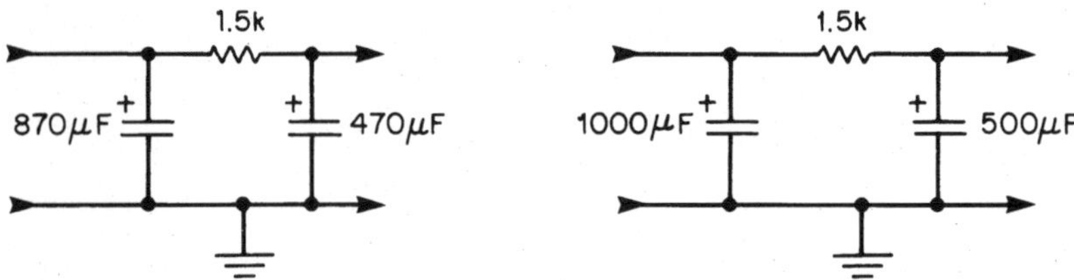

2-22 If an exact replacement for a 470- and 870-µF capacitor is not available, sub with a dual 500- and 1000-µF at 100 volts.

miniature electrolytic units. These capacitors have a very low impedance at high frequencies and very stable characteristics in the low temperature range. The capacitor tolerance is quite accurate at the higher frequencies.

High-frequency capacitors are found on the RF and oscillator circuits of shortwave and communication receivers operating within the high-frequency range. These capacitors will not change value or shift frequency as the different bands are turned on. The polarized high-frequency/high-temperature capacitors are widely used in computer and television power supplies. Likewise, the computer-grade radial nonpolarized high-frequency/high-temperature capacitors are found in the flyback circuits of the TV or monitor circuitry. These capacitors can operate above 125 degrees Centigrade, as they are less susceptible to heat stress.

When

Although these high-frequency capacitors very seldom break down in low-voltage circuits, they have been known to become leaky within the flyback and yoke circuits of the TV chassis (Fig. 2-23). Remove the suspected capacitor and check for leakage on the 10K-ohmmeter range. Test each capacitor in a capacitor tester for leakage or a change in resistance. Replace high-frequency/high-temperature capacitors with the original replacement part number.

High-voltage capacitors

What and where

The high-voltage capacitor operates within the TV or microwave oven and may operate from 200 to 3500 volts. The working voltage of the high-voltage capacitor found in the microwave oven may be from 2000 to 4000 peak voltage (Fig. 2-24). Although these high-voltage capacitors rate low in microfarads, the voltage rating is very dangerous. Always discharge the high-voltage capacitor before testing any component within the microwave oven.

The high-voltage capacitor is located in the secondary winding of the high-voltage transformer of a voltage-doubter circuit. An HV diode rectifies the ac voltage with the cathode terminal tied to the common ground. The output dc voltage may be from 1800

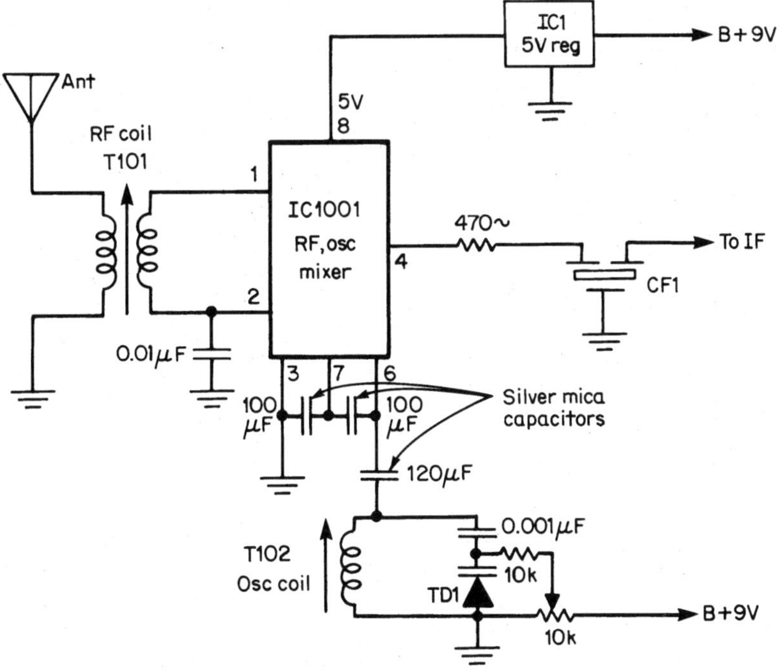

2-23 The high-frequency silver mica or monolithic capacitor are found in the RF and oscillator section of a shortwave receiver.

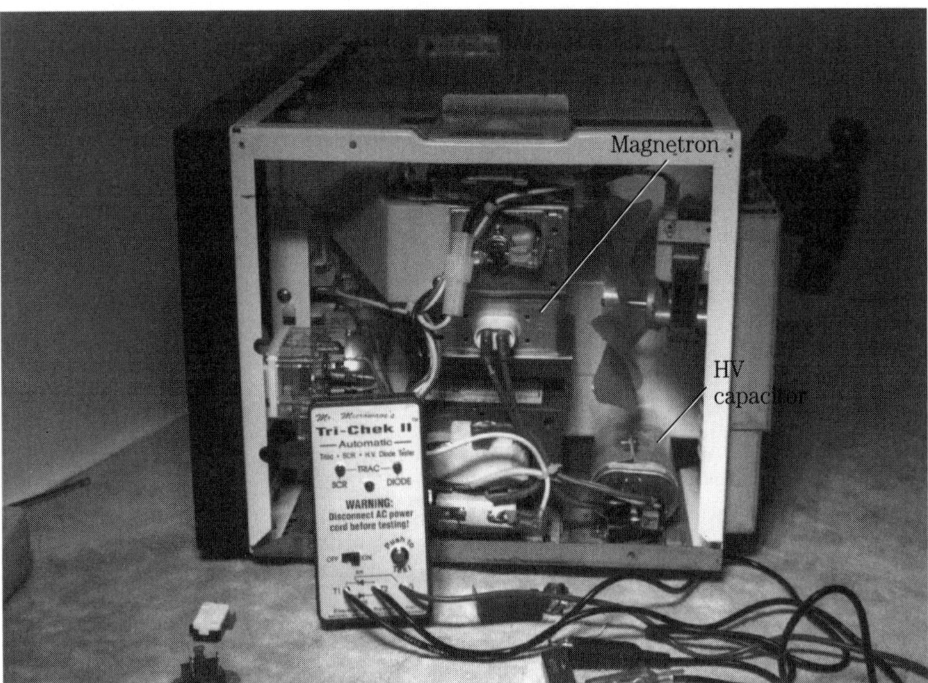

2-24 The high-voltage capacitor can be found in the microwave oven, within a voltage-doubler circuit.

to 4500 volts dc. In some microwave high-voltage circuits, a 10-megohm bleeder resistor discharges the high voltage to ground after the oven shuts down (Fig. 2-25). The HV is connected to the heater terminals of the magnetron tube.

When

When the microwave oven will not cook, blows fuses, or operates intermittently, suspect a defective HV capacitor magnetron or diode. Most oven high-voltage problems can be solved with a high-voltage test. The small VOM or DMM meters should never be used to check voltages within the oven. Keep all small meters out of the oven and use them for taking resistance-line voltage and continuity tests (Fig. 2-26). Do not attempt to use an analog meter (VOM), even if it measures up to 3500 volts. Use the Magnameter. It's safe and designed to measure the high voltage in the microwave oven (Fig. 2-27).

Testing

The Magnameter enables the technician to make both the HV and the plate current of the magnetron and HV circuit in one setup. Discharge the HV capacitor before connecting any test instrument into the circuit. You don't have to handle or change test leads during the voltage and current checks. When no HV or current is present on the meter, measure the resistance across the silicon diode to common ground. A low resistance-meter reading indicates a leaky capacitor, diode, or magnetron. Next take a resistance measurement across the HV capacitor. A low resistance measurement indicates a leaky or shorted HV diode (Fig. 2-28).

The Magnameter not only measures high voltage but also indicates how much current the magnetron is pulling. The current of most domestic microwave ovens will vary from 160 to 400 milliamperes, while commercial ovens might have a plate current from

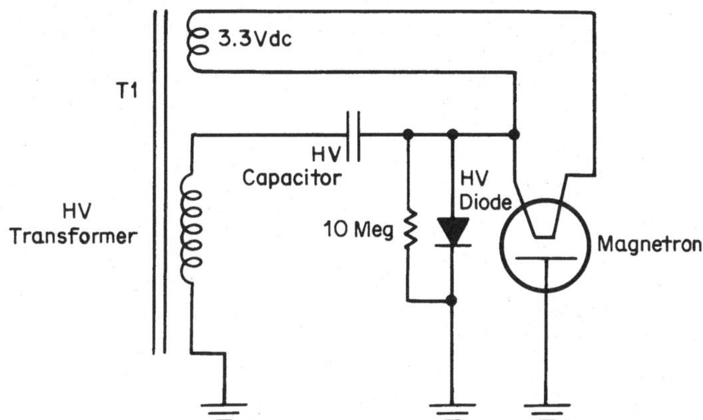

2-25 A 10-megohm resistor is found from the HV to ground to leak off charges from the HV capacitor after the oven shuts down.

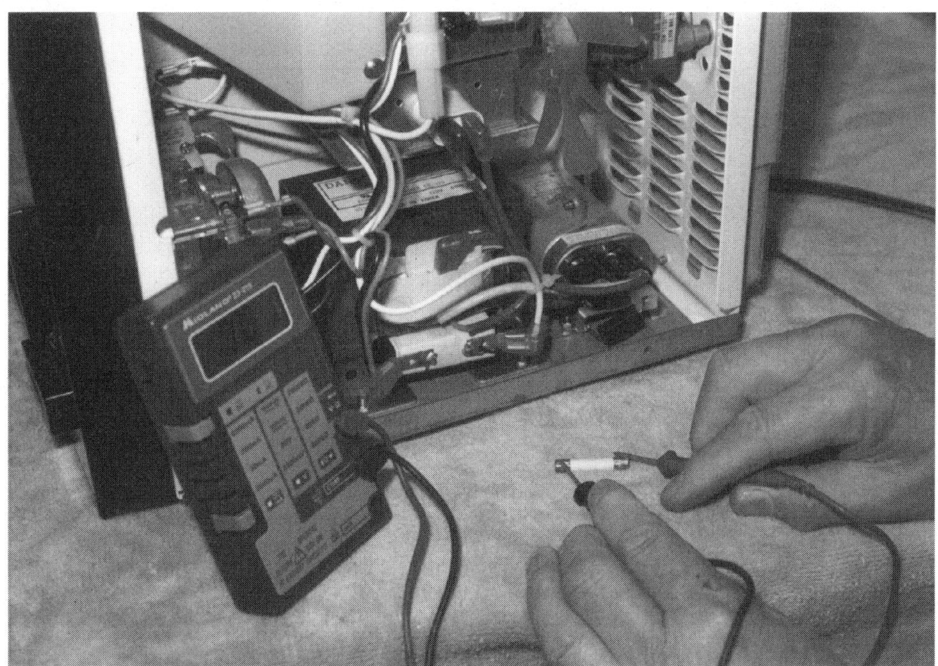

2-26 Keep those small meters out of the microwave oven and use them to check fuses and for continuity tests.

2-27 Use the Magnameter to measure high voltage within the magnetron circuits.

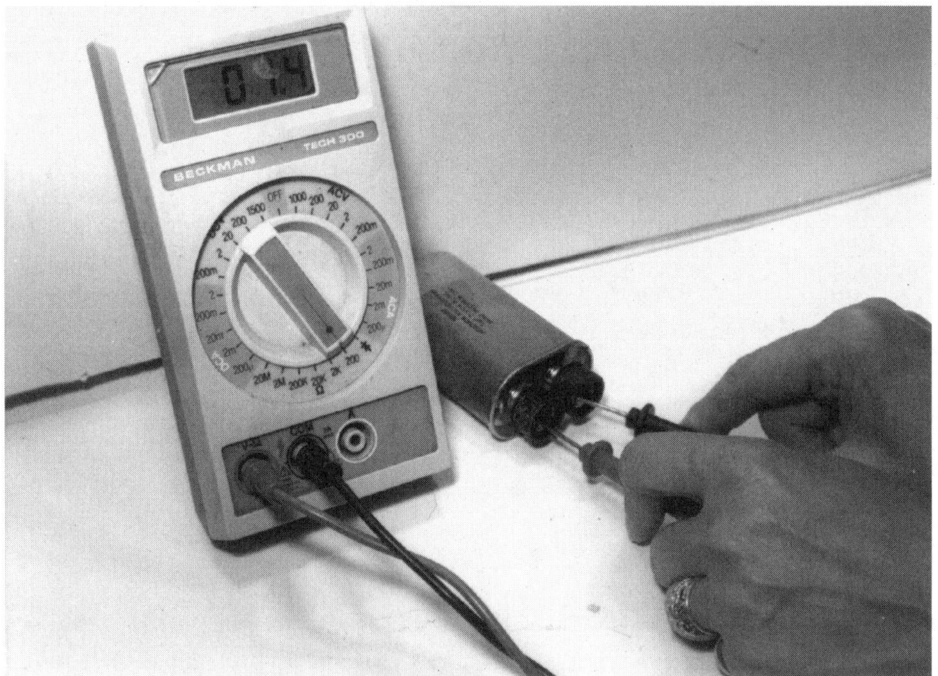

2-28 Here a microwave oven HV capacitor shows a leakage of 1.4 ohms.

200 to 750 milliamperes. The current meter is connected across a 10-ohm test resistor that is clipped in series with the HV diode. Simply remove the diode cathode end that connects to the common ground of the oven. Place the 10-ohm resistor in series with diode and ground connection (Fig. 2-29).

The Magnameter will quickly indicate problems in the high voltage or outside the magnetron circuit. Erratic meter movement will indicate an intermittent magnetron tube or poor filament connections. Low current reading might indicate a weak magnetron or low voltage. Both of these conditions may cause the oven to take too long to cook foods (Fig. 2-30).

TV safety capacitors

What and where

The safety or hold-down capacitors are located within the collector circuits of the horizontal output transistors or primary of the horizontal output transformer (flyback). These high-frequency/high-temperature capacitors can be mylar, ceramic, polyester, or disc-type capacitors. Since these capacitors have critical values to hold down the voltage generated by the flyback, they should be replaced with the exact part number, if possible (Fig. 2-31). These capacitors may vary from 0.0105 to 0.0068 at 1200 to 1600 volts. Try to replace the defective safety capacitor with exact capacity and a 1600-volt replacement. Most safety capacitors have axial terminals. In the older TV chassis, a four-legged safety capacitor is found in these circuits.

When

The high-voltage safety capacitor might become leaky, shorted, or open up. In the early ceramic types, the end of the capacitor would break the foil internally and

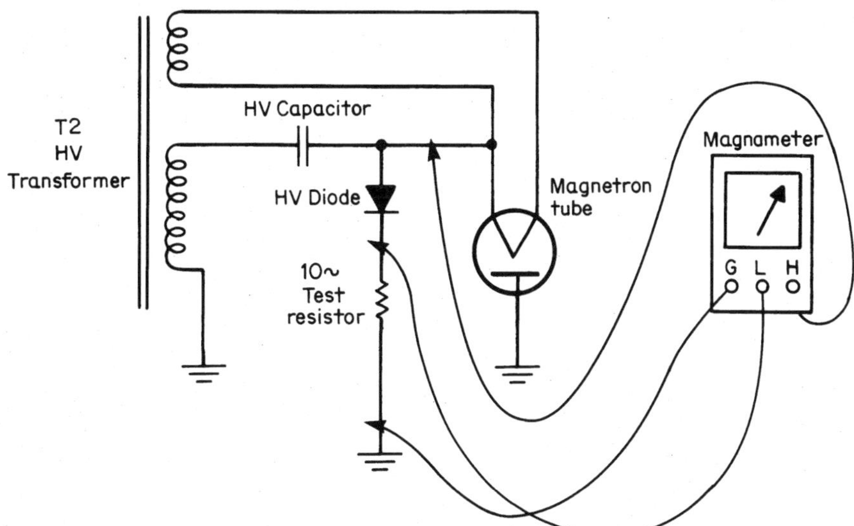

2-29 Insert a 10-ohm, 20-watt test resistor and connect the Magnameter to the HV circuit to test high voltage, current, and whether the magnetron is operating.

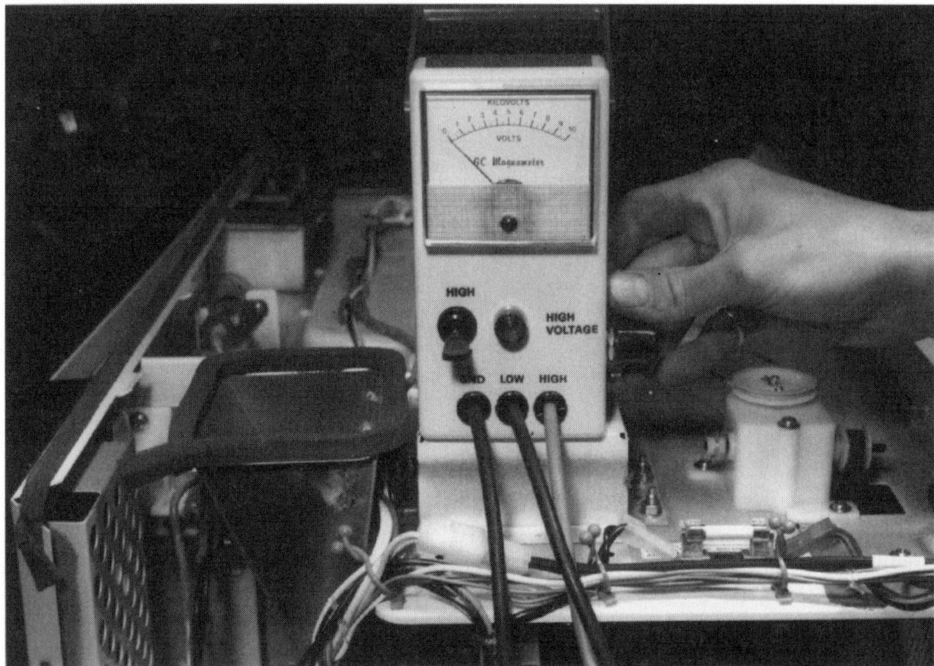

2-30 The Magnameter will spot a defective magnetron tube with low or no current measurement.

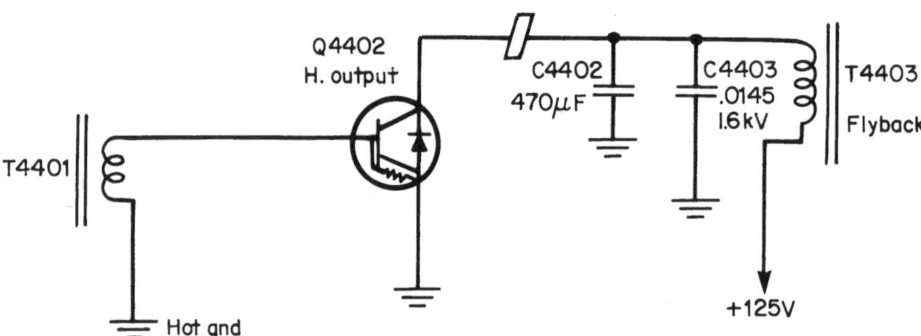

2-31 C4403 is the safety capacitor found in the horizontal output and primary winding of the flyback to hold down the high voltage.

open up. When the safety capacitor opens up, high voltage applied to the anode of the picture tube will go sky high. Excessive arcing or cracking is found at the picture tube, flyback, and anode cable. When the safety capacitor becomes shorted or leaky, the horizontal output transistor is damaged and the main fuse may blow (Fig. 2-32). The TV chassis will shut down. An intermittent safety capacitor can cause an intermittent cracking sound at the anode button of the CRT.

Testing

When excessive HV is indicated and the chassis shuts down, go directly to the safety capacitor. Plug the ac cord into a variable power line transformer with a voltage measurement at the anode of picture tube. Slowly raise the ac voltage and notice if the voltage has increased rapidly when about 85 volts of line voltage is applied to the chassis. Keep raising the voltage and notice at what line voltage the TV shuts down.

Check the safety capacitor for leakage when a fuse blows or the output transistor is damaged. Usually, the horizontal output transistor collector shorts to the emitter terminal. Remove the output transistor from the circuit for accurate resistance measurement. Test the capacitor for open tests on the capacitor tester. Simply remove and replace the safety capacitor when high voltage shuts down the chassis. Replace safety capacitor with the same value and at least 1600 volts (1.6 KV) rating.

Other fixed capacitors

Memory capacitors

What and where

Memory capacitors are found in computers, clock, and back-up circuits. These capacitors are found in the VCR and other electronic equipment to maintain clock and other presets when power is removed. The memory capacitor may have a capacitance of .047, 0.1, 0.22, 1, and 1.4 microfarads at 5.5 volts. The memory capacitor is short and squatty with radial terminals. A computer grade capacitor may have a capacity range from 5000 to 89,000 microfarads.

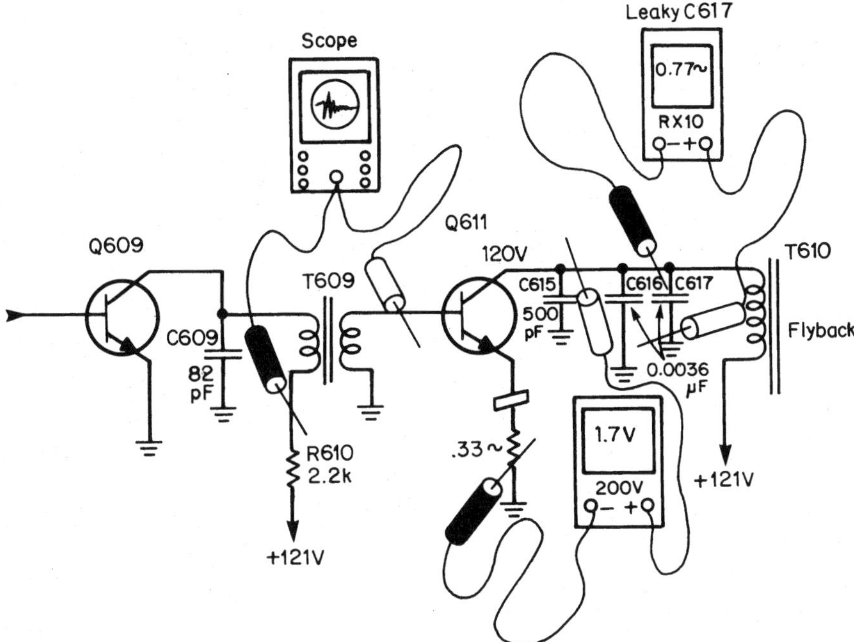

2-32 Leaky C617 (0036 µF) can destroy Q611 and may blow the secondary and line fuses.

Plastic film capacitors

What and where

The nonpolarized electrolytic capacitor is found in speaker crossover networks (Fig. 2-33). These capacitors can be constructed with mylar and ideal for custom crossover speaker connections. A plastic film or mylar capacitor has a mylar substance for a good dielectric. Crossover capacitors usually have axial type terminals.

The plastic film capacitor can also be made of polyester, polyethylene, and polystyrene. These plastic capacitors are used in audio and radio frequency circuits. The most used capacitor is from 0.001 to 10 µF.

Tantalum capacitors

What and where

The tantalum capacitor is another electrolytic capacitor made up of tantalum material instead of aluminum. These capacitors have high reliability and efficiency. They can be found in audio-frequency and digital circuits in place of aluminum electrolytics. Small plastic film and disc capacitors should be checked in the capacity tester for correct capacity. Check the capacitor for leakage with a 10X1 range. The open capacitor will not have any signs of charging or any indication on the capacity tester. Replace the capacitor if it is suspected of being intermittent.

Trimmer capacitors

What and where

The trimmer capacitor is a low-valued capacitor that can be adjusted to trim up the frequency, so to speak. A trimmer capacitor is usually operated in conjunction with a larger variable capacitor (Fig. 2-34). This small capacitor has only a few plates or elements and may be compressed with a screw into a ceramic base. The trimmer capacitor may vary from 1.5 to 50 pF.

When

Very seldom does the trimmer capacitor break down. Sometimes the dielectric insulating material will break down and fire through in higher-voltage circuits. When the mica insulator breaks off, the two metal plates may short against one another. When moisture or liquid gets between the plates and insulation, the tuning circuit is

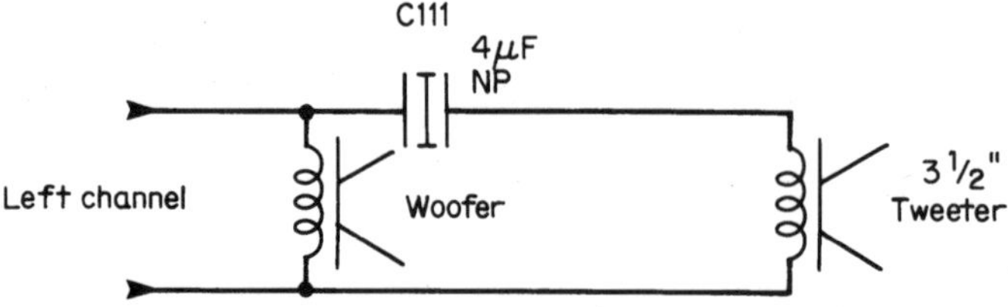

2-33 The nonpolarized electrolytic capacitor (C111) is wired between the 3-½-inch tweeter and main woofer speaker.

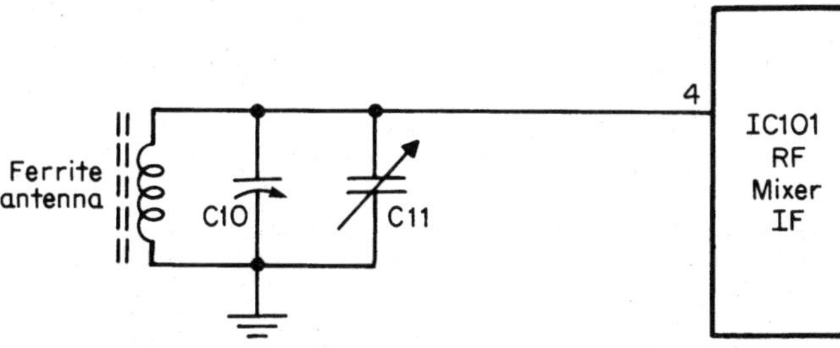

2-34 C10 trims up the RF antenna coil, mounted on top of C11, in the early radio chassis.

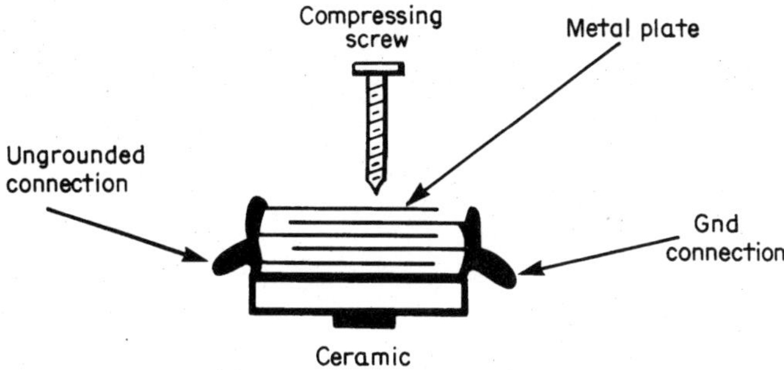

2-35 The trimmer capacitor found in the auto radio can cause weak reception when water enters the radio or antenna lead in.

unstable, like water dripping down into the car radio (Fig. 2-35). Do not try to repair trimmer capacitors. Simply replace with a new one.

Variable capacitors

What and where

A variable capacitor can be changed from a low value of capacity to a higher one by rotating the plates into the stator section. The insulation of variable capacitors is air, and sometimes they are called air-tuning capacitors (Fig. 2-36). The rotor plates are rotated into the stator plates, increasing the capacity in picofarads (pF). Several different sections of plates may be connected to the same dial shaft and all sections turned at the same time. These gauged variable capacitors were found in the early radio and shortwave receivers to tune in the various broadcast stations. Today, the variable capacitor has been replaced with varactor diode tuning. (See variable diode, Chapter 5.)

When

The variable capacitor may collect dust and dirt between plates, shorting out the capacitor. With dirt and dust between the stator and rotor plates, suspect noisy tuning.

2-36 The rotor plates are rotated into the stator plates to tune the broadcast band in early radio receivers.

A rotating noise can be caused by a dry bearing. Clean it up with cleaning spray. Blow out the dust with a hair dryer or spray can to eliminate noisy tuning. When one or more rotor plates begin to touch one another, there will be an intermittent signal, or no signal can be tuned in.

Testing

Check the capacitor by connecting a RX1 range VOM or FET-VOM across the stator and rotor terminals. Remove the stator wire from the capacitor section before testing. Now rotate the capacitor and watch the meter hand (Fig. 2-37). If the hand moves or shows that the plates are touching, a very low-resistance measurement is made. The variable capacitor can also be checked on the capacity tests in the pF range. Try to straighten up and clean the plates for infinite resistance measurement. A good air cleanup with lubricating light oil in each stator bearing can solve most variable-capacitor problems.

SMD capacitors

What and where

The surface-mounted device (SMD) or capacitor is found in the same circuits as other capacitors. This surface-mounted capacitor is soldered directly on top of the PC wiring. These miniature capacitors are designed to fit into compact areas. The combination portable AM-FM radio and tape player might have surface-mounted capacitors in just about every radio circuit. A ceramic disc or electrolytic capacitor is the most commonly used SMD capacitor (Fig. 2-38). Surface-mounted replacement or universal capacitors can be purchased in 5 or 10 per bag.

The SMD case multilayer chip capacitors can be obtained in 0.01-, 0.022-, 0.1-, 0.22-, and 1000-pF values. Tantalum chip electrolytics can be purchased from 1 µF

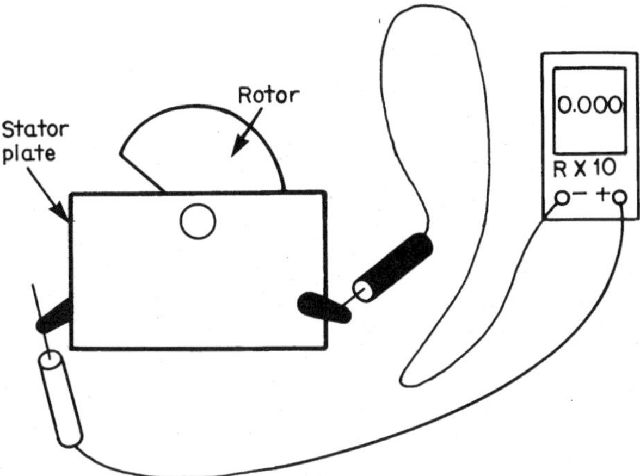

2-37 Connect the ohmmeter across the stator and rotor terminals. Rotate the tuning capacitor and notice if any reading indicates that the plates are touching one another.

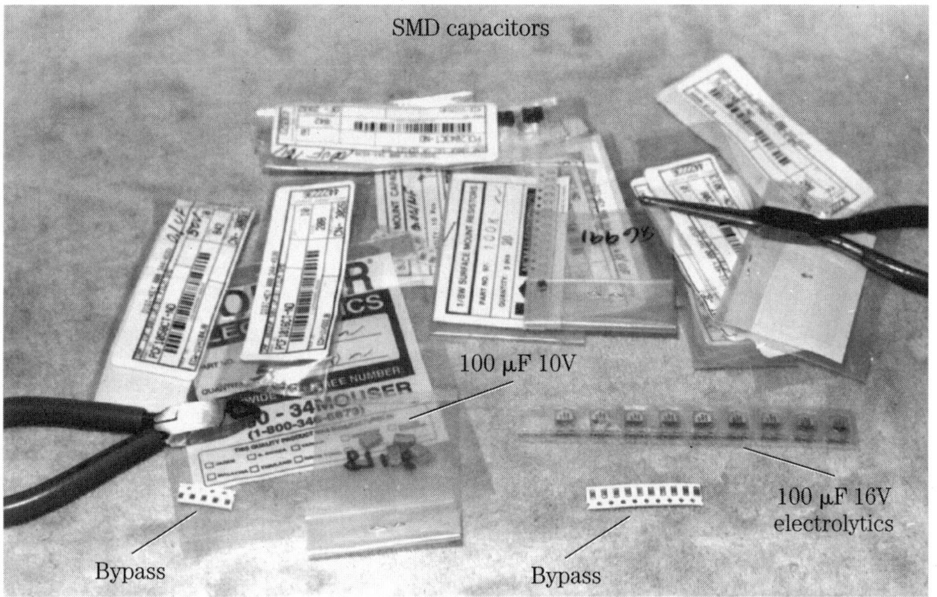

2-38 The bypass and electrolytic SMD capacitors are tiny.

to 47 µF at 16 or 50 volts. Surface-mounted electrolytic capacitors can be selected from 1 µF to 100 µF in 10- and 16-volt ranges (Fig. 2-39). Always choose the highest voltage range for replacement. You may find that a 10-volt electrolytic SMD capacitor will break down with an operating voltage of 6 volts. Always replace SMD capacitors with the manµfacturer's part, if possible.

When

The SMD capacitor has a tendency to become intermittent with poorly soldered or broken PC connections. Inspect the soldered ends and PC wiring right where the end is soldered to the capacitor. SMD capacitors can short, become leaky, and appear open.

Testing

These capacitors can be tested with voltage, resistance, and capacity testers. Do not apply too much heat to the end of the capacitor, or the whole end contact will come loose from the multilayer capacitor. Make sure each end is covered with solder and connected to PC wiring. Test each end with a RX1 resistance measurement. Use sharp pointed test probes. Check the soldered connection under a lighted magnifying glass.

2-39 The SMD electrolytic capacitors come in 10- and 16-volt types that solder directly to the PC wiring.

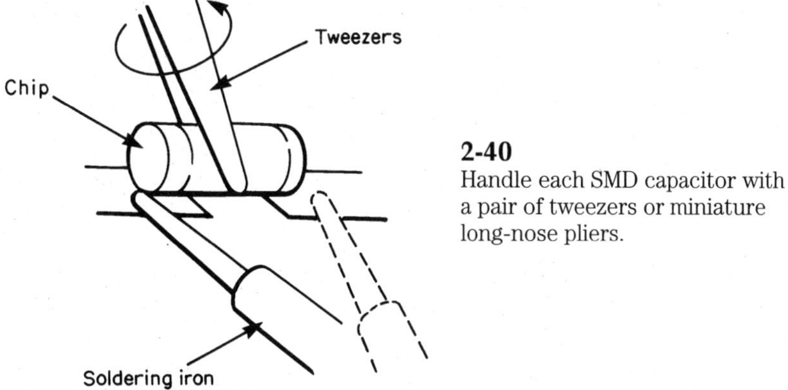

2-40
Handle each SMD capacitor with a pair of tweezers or miniature long-nose pliers.

Handle each SMD capacitor with a pair of tweezers or miniature long-nose pliers (Fig. 2-40). Remove only one SMD part from the package and lay it on a piece of white paper so you can see it. These small SMD capacitors can be easily lost. Place the SMD capacitor on the soldering pad and solder up each end with a low wattage or controlled soldering iron. Do not press down on the SMD, as you can damage or break capacitor end connections.

Conclusion

All fixed capacitors can be tested with voltage, resistance, signal, charging, and capacity checkers (Fig. 2-41). Even the SMD capacitor can be tested with the charging test. Remove one end of the capacitor for a correct in-circuit test. Remove radial-type capacitors from the circuit to be checked on the capacitance tester. Sub the suspected electrolytic capacitor when you hear hum in the sound or hum bars are found in the picture tube. When an intermittent capacitor is suspected, remove and replace it with the same or higher value in capacity and working voltage.

The fixed mica capacitor may have a color-coding dot marking found on the body of the capacitor (Table 2-1). Usually the first dot (A) is either white or black, indicating a mica capacitor. Dot B indicates the first number of the capacitance value, and dot C is the second figure. The multiplier is given by dot F. These numbers are given in picofarads and to convert to microfarads, divide the number by 1,000,000. Dot D represents the temperature coefficient, while dot E is the tolerance or + and − charge of the value.

If you have located a leaky mica capacitor with dot A as black, dot B as brown, dot C green, dot E silver, and dot F brown, the mica capacitor value is 15 × 10 equals a 150-pF capacitor at 10% tolerance. Of course, the leaky capacitor will indicate leakage on the capacity tester and no actual value except the dot color-coding.

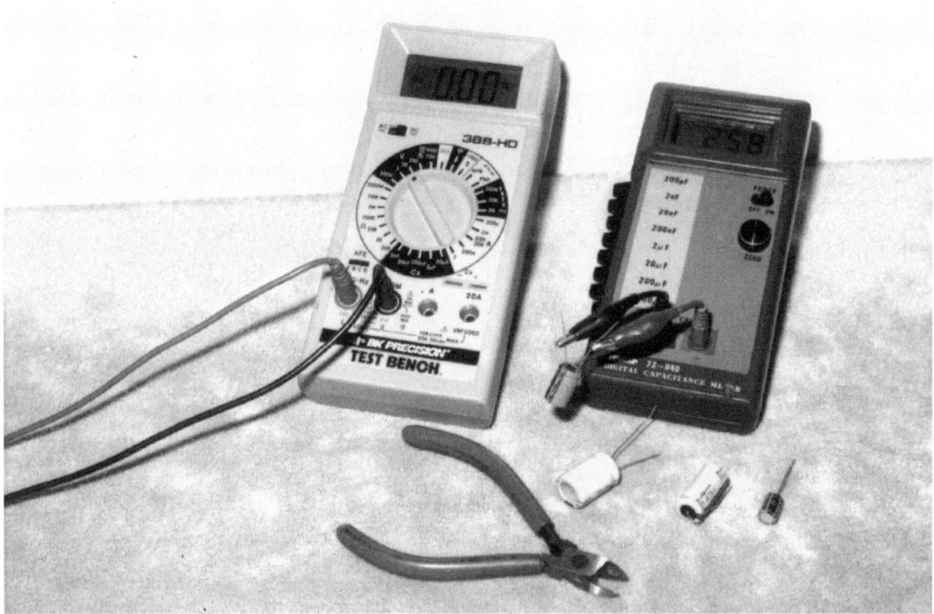

2-41 Here small coupling and decoupling electrolytic capacitors are tested in a capacitor checker.

Table 2-1. The fixed mica capacitors are color-coded with dots, indicating the value in capacity and tolerance

Color	Dot A	Dot B	Dot C	Dot D	Dot E	Dot F
Black	Mica	0	0	± 1000	± 20 %	×1
Brown	"	1	1	± 500	± 1%	×10
Red	"	2	2	± 200	± 2%	×100
Orange	"	3	3	± 100	± 3%	×1000
Yellow	"	4	4	± 20	±	
Green	"	5	5	± 10		
Blue	"	6	6			
Violet	"	7	7			
Gray	"	8	8			
White	"	9	9			

When repairing and servicing foreign products, the circuit diagram may have a different marking on the capacitors (Fig. 2-42). The different parts and components might have different symbols than found in American diagrams. The foreign publication and schematics might have the capacitors marked in nanofarads (nF).

Simply translate nanofarads to microfarads when replacing a defective capacitor within the electronic chassis. Likewise, when constructing a foreign electronic project with different capacity symbols, follow the capacitor chart in Table 2-2. Now see how the various components are tied together in the schematic.

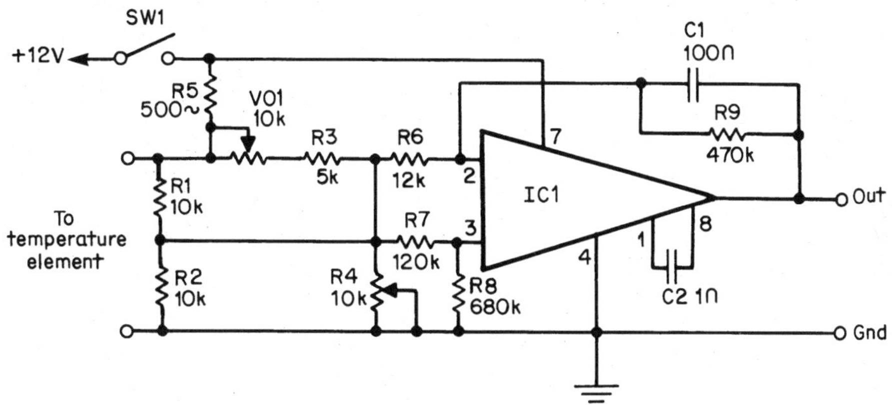

2-42 Foreign schematics may have a different capacitor symbol.

Table 2-2. A foreign capacitor chart

nF	μF
1.0	0.001
10.0	0.01
100.0	0.1
1000.0	1.0
10,000	10
100,000	100.0
1,000,000.0	1000.0

3
CHAPTER

Testing electronic components C

(Cartridges through circuit breakers)

The defective phono cartridge can produce distorted music, while the dynamic cartridge needs another stage of amplification before it can be heard. Today, the ceramic filter is found in radio and TV IF circuits. RF chokes prevent RF signal, and audio and filter chokes smooth out the ac ripple in the low-voltage power supply. There are many different circuits in the cassette player, radio, and TV chassis. In the early TV chassis, the circuit breaker replaced the power-line fuse. All of the above is discussed in this chapter.

Cartridge

Phono-cartridge

What and where

The crystal cartridge or pickup is found within the portable or combination AM-FM-MPX receiver that has a record player (Fig. 3-1). The crystal is attached mechanically to a stylus or needle, whose movement follows the groove of the record. By twisting or moving the crystal element, voltage is developed and applied to an audio amplifier so one can hear the music recorded upon the record. The piezoelectric crystal can be damaged if the pickup arm is accidentally dropped upon a hard surface.

3-1 The phonograph might have a crystal or magnetic cartridge.

The piezoelectric crystal can be made of quartz, Rochelle salt, or other materials that produce a voltage when force is applied. The crystal cartridge output voltage may be as high as 3 volts. A phono-crystal is located at the end of a phonograph pickup arm and is usually mounted with two mounting screws. The defective pickup crystal may produce distorted, intermittent music or no sound. These crystal cartridges can melt down and produce distorted music under 130 degrees. The portable phonograph should not be placed directly where the hot sun shines upon it.

When

The crystal cartridge can become damaged in three ways: dropping the pickup arm, treating the arm roughly, or leaving it in too much heat. A dirty stylus or needle can cause distortion or fuzzy music (Fig. 3-2). Excessive dirt between the stylus and cartridge might produce tinny and weak sounds. First clean around the stylus with a small brush. Look at the stylus under a magnifying glass to determine if the point is worn or chipped. A chipped diamond stylus can cause scratchy music and might gouge out the recording.

Testing

Check the crystal cartridge by connecting it to an external amplifier. You may find one stereo channel distorted and damaged with the other channel normal. Suspect a weak crystal cartridge with very low or no volume and music. The crystal cartridge can only be checked for leakage with the ohmmeter. No continuity should be measured with a normal cartridge (Fig. 3-3). Excessive motorboating sounds might be heard with a defective cartridge.

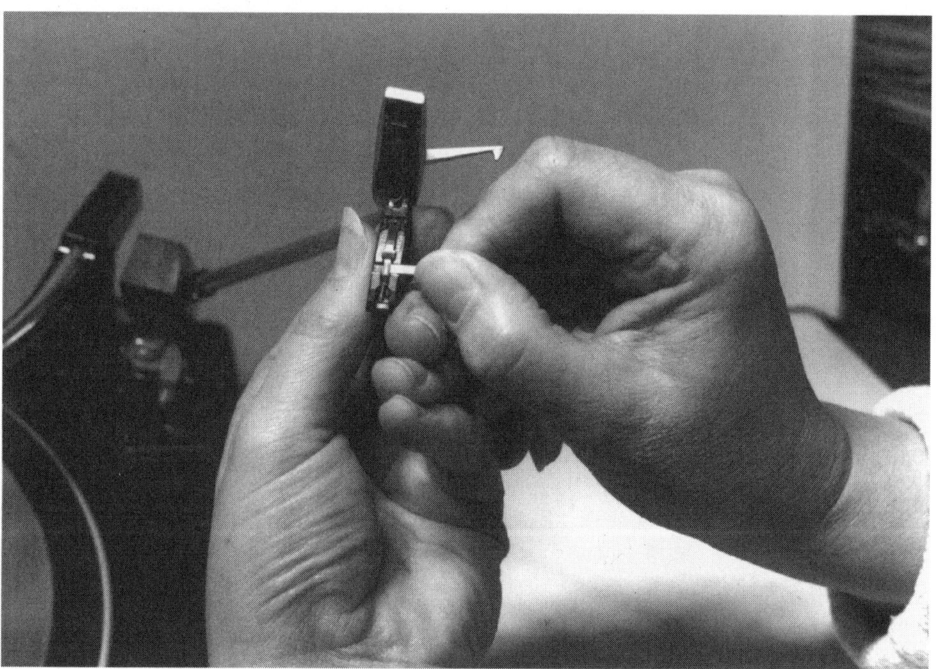

3-2 Check the stylus for excessive dirt and dust, which cause mushy and distorted music.

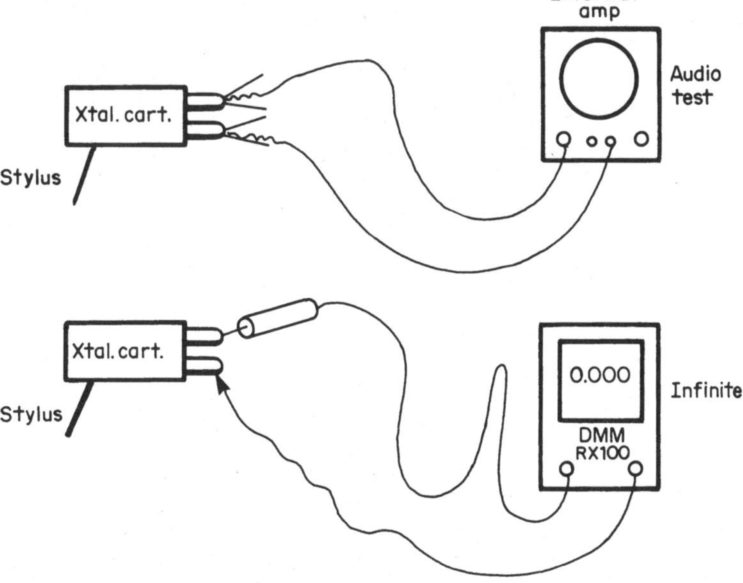

3-3 Test the crystal cartridge with the external amp and resistance measurements.

Replace the cartridge if in doubt. Check the voltage and universal replacement in a crystal replacement manual. Be very careful replacing the cartridge or the stylus. Some needles are removed by pulling straight out. Sometimes the stylus might slip under a metal keeper to hold it in position. When removing the needle, be careful that the small saddle attached to the crystal or moving metal vane is not broken. Replace with original part number when possible.

Dynamic cartridge

What and where

A defective dynamic cartridge may cause the same symptoms as the crystal cartridge. The magnetic cartridge may be called a variable-reluctance pickup. The dynamic cartridge has two moving vanes with coils in the latest phonographs (Fig. 3-4). The metal vane is attached to the stylus. When the metal vane moves within the record groove, it changes the magnetic field of the nearby coils. Two separate coils are found in the stereo magnetic pickup. The coil leads are attached to the magnetic cartridge and to the amplifier. Another stage of amplification is needed for the magnetic cartridge since the pickup signal is very weak compared to a regular crystal cartridge.

Testing

Take a continuity test of each winding when the magnetic cartridge is suspected of causing distortion or weak sound (Fig. 3-5). First brush out all dirt and

3-4 The dynamic or magnetic cartridge has coils that are found in the latest phonographs.

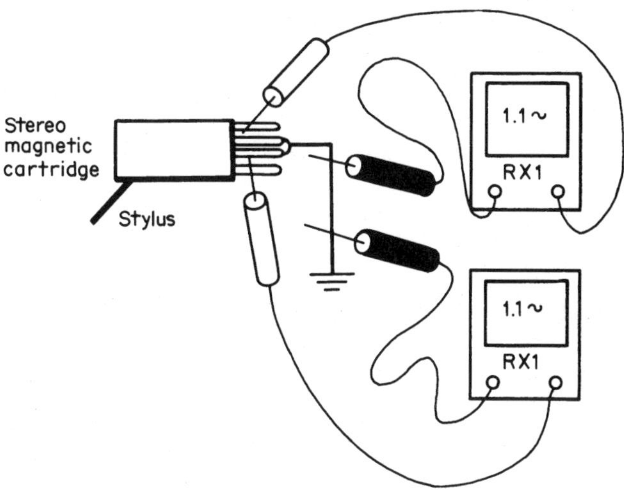

3-5 Take a continuity or resistance measurement when suspecting a defective magnetic cartridge.

dust with a small artist's brush. Check the resistance of each coil with the RX1 range of the DMM, if the signal is still distorted. This resistance should be very low and comparable to the other channel. Replace the cartridge if one side measures 2 ohms and the other 100 ohms. Both stereo cartridge resistances should be the same or comparable in resistance. Push up and down on the tone arm and notice if the resistance cuts up and down upon the meter (Fig. 3-6).

Intermittent or erratic phono reception might be caused by broken wires or loose clips on the cartridge. Check each connection at the cartridge terminals. Hum might be present if there is a broken negative or black ground wire. Measure the continuity of each channel with the ohmmeter where the female phono plugs are found on the turntable. This test will detect an open magnetic cartridge or broken wire lead. Remember, crystal cartridges should have infinite resistance while a magnetic cartridge will have a very low resistance measurement.

Ceramic filter

The ceramic filter may take the place of a set of RF-video coils in the IF stage of the TV set. Today, the ceramic filter network takes the place of an intermediate frequency (IF) transformer within the radio or shortwave receiver (Fig. 3-7). Usually, the ceramic filter causes very few problems and failures. The ceramic filter contains a fixed frequency made up of piezoelectric ceramic material.

Radio IF sound ceramic filter

What and where

The ceramic filter found in the auto, clock, and shortwave radios is usually found in the AM-FM IF circuits. The ceramic filter may follow a regular IF transformer or is

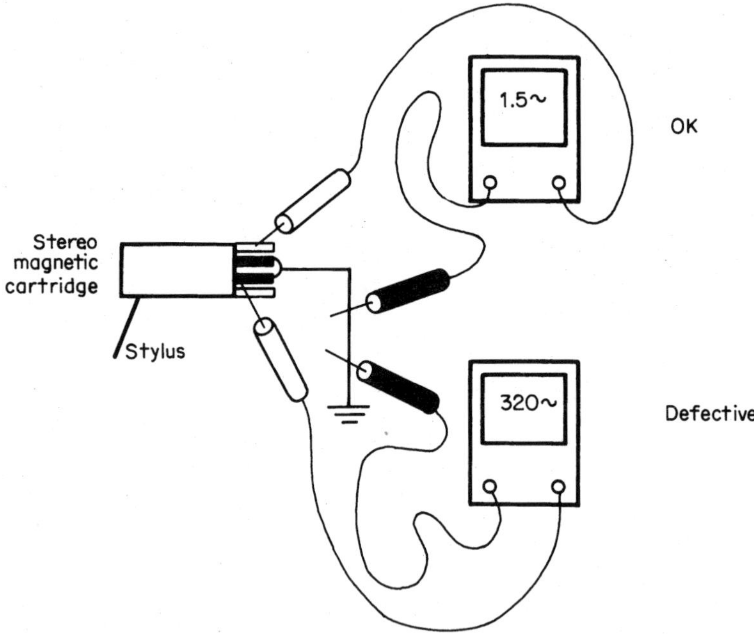

3-6 Both magnetic coils in the stereo channels should have the same resistance.

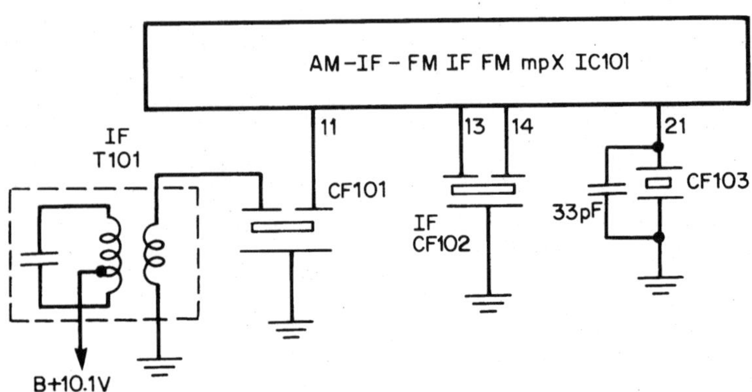

3-7 Instead of IF transformers, ceramic filters are found in present-day radio receivers.

mounted in a separate IF circuit. The ceramic filter is cheaper to manufacture and takes up less mounting space. The ceramic filter found in the AM-IF circuit is cut at 455 kHz frequency, while in the FM-IF circuits the frequency is 10.7 MHz (Fig. 3-8). The ceramic filter has no adjustments as the IF transformer.

Testing

Check the ceramic filter with an RX10 resistance range. Infinite resistance should be read from input terminal and output terminal to common ground. Remove the input and output terminals from the circuit for correct resistance measurements. The input and output terminals can be tested upon a crystal checker on a weak, good, or bad scale.

TV video IF and sound ceramic filters

What and where

The ceramic IF filter within the video circuits of the TV chassis is located after or before a transistor IF preamp circuit. The ceramic filter may be called a saw filter in some TV chassis (Fig. 3-9). The ceramic filter is also found in some TV chassis within the sound IF-IC circuits. For instance, within the RCA CTC159 chassis, a saw ceramic filter is found between IF preamp transistor (Q2301) and input of IF-IC U1001, CF1201 is located in the sound circuit of pins 40 and 35 of U1001, and a ceramic comb filter (CF2601) is found between video switch U1401 and pin 49 of IF, chroma/luma processor U1001.

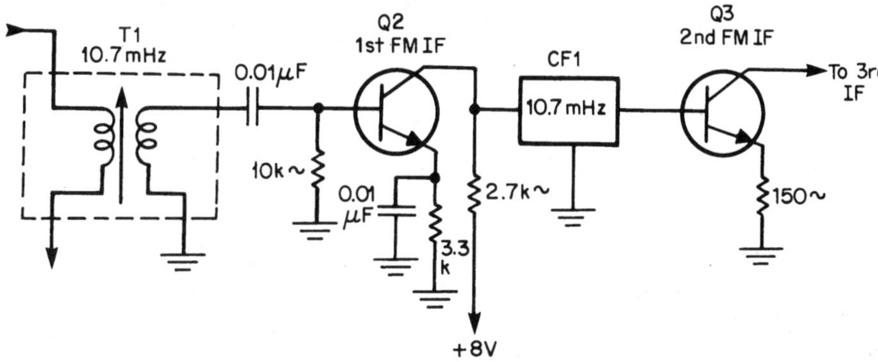

3-8 The FM-IF ceramic filters found in the clock or portable radio.

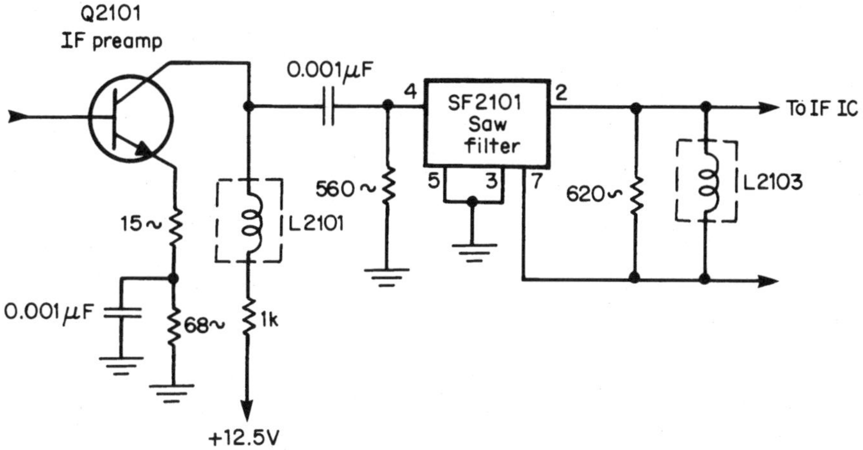

3-9 A saw filter network (ceramic) found in the IF circuit of a TV chassis.

Testing

Take critical in-circuit resistance tests between input and ground, and output and ground. If a low resistance measurement is found, remove one end from the circuit with solder wick and take another measurement. If in doubt, test the ceramic filter out of the chassis with a crystal tester (Fig. 3-10).

Ceramic resonator

What and where

Ceramic resonators consist of the piezoelectric material converting a mechanical stress into electrical energy or vice versa. The ceramic resonators are made from thin slices of quartz, and when current is applied a vibration results. Usually, ceramic resonators have only two electrodes.

Ceramic resonators may be found as clock generators for various microprocessor circuits, in telecommunication equipment, and in personal computers. High accuracy and stabilization result with a frequency shift ±3%. A Colpitts oscillator can be constructed from an IC4069, 1 megohm resistor, two 470-pF capacitors, and a ceramic resonator (Fig. 3-11). The output signal can be checked with a frequency counter or oscilloscope.

Testing

Ceramic resonators can be checked with a crystal checker and for leakage on the RX100 scale. Remove resonator out of the circuit for leakage or crystal checker

3-10 Check the suspected ceramic filter with a crystal checker.

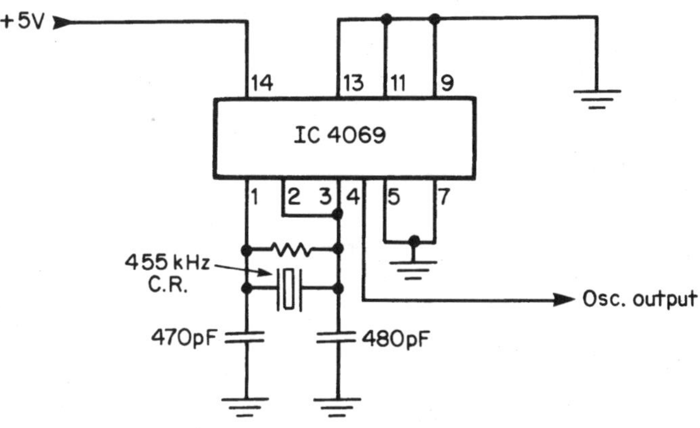

3-11 A 455-kHz ceramic filter found in a ceramic resonator Colpitts oscillator circuit.

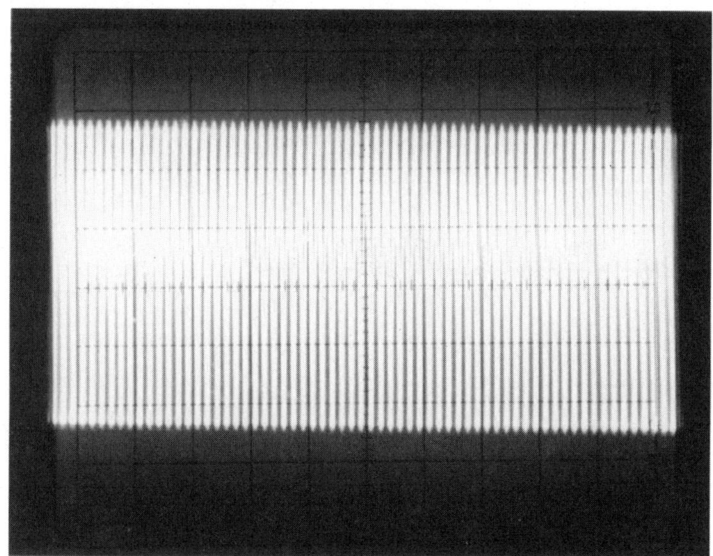

3-12 Check the output waveform of a crystal oscillator circuit.

tests. The oscillator frequency can be tested with a frequency counter or oscilloscope to see if the ceramic resonator is oscillating (Fig. 3-12).

Chokes

A choke coil provides high impedance to alternating current (ac) while passing direct current with practically no opposition. Choke coils are found on audio, radio, RF, and filter circuits. The RF choke coil prevents RF from entering the audio circuits and might pass DC current at the same time. A filter choke coil or transformer usually has a large inductance and smoothes out the pulsating current within the low-voltage power supplies (Fig. 3-13). Choke coils may have air, ferrite, iron core, and several laminations as in the filter-choke transformer.

Hi Q and RF choke coils

What and where

The hi Q choke coil has a high value for the rates of reactance to resistance (X/R). Usually, high Q and RF chokes are found in radio frequency (RF) circuits. The RF chokes may be molded types of electromagnetically shielded, subminiature iron core, miniature epoxy-coated chokes, micro-mini molded, ferrite and ferrite beads, and a three-section pi-wound portion. The RF wound choke coil can be substituted for RF coils in shortwave receivers (Fig. 3-14). The RF choke coils may start at 0.10 microhenries (μH) to 22 millihenries. A three-section 2.5-mH choke coil was used in the early regenerative receiver circuits (Fig. 3-15).

Testing

The RF choke may have a broken terminal lead, coil wire breakage where it connects to the terminals or becomes entirely open. Check the RF coil right in the circuit with an ohmmeter continuity test (Fig. 3-16). The high frequency RF choke coils may have less than 1-ohm resistance while the filter choke has several ohms of resistance. Finding the open winding right at the connection terminal wire can repair most open

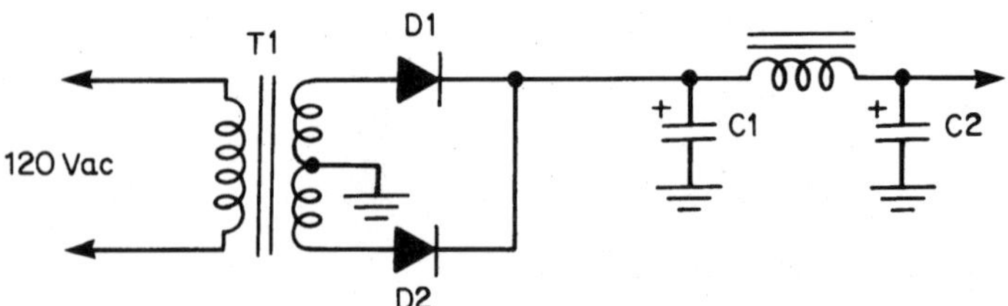

3-13 A filter choke is found in a fullwave rectifier circuit for smoothing out the ac ripple.

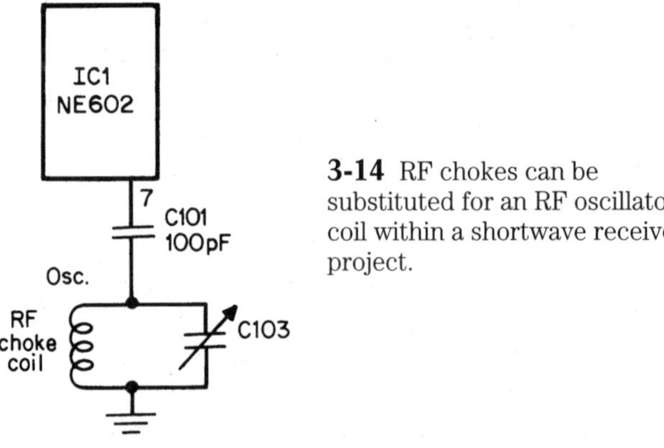

3-14 RF chokes can be substituted for an RF oscillator coil within a shortwave receiver project.

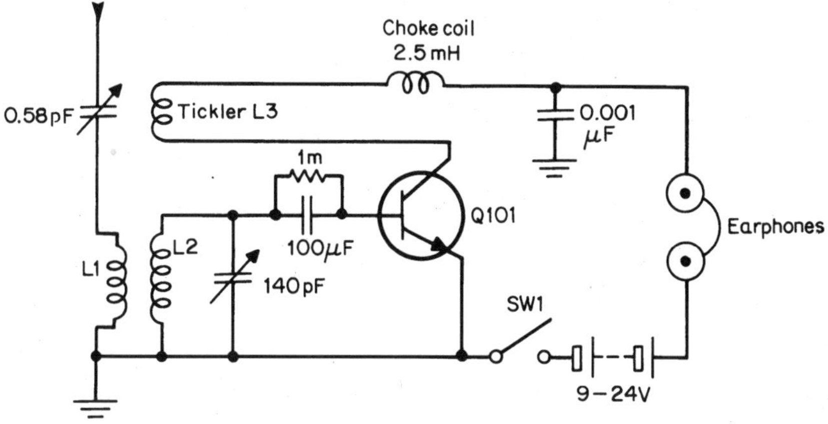

3-15 The RF (2.5-mH) choke coil was found in the early regenerative shortwave receiver.

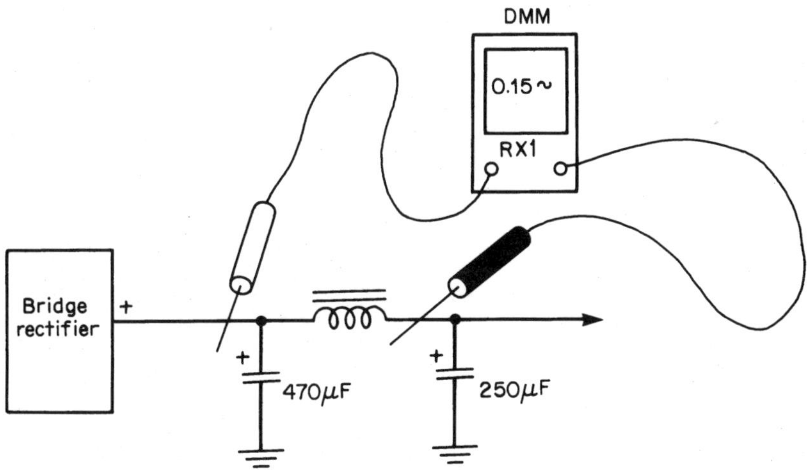

3-16 Check the RF choke coil for continuity with the RX1 scale of the DMM.

RF choke coils. Find the loose end and unwind one turn, scrape back the enamel or cotton cover, and resolder to the fixed terminal lead.

Heavy-duty hash chokes

What and where

The RF1 suppression coils or chokes might have an inductance from 1 µH to 10,000 µH. A high-current suppression coil or choke may have an inductance from 3.9 µH to 10,000 µH with a high dc current capacity from 0.30 amps to 12 amps. These hash or high-suppression choke coils might be wound over ferrite or iron oxide cores.

Heavy-duty hash chokes might be wound with axial terminal leads, and high-current chokes are wound on a ferrite or iron-core toroid form. Often the resistance will vary from 0.019 to 30 ohms. The hash and heavy-current filter chokes help eliminate electrical noise in power supplies, audio radio dc circuits, chargers, motors, tubes, and mercury-vapor tube circuits.

Testing

Check all choke windings with RX1 ohmmeter range of a DMM or VOM. Resolder poor choke terminals (Fig. 3-17).

Filter chokes

The filter choke or transformer is an inductor that provides inductive reactance in the power supply while passing direct current. A filter capacitor and choke smooth out the ripple found in a full-wave or half-wave rectifier circuit (Fig. 3-18). Half-wave rectification provides more ripple than full-wave. The half-wave ripple is 60Hz while the full-wave is 120Hz. Often the large filter choke is rated in henrys and carries heavy current. Large high-current filter chokes are wired with larger diameter wire.

What and where

The filter choke may be inserted after or before a filter capacitor. Two electrolytic filter capacitors can be connected at the input and output of a large filter choke for minimum ripple effect. This filter current may be also called a Pi-filter (Fig. 3-19). An

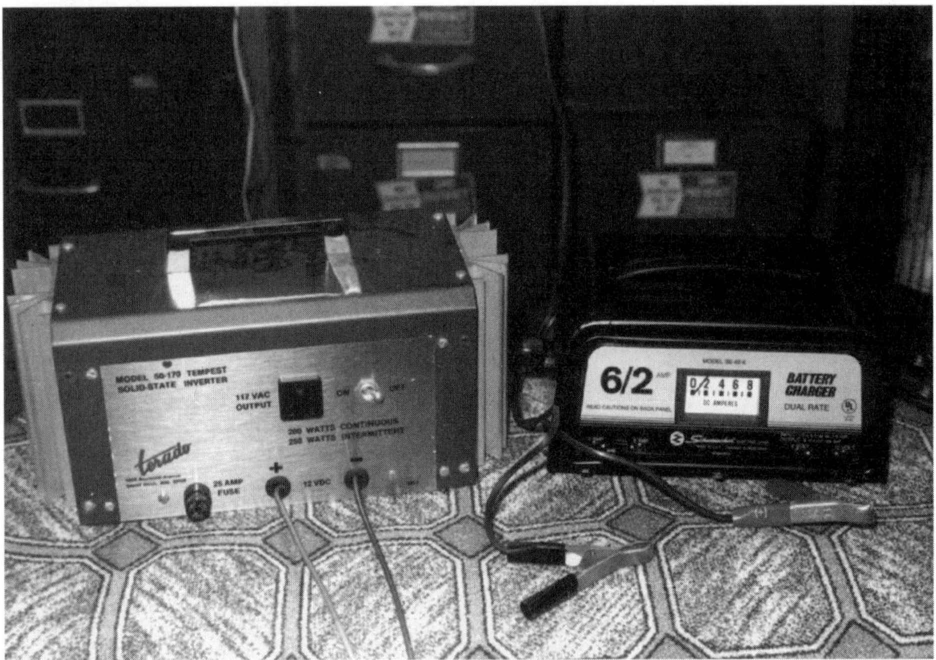

3-17 A large choke coil found in the auto battery charger.

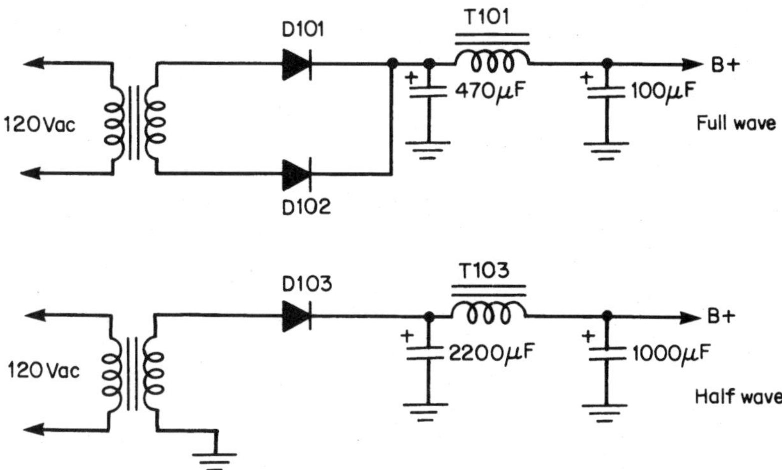

3-18 The filter choke can smooth out the ac ripple found in the full-wave and half-wave power supplies.

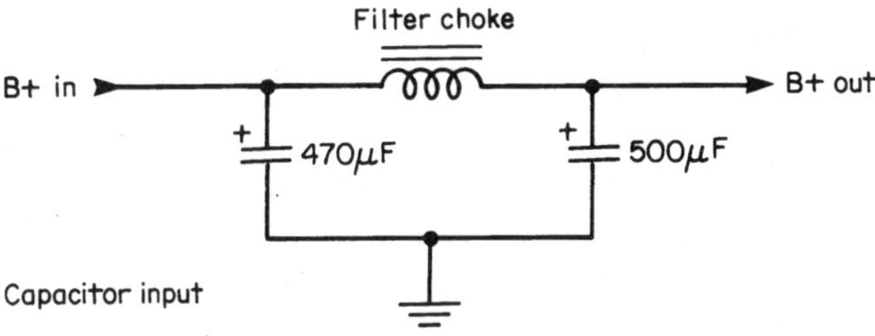

3-19 A pi-filter network with electrolytic capacitors on each side of a filter choke in the power supply.

electrolytic capacitor is mounted ahead of the filter choke in a capacitor-input circuit. When the choke is mounted ahead of the electrolytic capacitor, it is called a choke-input filter network (Fig. 3-20). You will find the output dc voltage is higher with capacitor-input than in a choke-input filter network.

In the early radio circuits, the field coil of an electrodynamic speaker served as a filter choke coil. This coil was wound around a metal core. When heavy current passed through the winding, the winding served as a choke coil and provided a magnetic field for the speaker (Fig. 3-21). The dynamic speaker voice coil is mounted over the same metal piece that the field coil was wound on. Today the dynamic speaker has a permanent magnet instead of a field coil.

When

The defective filter choke or transformer can produce 60 or 120Hz hum bars on a TV screen and loud hum in the audio. When the filter choke goes open, high voltage is found at the main filter capacitor and no voltage at the dc converted circuits (Fig. 3-22).

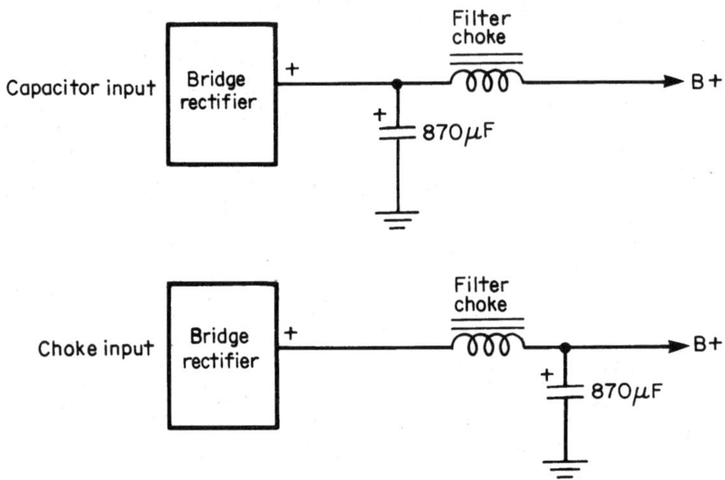

3-20 Capacitor input and choke input filter networks.

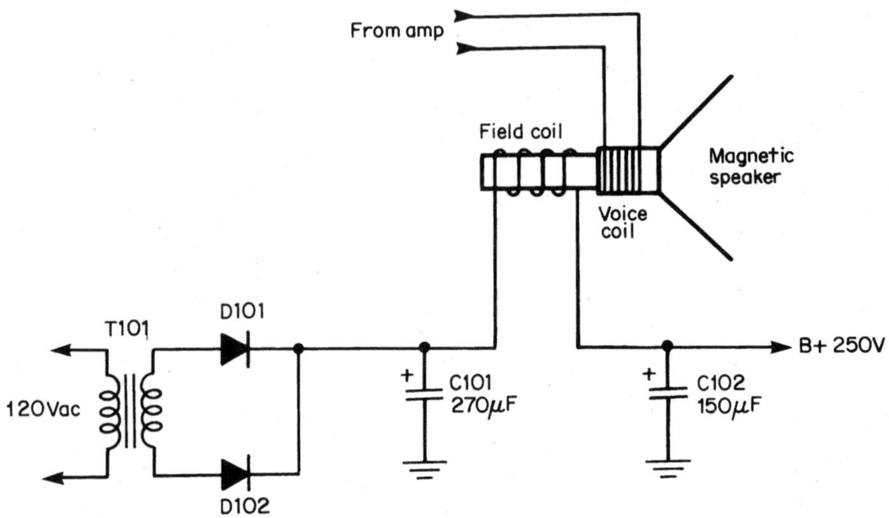

3-21 The early field coil in a magnetic speaker provided filtering action in the early radios.

If a large filter capacitor or component is shorted after the filter choke, the coil winding can be damaged and become shorted internally. Often, heavy brown or burned marks are found on the covered windings. Very poor filtering action results with a burned choke transformer.

Testing

After shunting the electrolytic capacitor within the low voltage power supply, suspect a defective choke coil with no or low output voltage. Determine if the input filter has higher dc voltage with real low-output voltage. Check the dc voltage at both ends of the filter choke transformer.

Check the ripple input and output waveform with the oscilloscope. The ripple waveform should be flatter or smoothed out at the output choke-transformer connection. These measurements can be taken across the main filter electrolytic capacitors. If the ripple waveform is higher and real jagged, compare to the input, suspect a defective choke transformer or electrolytic capacitor. Clip another known capacitor across the suspected one and take another waveform with the scope.

Suspect either an overloaded circuit connected to the filter choke or a shorted choke coil with low voltage. Measure the resistance across the choke transformer winding (Fig. 3-23). Replace the defective choke transformer since it cannot be repaired. Now check the input ripple in and out of the choke transformer with the scope.

Circuits

An electronic circuit consists of several electronic components interconnected. The diagram or schematic is a drawing of an electronic circuit. The circuit diagram contains symbols and lines representing the components within an electronic circuit. Figure 3-24 shows a drawing of each component and how they are tied together. The cassette tape head, radio RF, TV horizontal driver circuit, and TV vertical output circuits are given with examples and troubleshooting procedures:

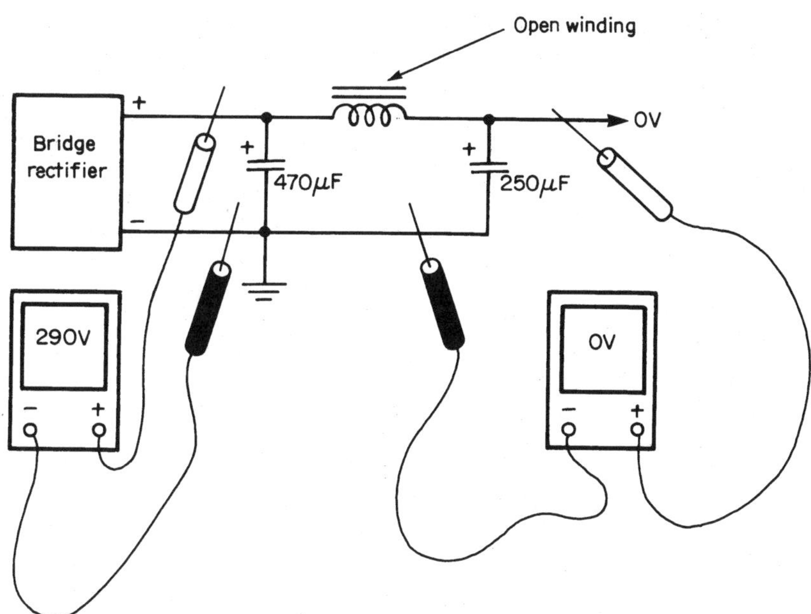

3-22 Check the voltage at the main filter capacitor and on both sides of the filter choke for correct voltage.

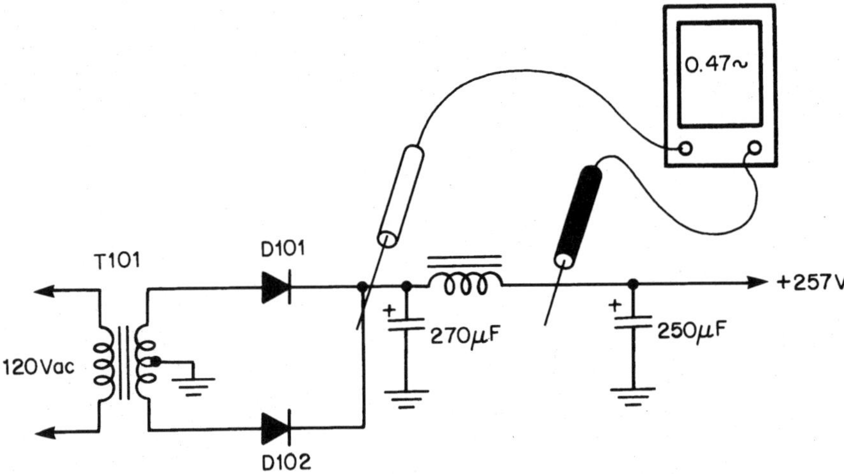

3-23 With excessive hum in the speaker, measure the resistance across the suspected filter choke.

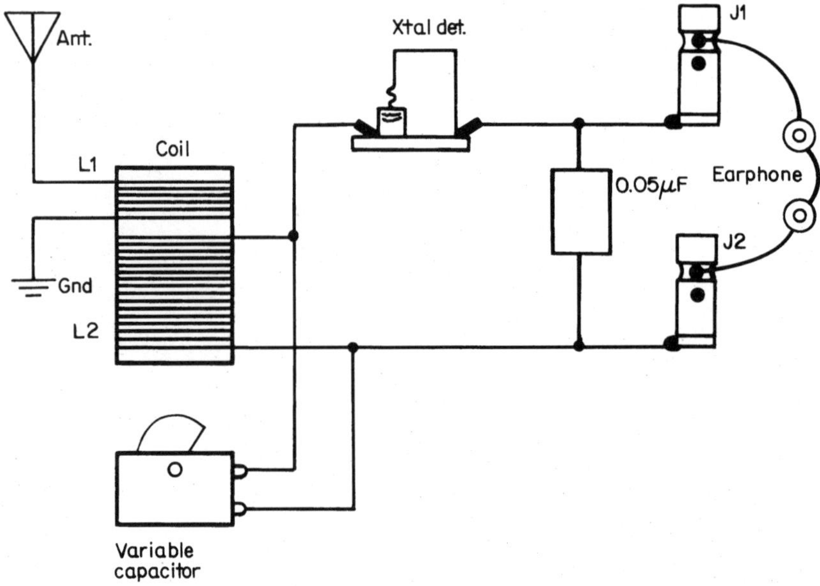

3-24 A diagram of a simple crystal receiver.

The cassette tape head circuit

What and where

The tape head has a magnetic core with many windings of fine wire for picking up the magnetic recording found on the cassette tape. The cassette tape head may consist of one monaural winding or two separate windings for stereo operation. The tape head input circuit may consist of a tape head winding, electrolytic coupling capacitor and preamp transistor, or IC amplifier. The amplified music is then switched into the AF and audio output circuits (Fig. 3-25). The cassette tape head is located in tag-alongs, portables, boom-boxes, AM-FM-MPX receivers, microcassette recorders, professional recorders, car stereos, compacts, and cassette players and decks.

When

Besides collecting oxide dust, the R/P tape head can become open, intermittent, or cause distortion in the speakers. The worn tape head can cause a loss of high frequencies. A magnetized tape head can cause extra noise in recording. An open head winding will produce a dead channel. Poorly soldered connections or broken terminal connections can cause intermittent music. Check the open or high-resistance head winding with the ohmmeter. Inspect the cable wires and resolder for intermittent connections. Make sure that one screw is not loose, which would let the head swing out of line. The typical tape head resistance is from 200 to 830 ohms.

Testing

The open tape head winding or broken wire connection can produce dead or intermittent reception. Inspect the tape head connection with no head pickup. Check the head resistance with an ohmmeter (Fig. 3-26). Make sure the moving tape is pressed against the tape head. Always, clean off the excess oxide with a cleaning stick and alcohol. Degauss the tape head for noisy recordings. Clean up the record/play switching contacts.

Signal trace the tape head and circuit for a weak, dead, or low audio signal. Insert a test cassette and signal-trace the audio signal from tape head to the preamp tape head output circuits. Connect an audio signal generator at the tape head for weak or distorted audio. Signal-trace the tape head circuit with the scope or external audio

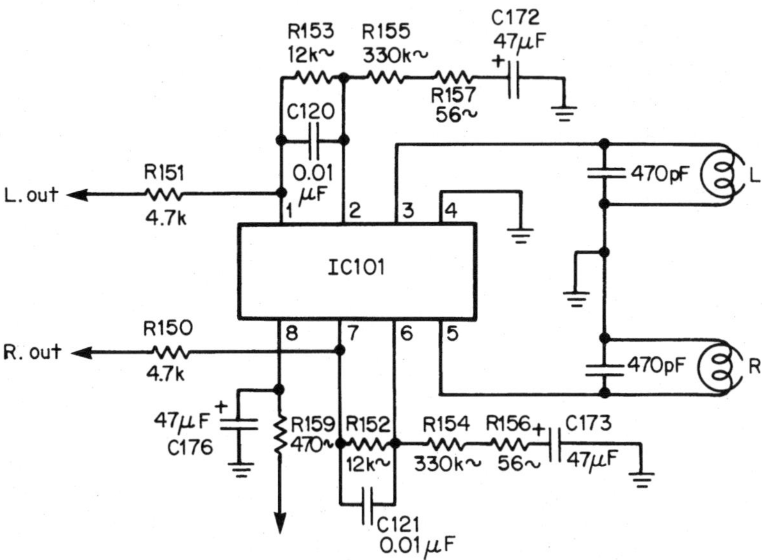

3-25 A stereo cassette tape head connected to a dual IC preamp circuit.

3-26 Check the tape head resistance with the RX1K ohm range of the DMM.

amplifier. When the signal stops, the defective stage is nearby. Then, take critical voltage and resistance measurements.

To prevent cross-talk, the tape head must be properly lined horizontally for optimum sound reproduction. Usually, one side of the tape head is fastened with a small screw and the other side with an adjustable spring. Improper head azimuth adjustment can cause cross-talk, distortion, and loss of high frequencies. You can adjust the azimuth screw by playing a recorded cassette of violins or high-pitched music to maximum into speakers.

Accurate azimuth adjustment can be made with a 3-, 6.3-, or 10-KHz test cassette with 8-ohm load instead of speakers. Connect the low ac range of the DMM across the 8-ohm resistor (Fig. 3-27). Play the recorded cassette and adjust the azimuth screw for maximum on the DMM or VOM. The oscilloscope and frequency meter can also be used as an indicator.

Signal-trace the audio preamp transistors or IC with a test cassette and oscilloscope. You can also trace the audio signal from the tape head winding, through the preamp circuit with an external audio amplifier. Compare the signal gain and performance to the other stereo channel. Take an in-circuit test of the preamp transistors. Remove from the circuit if in doubt. Test the input and output signal of the preamp IC with the scope or external amp.

Radio RF circuit

The radio frequency (RF) amplifier amplifies the radio frequency signals. The RF transistor amp in the table radio picks up the signal from the antenna and couples the signal through an RF coil. This RF coil might be a coil winding wound on a ferrite core or form called the ferrite antenna. Often, the RF amplifier of the FM band consists of a few copper wire turns connected to an RF FET transistor. The RF stage is found in large and expensive AM-FM-MPX receivers.

What and where

The RF coil in the FM receiver contains only a few turns of wire while the AM-RF coil has many turns. Most FM receivers use an FET transistor in the varactor or conventional capacitor RF stage. Diode D3 is the varactor diode in the RF stage of the varactor tuner in Fig. 3-28. Here the tuning voltage applied to D3 is supplied through R4 (33K ohms). The FM coil L4 is tuned by inductor D4. The same tuning

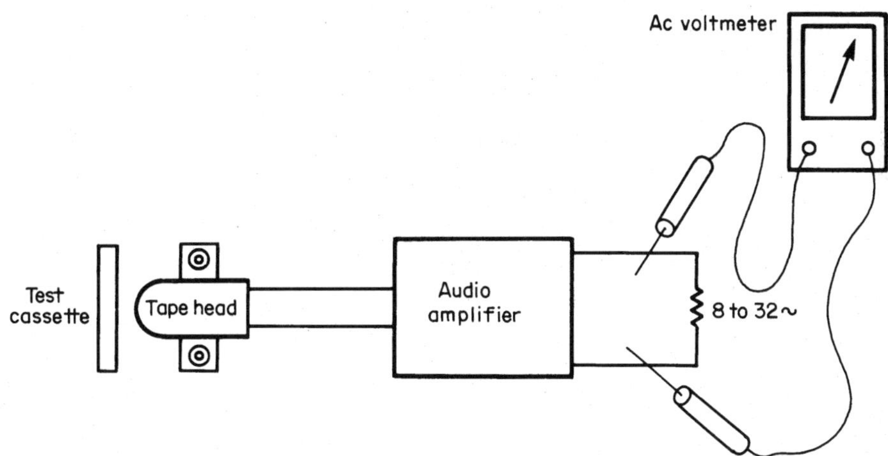

3-27 Correct test setup for tape head alignment.

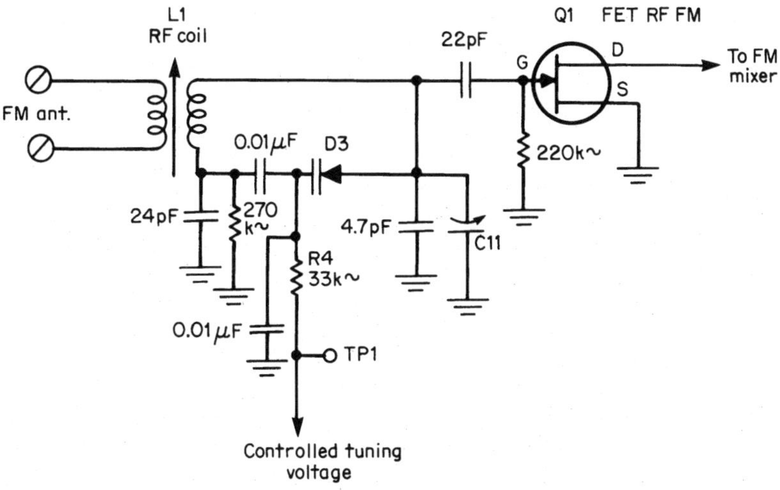

3-28 The RF coil (L1) is coupled to the gate of the FET-RF-FM transistor.

source occurs at R7 from the same tuning-voltage source. TP1 is a test point to check the varying voltage supplied by the controller or pushbutton voltage selection.

You might find that the AM varactor tuner consists of a separate RF FET transistorized stage with a separate IC for the converter and IF circuit (Fig. 3-29). A varactor tuning diode is found in the RF and converter stage. The same tuning voltage is fed to both tuning circuits. The IF output is fed to multiplex IC circuits. In large deluxe models, you might find entirely separate AM and FM RF circuits.

The FM RF and oscillator coils are easily located with only a few turns of bare copper wire. The AM RF coil can be located with a ferrite core or with an outside antenna connection. Sometimes the AM RF and FM RF coils are shielded. The coil with a lot of coil turns represents the AM RF band.

When

A defective RF AM stage can cause a weak, intermittent, or local-only signal. The dead RF circuit can result from a shorted or leaky RF transistor. Improper or no voltage applied to the RF stage might cause weak or no reception. Inspect the AM antenna coil for a broken core or wire connection. The open RF transistor will have a dead radio symptom. A noisy RF transistor or IC can produce a frying noise in the sound.

You may find a no FM-AM symptom, caused by an open or shorted FM RF transistor. Suspect a defective regulator transistor with low or no voltage at the drain element of FET RF transistor. A leaky or open FET RF transistor might tune in one local FM station. The intermittent AM-FM reception can be caused by a dirty AM-FM switch. Weak or intermittent RF can result from lightning damage.

Testing

With the diode-transistor test of the DMM or transistor tester (Fig. 3-30), check the RF transistor in a circuit with weak or dead front-end symptoms. Check for accurate forward bias between base and emitter terminals. Measure the voltage on the collector

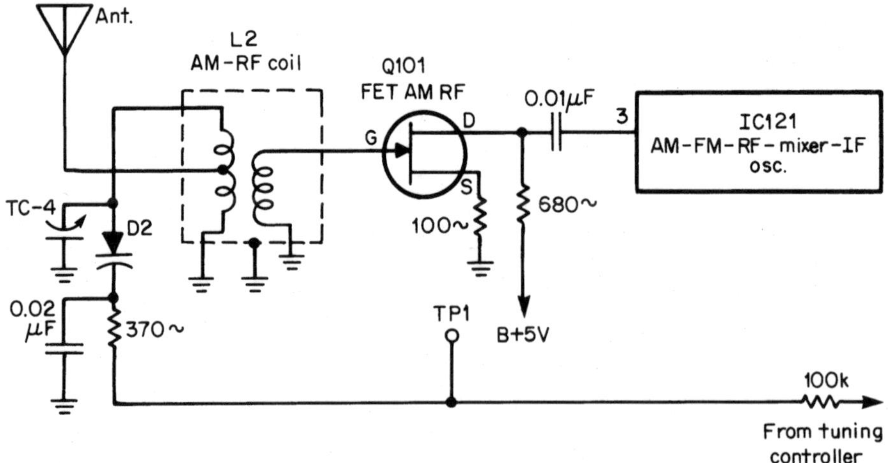

3-29 An AM-RF coil tied directly to the gate terminal of the FET-AM-RF transistor.

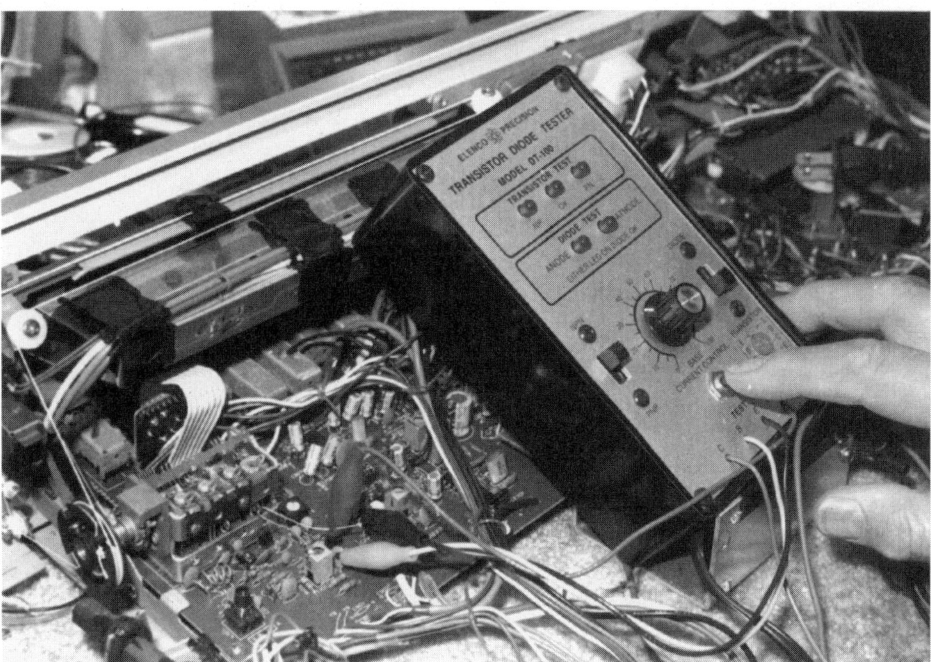

3-30 Check the RF transistor in the circuit with a transistor tester or the transistor test of the DMM.

or drain terminal of the RF transistor. Very low-drain or collector voltage indicates a leaky transistor or improper voltage source. High-drain or collector voltage indicates an open RF transistor or emitter resistor.

Take a resistance measurement of the antenna or RF coil. A broken antenna coil lead can shut the RF reception down. Make sure the antenna core is not cracked into, creating weak reception. Inspect all wiring connections from coil to the PC board circuits. Short out the base and emitter terminals of the RF transistor to find if the transistor is noisy. Move components around with a plastic tool to determine if a bad connection exists causing intermittent reception.

Replace the AM and FM RF transistor with the exact part number or universal components. Cut all transistor leads the same length. Replace the FM RF transistor in the same spot with the same lead length. Make sure the drain terminal is connected to the highest voltage and the source terminal to the ground of an RF FET transistor.

TV horizontal driver circuit

Within the present-day TV chassis, the horizontal driver circuit receives a pulse or waveform from the countdown or oscillator circuit. When the waveform or drive signal is missing, the driver transistor becomes very warm and might destroy the driver transformer or power resistor. The horizontal driver transistor greatly amplifies the drive signal at the horizontal output transistor, through a driver transformer to the base terminal of the output transistor. Besides damaging components within the driver circuits, no drive voltage on the horizontal output transistor might be destroyed with a blown line fuse.

What and where

The horizontal driver transistor circuits are practically the same in every TV chassis. This horizontal drive circuit takes the signal from the sweep or countdown IC and amplifies the signal to drive the horizontal output transistor (Fig. 3-31). The horizontal drive waveform can be scoped from the sweep output IC to the base and collector terminals of the driver transistor. When no or improper horizontal drive waveform is found on the base of the horizontal output transistor, suspect a defective sweep IC or driver transistor circuit. Usually, the TV chassis is shut down and will not start up.

When

Go directly to the sweep IC when no drive signal is found at the horizontal output transistor. Often, the horizontal output transistor and line fuse are destroyed if the chassis does not shut down at once. Likewise, the horizontal output and flyback circuits must operate before a run voltage is applied from the scan-derived voltage circuit of the flyback.

Check to see if the supply-pin voltage of the sweep IC is fed from the secondary winding of the flyback or low-voltage power supply. Remove the power plug. Insert a dc voltage at the supply pin of the sweep IC from an external voltage power supply. Now take a sweep waveform at the horizontal output drive pin, tied to the horizontal driver transistor, to determine if the sweep IC is operating.

Testing

When the horizontal sweep IC has a drive waveform and no driver waveform exists, check the input and output drive waveforms with the scope. If a drive waveform is found at the base of the driver transistor and no amplified waveform, take critical voltage measurements on the driver transistor. The driver transistor will overheat and run very warm with no drive pulse. If left on too long, the transistor may become leaky and overheat the primary winding, burning an isolation resistor in series with the voltage source and transformer. Remember, this voltage source applied to the primary winding of the driver transformer is the same as found on the primary winding of the flyback.

Take a critical resistance measurement of the primary winding to determine if the winding resistance has changed (Fig. 3-32). Compare the measurement to the schematic, if one is handy. Take a resistance measurement of the overheated resistor and replace if changed. Test the driver transistor out of the circuit.

Resolder all driver transformer connections on the PC board for intermittent or no start-up symptoms. Replace the small electrolytic capacitor (4.7 to 10 μF) within the supply-source circuit when the horizontal output transistor runs warm or is too hot to touch. Install a new driver transformer when the horizontal output transistor runs excessively hot, after the driver transistor, resistor and electrolytic capacitor have been replaced. Now check the drive waveform at the base terminal of the horizontal output transistor.

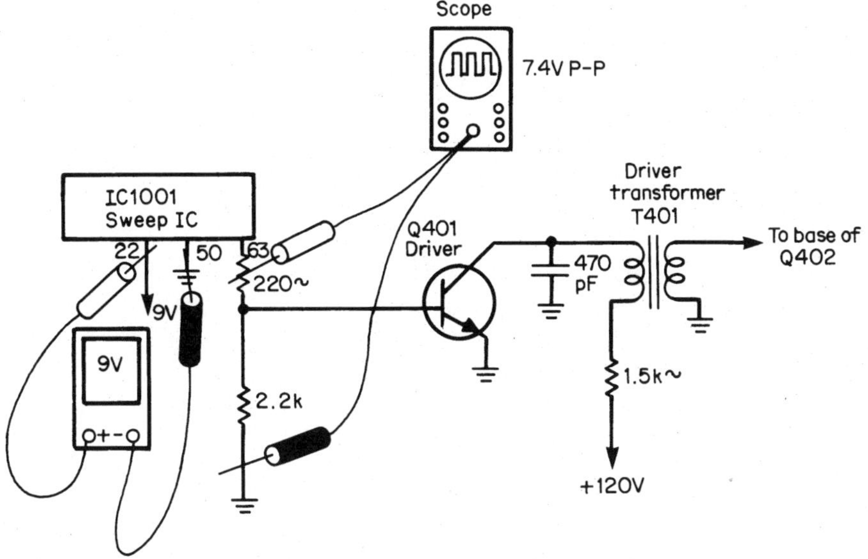

3-31 Inject a dc voltage source into the sweep IC with the scope probe at the horizontal drive terminal to determine if the sweep oscillator circuits are normal.

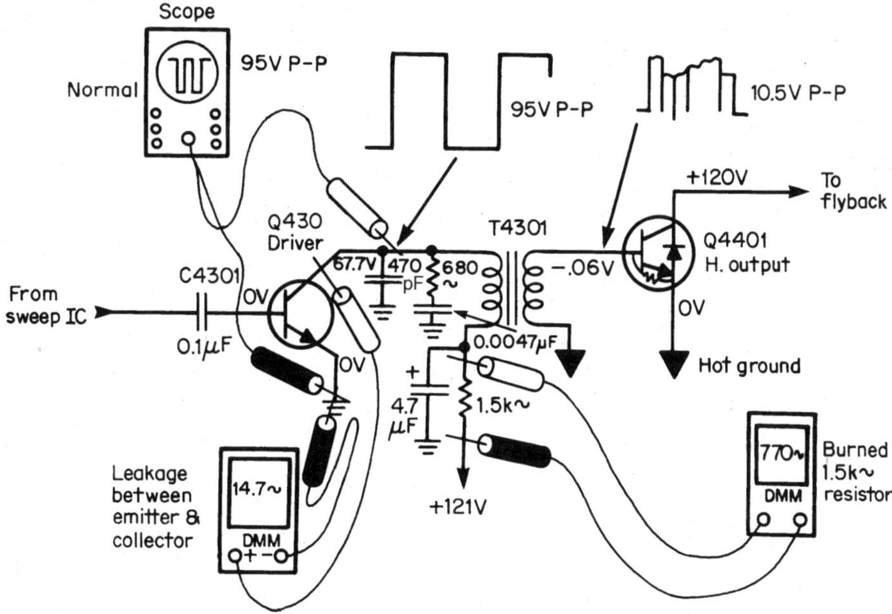

3-32 Take critical scope, voltage, and resistance measurements to locate defective components in the horizontal driver circuits.

TV vertical output circuit

Like the horizontal drive circuits, the vertical drive is taken from the same sweep IC in the latest TV chassis. The vertical drive pulse might be applied to the base of a drive transistor or IC. The vertical drive transistor amplifies the vertical pulse or waveform and drives the vertical output transistor or IC. The output transistors or IC provide sweep to the vertical deflection yoke (Fig. 3-33).

What and where

The vertical output circuit may contain transistors or IC components. A directly-coupled resistor is found between the sweep output IC and the vertical driver transistor. Likewise, the output pulse is directly connected to the input terminal of the output IC. Usually, the vertical yoke winding is capacity-coupled to the output IC. You will find a return resistor and electrolytic capacitor found in the yoke return path to ground. Today most vertical circuits use either transistors or have an IC within the vertical output circuits.

When

The vertical circuits might break down with a white horizontal line or no vertical sweep, insufficient sweep, bunching vertical lines, lines at the top, intermittent sweep, vertical crawling, poor vertical linearity, vertical rolling, and vertical foldover. The horizontal white line can be caused by just about any component within the vertical circuit. An open yoke winding, open coupling capacitor, transistors, and IC can cause a white horizontal line.

Insufficient vertical sweep or black at the top and bottom result from a defective vertical output circuit. Improper drive voltage can cause insufficient vertical sweep. A leaky top or bottom transistor can cause poor vertical sweep. The leaky vertical IC or component tied to the IC can produce improper sweep. An improper voltage source can create insufficient vertical sweep.

Intermittent vertical sweep is the most difficult to locate and repair. Check for transistor or IC breakdown, poor board connections, a loose output transistor mounting screw, or broken board connections. Suspect a poor eyelet or griplet board connection. Try to narrow the intermittent to a certain section of the board. Intermittent transistors and output ICs can be located by applying several coats of coolant. Monitor the vertical circuit with a scope and a DMM to help locate the intermittent component.

Check for leaky output transistors and improper bias resistors when vertical foldover occurs. Most vertical foldover problems occur in the vertical output circuits. Foldover might occur at the top or bottom of the picture. Replace both output transistors when one is found leaky. When vertical foldover and linearity is found, check the feedback circuits within vertical output circuits for defective electrolytic capacitors (Fig. 3-34). Replace both output transistors when one is found leaky.

Scanning lines slowly moving up the picture with a dark section is called vertical crawling. You may find a dark bar at the top or bottom of the raster. Improper filtering in the low voltage source produces most vertical crawling symptoms. Clip another capacitor across the main filter to locate the dried-up filter capacitor. Improper voltage applied to the vertical circuits can cause vertical crawling. Bunching lines at

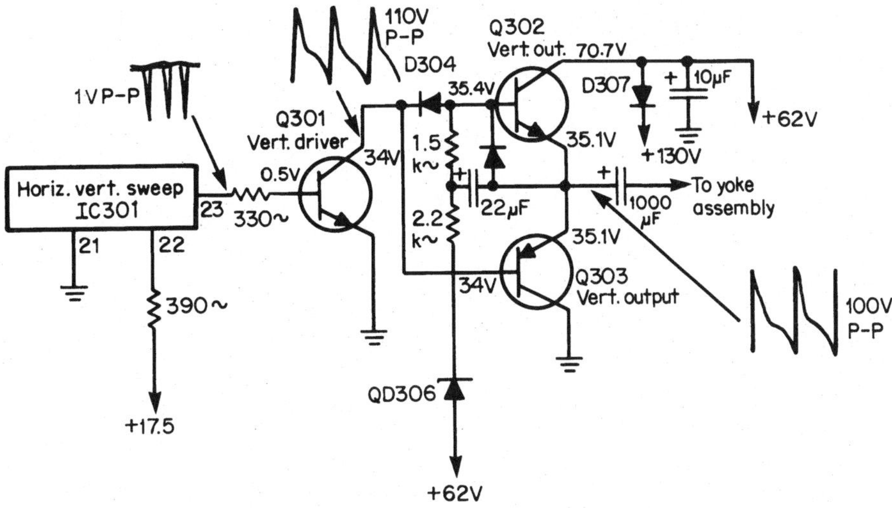

3-33 Take critical vertical waveforms of the transistor output circuits that feed the vertical sweep to the yoke coils.

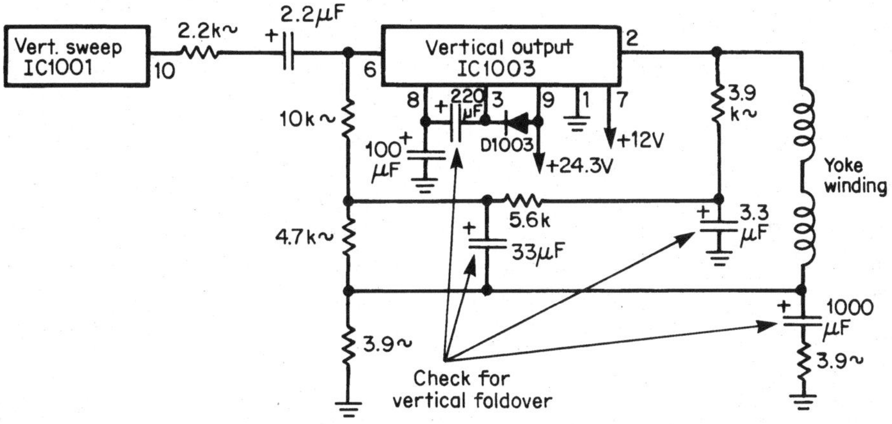

3-34 Check all electrolytic capacitors in the vertical output and feedback circuits for poor vertical linearity and foldover.

the top of the picture might result from defective transistors, bias resistors, and vertical output coupling capacitors.

Testing

Scope the vertical circuits by the number (Fig. 3-35). If the symptom is a white horizontal line, start at the output terminal of the sweep IC. Proceed to the input terminal of the vertical output IC. Check the waveform at the output IC terminal to the deflection yoke. If the correct waveform is found at the output IC terminal and no sweep exists, check the yoke winding, electrolytic coupling capacitor, and return resistor. Remember the scope waveforms are very unstable in the vertical output circuits.

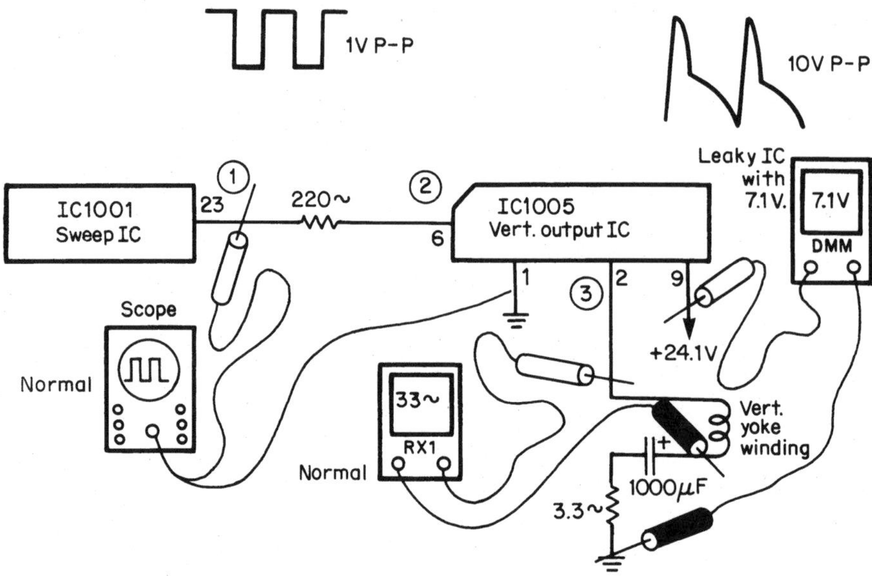

3-35 Scope the vertical circuits by the number on a vertical output IC.

When the waveform stops or is not located, take correct voltage and resistance measurements. If there is a correct waveform at the input terminal and there is no output sweep, check the vertical output IC circuits for defects. Take a voltage measurement on each IC terminal. Measure the resistance from each terminal
to the ground. Suspect a defective IC when voltage and resistance measurements are fairly close to the schematic. Do not overlook bad or poor wiring connections to the output IC. Double-check the pin-cushion circuits for vertical sweep problems.

Circuit breaker

A circuit breaker is a resettable device protecting circuits that might be overloaded. The circuit breaker takes the place of a fuse (Fig. 3-36). It can be used over and over again. The blown fuse must be replaced each time. You do not have to replace unless it is damaged or will not hold its reset value. Check the circuit for overloading conditions when the breaker kicks out at once.

What and where

The circuit breaker can be found in the home fuse box instead of fuses. This circuit breaker protects any product connected to the fusible power line. Circuit breakers were found in the early ac-dc power-line TV sets (Fig. 3-37). The circuit breaker worked in place of the main fuse and the horizontal output tube circuits. A circuit breaker might be found within the speaker circuits to protect the speakers from excess voltage or too much volume. The ac motor might be protected by an overloaded circuit breaker. Today, the circuit breaker is located in many commercial home appliances.

When

The circuit breaker must be replaced when it keeps kicking out with a normal load. Circuit breakers can be damaged when they are reset and held with overloaded conditions. The circuit breaker can be extensively damaged by lightning or power outages.

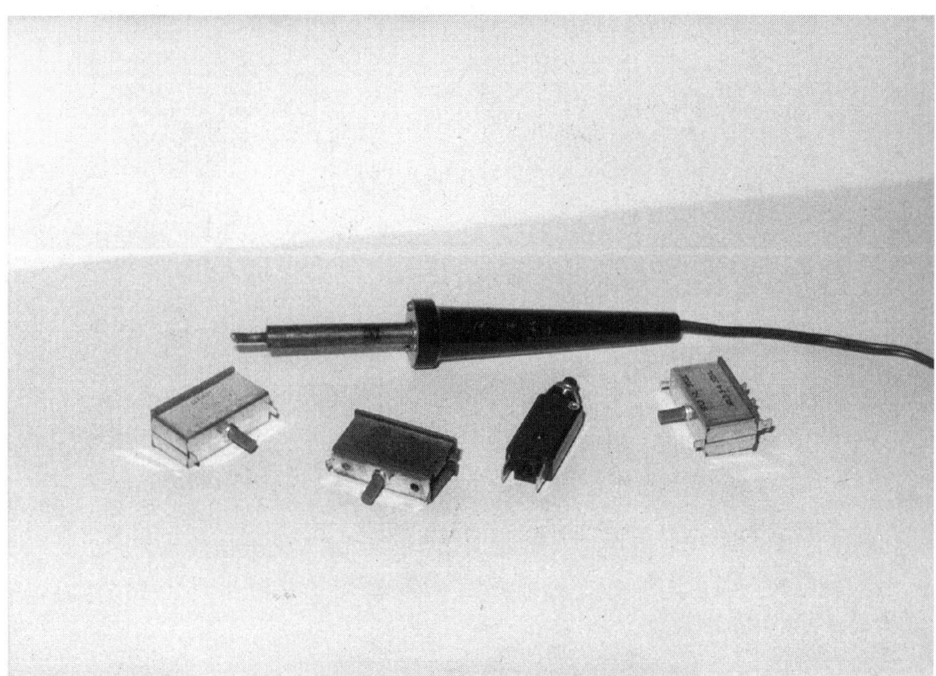

3-36 The circuit breaker can be found in motors, speakers, and early TV circuits.

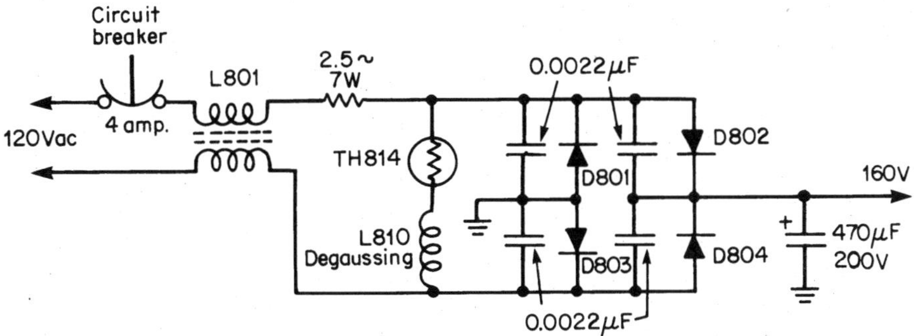

3-37 A circuit breaker was found as the main fuse protector in the early TV power supply circuits.

Testing

Check the continuity of the circuit breaker with the RX1 scale of a DMM or VOM. If the circuit breaker will not reset, replace it. When the circuit breaker will not kick out with an overload, install a new one. Check the circuit for correct amperage and the amperage of the circuit breaker listed on the schematic. Do not install a 1.5-amp circuit breaker with a 3-amp type. For best protection, install a 2-amp circuit breaker, when a 1.5-amp is not available.

4
CHAPTER

Testing electronic components C

(Coils to crystals)

An open oscillator coil can prevent oscillations from occurring in the table radio, while an open peaking coil might produce a smeared picture in the TV. The defective TV horizontal yoke might appear open or become leaky, causing no vertical sweep and a trapezoid raster. Hum and buzz in the sound of the TV chassis might be caused by a defective sound coil (Fig. 4-1). A defective CRT might result in an intermittent signal, lines in the picture, or no raster. The defective crystal can cause many color problems within the TV chassis, and on it goes.

Coils

Induction coils

What and where

The symbol of inductance is L and is rated in henries. The inductance is the inertial property of an electronic device that opposes the flow of current and a change in current. Large inductance might be rated in henries while small coils found in electronic products are in millihenries (mH) and microhenries (μH).

When

Induction coils can be found in radios, cassettes and CD players, VCRs, camcorders, and TV sets. The defective coil can appear open, have poor coil connections, and have shorted turns. Most defective coils are either open or have poor

4-1 Hum and extreme distortion can result from a misaligned discriminator coil in the TV chassis.

connections. Sometimes a shielded coil might short against the outside metal shield and cause a dead circuit.

Testing

Check the suspected coil with a continuity or resistance measurement of the VOM or DMM (Fig. 4-2). Actual resistance measurements should be tested with a digital multimeter. Resolder all coil leads with intermittent operations. Replace the entire coil with corroded green areas of a suspected coil. In many cases, the coil can be tested with a voltage check within the same circuit. For a correct inductance test, test the coil with an LCR meter or inductance range found on the DMM.

Radio RF and oscillator coils

What and where

The RF coil found in clock radios may consist of an antenna coil wound upon a ferrite coil form or a separate RF coil fed into an FET-RF transistor (Fig. 4-3). An oscillator coil found in the broadcast receiver has many turns of wire, while the shortwave or higher range of frequencies might consist of several turns. The oscillator coil found in the FM oscillator stage might consist of only a few turns of solid wire.

When

The oscillator coil within the table radio can go open or become intermittent producing no music or intermittent sound. Often, coils cause minor problems within

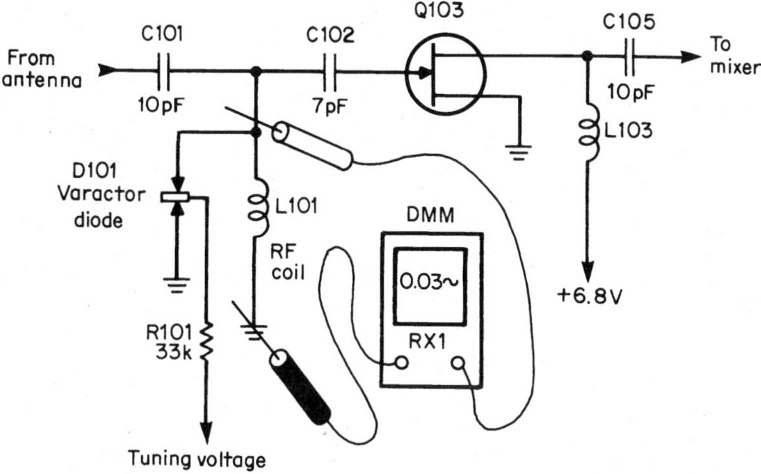

4-2 Take a quick continuity test of the suspected RF coil with the low-ohm range of the ohmmeter.

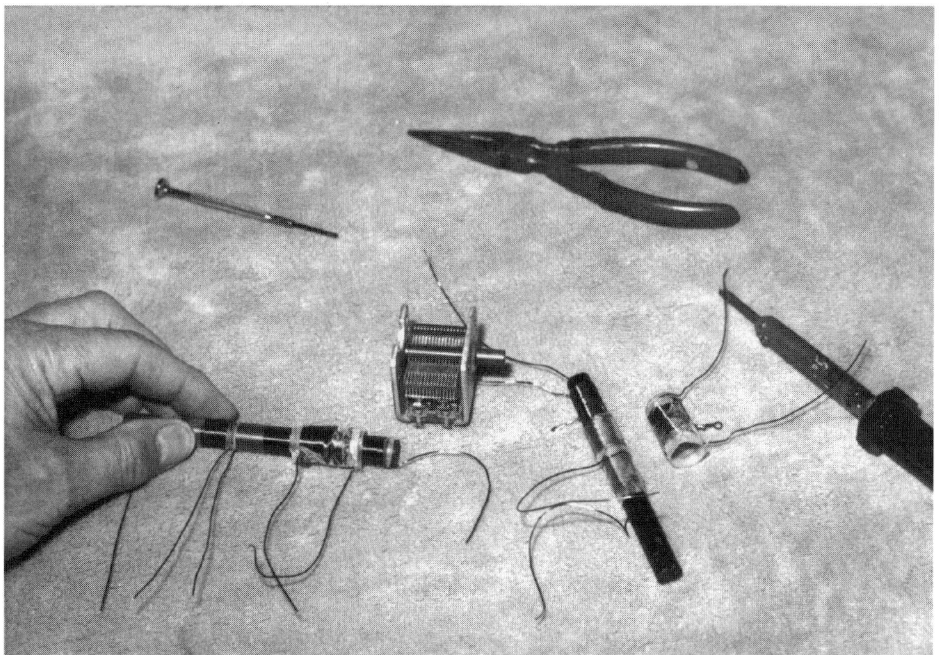

4-3 The antenna RF broadcast band coil might be wound on a long ferrite form.

the radio receiver. When no stations can be heard with only a local station, suspect the RF coil or transistor. Suspect a drifting oscillator coil or stage when you must keep adjusting the tuning capacitor. If a loud rushing noise is heard and no stations, check the oscillator or mixer circuits. For intermittent radio reception, inspect and resolder the coil terminals.

Testing

Check the coil winding with a quick continuity test or ohmmeter measurement. The oscillator stage can be checked with another radio near the defective receiver. Simply rotate the tuning coil of the suspected receiver and listen for a beat frequency as the dial is rotated (Fig. 4-4). A squeal can be heard in the normal receiver if the oscillator stage is operating.

The oscillator circuit can be tested with a calibrated signal generator and scope. Connect the scope probes to the oscillator-tuned circuit and adjust for only a few cycles. Now connect scope probe to the generator and turn generator until you have the same number of cycles. This will indicate the frequency of the oscillator circuit. Notice if the oscillator circuit drifts off frequency when the number of cycles changes. The oscillator output frequency can also be tested with a frequency counter test instrument (Fig. 4-5). Critical voltage measurements upon the base and collector terminals can indicate if the oscillator circuits are functioning. Check the oscillator frequency with an LCR meter, if handy.

The oscillator, RF, and IF stages can be tested with a signal generator. Connect the output of signal generator with four to six loops of wire wound around ferrite antenna coil. Set the signal generator to 1400 kHz frequency. Rotate the receiver to 1400 kHz. Now you should hear a modulated signal if the oscillator and RF stages are normal. Check the lost signal with an RF probe for scope or external audio amp at the IF stages when no sound is heard in the speaker.

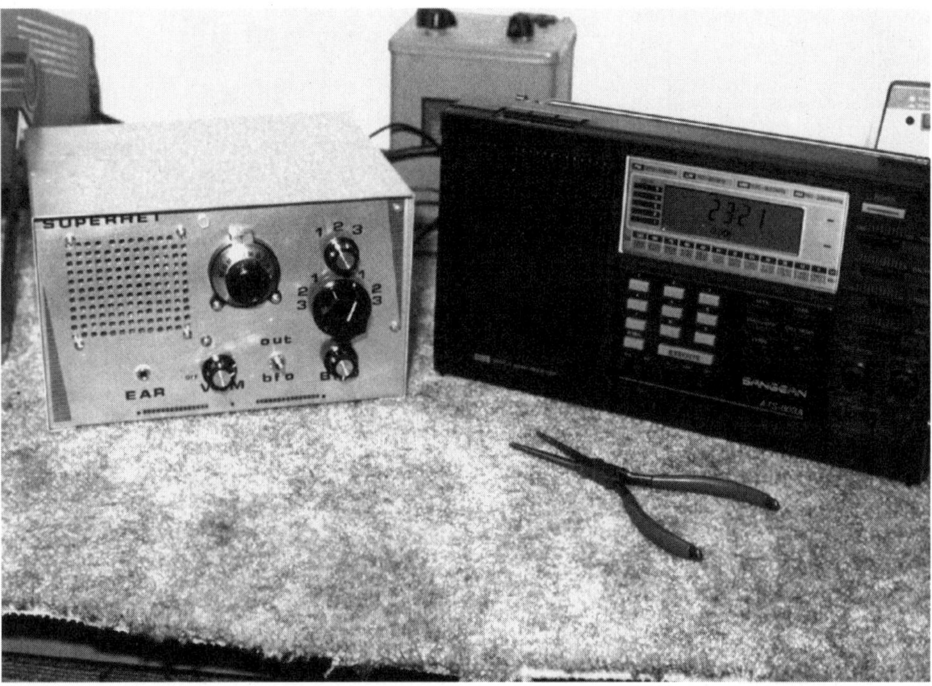

4-4 Check the oscillator stage by tuning across the same band with another radio, and listen for a whistle or squealing noise.

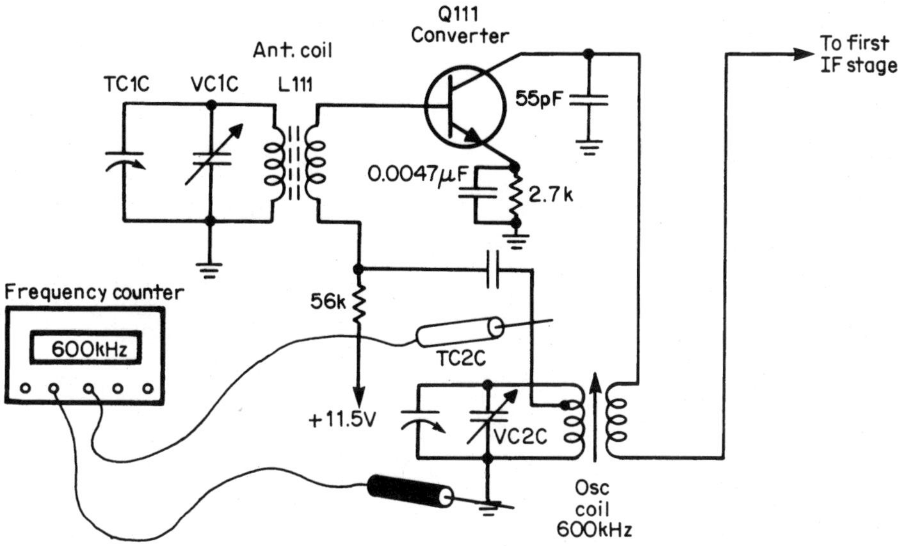

4-5 Check the oscillator frequency with a frequency counter.

Toroid coils

What and where

A toroid coil is a doughnut-shaped coil. The coil is wound in and out of the center hole area. Most toroid cores are made from powdered iron and ferrite material (Fig. 4-6). These toroid coils are found in ac and very high frequencies (VHF) circuits. The toroid coil is very compact compared to an open-wound coil. A higher Q factor can be obtained with the toroid coil than with an open-wound coil.

The powdered-iron toroid coil is found in RF circuits to medium VHF frequencies. You may find that the ferrite cores are found in audio, low RF circuits, and power supply circuits. Large toroid coils are found in the high-powered automatic sound system. The ferrite core is found with coils up to 10 MHz. As a rule, the powdered-iron toroid's coils are found in electronic products above 10 MHz.

Testing

The toroid coil can be tested with a low resistance measurement or with an LCR meter. You may find some digital multimeters may have a combination inductance range of 7 mH to 20 henries, while the LCR meter measures from 1 mH to 10,000 henries (Fig. 4-7).

TV peaking coils

What and where

The TV peaking coils are usually located in the video and detector output circuits. The small inductor compensates the frequency response of a circuit. You may

4-6 Toroid coils are wound around a donut-shaped ferrite or powdered-iron coil form.

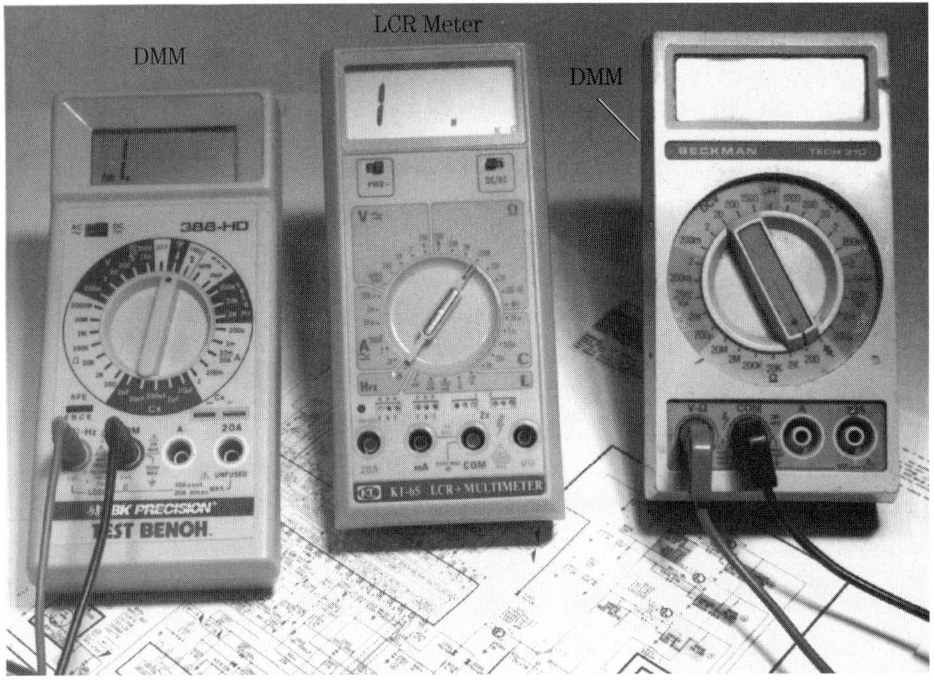

4-7 Check the coil inductance on an LCR digital meter.

find a series or shunt-type inductance. These peaking coils were found in the early tube chassis. The video peaking or inductance coil might be found in the collector circuit of the video amp after the delay line (Fig. 4-8). A peaking coil might be found in series with the base terminal of video or buffer amp.

When

An open or shorted peaking coil can cause a smearing TV picture. Remember a defective delay line can also cause smeary pictures. Sometimes the smeary condition can be confused with a poor focused picture.

Testing

Check the resistance of the peaking coil. Often the peaking coil is found open. The peaking coils in some Sams Photo-Facts are rated in microhenries (µH). Test for correct inductance with the LCR meter (Fig. 4-9).

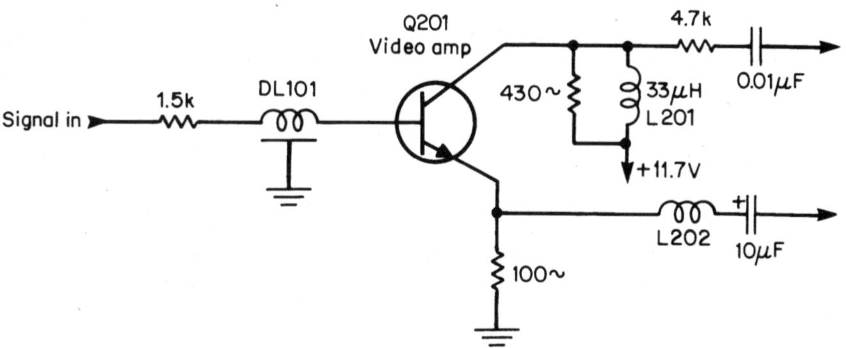

4-8 Here L201, a peaking or loading coil, is found in the video amp output circuit.

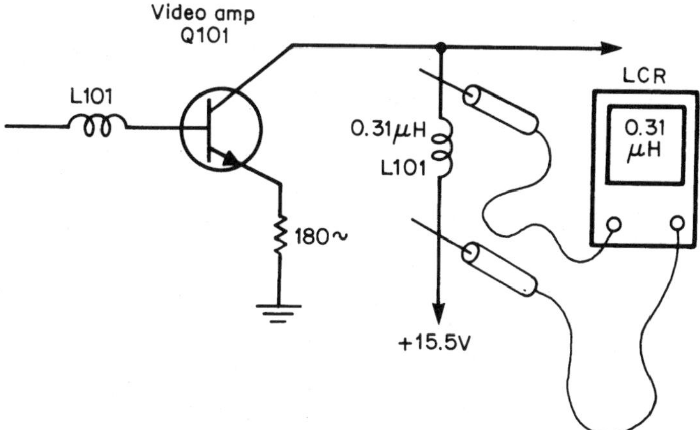

4-9 Check the inductance of L101 with an LCR digital multimeter.

The horizontal yoke

What and where

The deflection yoke found in the TV receiver, monitor, magnetic-deflected oscilloscope, or computer, consists of two different sweep coils that provide vertical and horizontal sweep. This yoke has two different horizontal and two different vertical coils that are connected in series. The horizontal output transformer circuits feed the horizontal deflection coils, while the vertical output transistors or IC, provide sweep to the vertical circuits (Fig. 4-10).

The deflection yoke is mounted upon the rear bell of the picture tube. This yoke must be adjusted back and forth, up and down in picture-tube purity alignment. Rubber wedges are found, in some sets, behind the yoke assembly to help keep the yoke assembly in position, after purity adjustment (Fig. 4-11). On some CRTs the deflection yoke is glued to the tube. A new tube with a yoke must be ordered when either one is defective.

When

Rotate the yoke assembly to the left or right to level the picture within the CRT. The yoke assembly might move when changed around in the room or hauled to the shop. Make sure the level adjustment does not upset the picture-tube purity alignment. Then tighten the yoke assembly snug-up screw at the rear of the yoke assembly (Fig. 4-12).

Look for a trapezoid pattern when the vertical and horizontal windings arc over between windings. Often, smoke will curl up from the yoke when arcing occurs between the two windings. Water, coke, or liquid spilled down in the rear of the TV set can cause

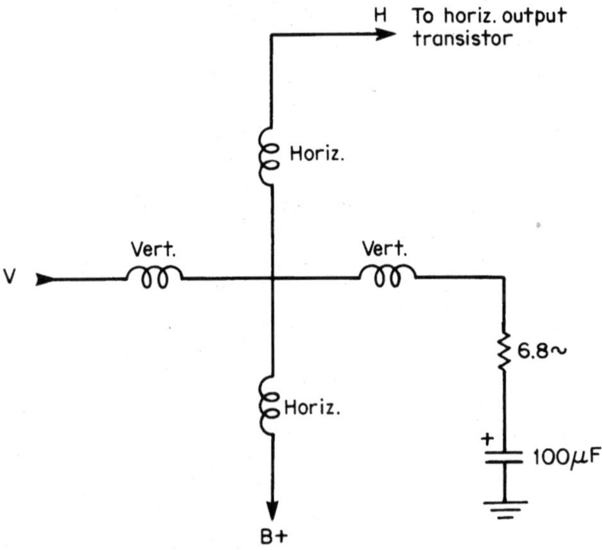

4-10 The vertical output sweep IC feeds to the vertical winding of the deflection yoke in the TV chassis.

4-11 The rubber wedges in front of the yoke assembly provide correct purity adjustment.

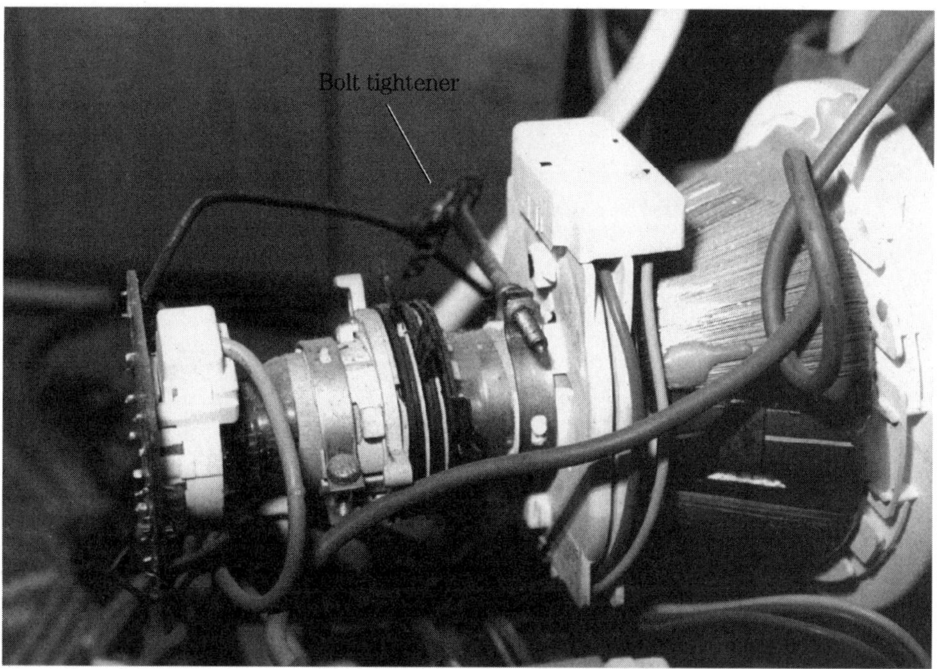

4-12 Tighten the yoke assembly bolt-strap after leveling the yoke on the neck of the CRT.

the winding to arc over (Fig. 4-13). The vertical or horizontal series coils can become shorted between each coil winding.

Testing

In the early black-and-white TV chassis, the yoke assembly was tested by pulling the yoke and feeling the inside of the yoke to determine a hot spot. Check the yoke for open conditions on the low range of a DMM (Table 4-1). Although you cannot tell if several turns of the winding have shorted together, the resistance measurement will indicate if the winding is open. The random yoke resistance chart (Table 4-1) shows the various resistances of yokes found in the latest TV chassis. The larger the screen size, the less the resistance of the yoke windings.

Check the suspected yoke with a yoke tester or sub another yoke. Remove the red wire from the yoke winding and notice if the high voltage increases. Of course, the high voltage will be about half the original measurement. By removing the yoke from the circuit, it will indicate the yoke or parts tied to the horizontal winding are defective and are pulling down the high voltage. Take a waveform test with the scope probe near the flyback (Fig. 4-14).

TV width coil

Check the high-voltage regulator circuits for poor width in the latest TV chassis. The high-voltage regulator transistor, SCRs, and zener diodes in the regulator circuits can produce insufficient width. Poorly soldered connections at pincush-

4-13 The yoke assembly may arc over after water or liquid has been spilled through the back of the TV cabinet.

Table 4-1 A Random Yoke Resistance Chart of Several Different TV Manufacturers

Manufacturer	Model #	Horiz. yoke	Vert. yoke
Goldstar	CMT 2612	1.6 ohms	8.7 ohms
Goldstar	CMS 4841N	3.3 ohms	12.9 ohms
RCA	CTC 167	1.4 ohms	5.6 ohms
RCA	CTC166	1.8 ohms	5.9 ohms
Samsung	CT 339 FB	10.5 ohms	12.8 ohms
Sharp	19SB60	3.5 ohms	15.9 ohms

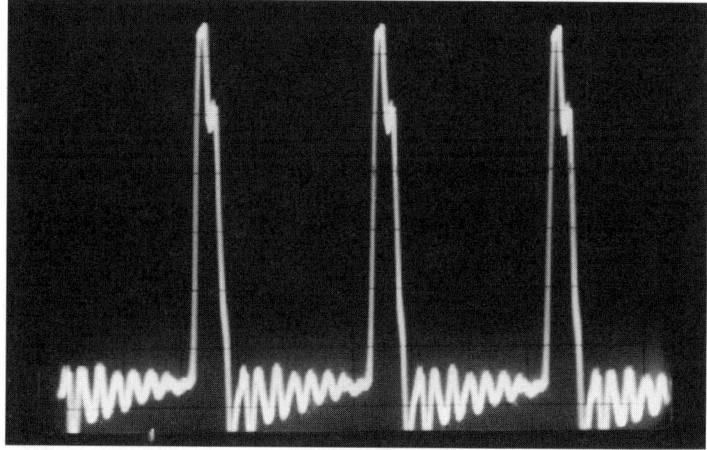

4-14 Check the horizontal sweep circuits with the scope probe held near the flyback.

ion, regulator, and driver transformers can result in poor width. Low voltage applied to the horizontal output circuits can cause poor width (Fig. 4-15). Open bypass or coupling capacitors in the horizontal and high-voltage circuits also can cause poor width.

What and where

In the early TV chassis, the width was adjusted with a variable-resistance control. Today, the width control might not be found in the same horizontal yoke circuits. You might locate a width coil in the bottom leg of the horizontal yoke winding. A magnet slides in or out for correct horizontal width (Fig. 4-16).

Testing

The yoke coil might appear open or have a poorly soldered connection. Check the yoke coil with the low ohm scale of an ohmmeter. If the coil continuity is normal with soldered connections, and the width is pulled inward, suspect other components and lower voltage in the horizontal output circuits.

4-15 Check the horizontal output circuits for low voltage and leaky output transistors, and check the driver transformer for poor width.

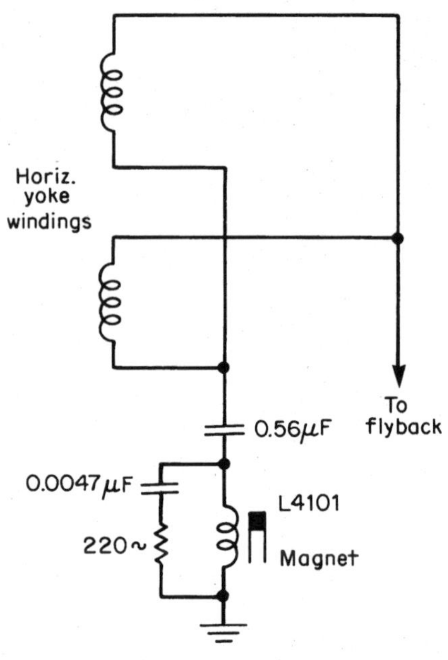

4-16
You may find a width coil with an adjustable magnet to adjust the width in the latest TV chassis.

TV pincushion

When and where

The pincushion circuits prevent the raster from sagging towards the center of the raster in large TV screens (Fig. 4-17). Sometimes pincushion magnets are placed on the outside area of the picture tube to help cure the pincushion effect. The pincushion circuit might consist of transistors, coils, and a transformer in a parabola

4-17 Suspect a pincushion output transistor when the sides are curved in on larger-screen TVs.

generator circuit to provide linearity to the ends and width of the TV raster. Only a pincushion coil or transformer might be found in the lower-priced and smaller-sized picture tubes (Fig. 4-18). The pincushion coil and circuits are found in the bottom leg of the horizontal yoke winding.

Testing

Test the coil continuity with the low-ohm range of the DMM. Resolder both coil terminals. Suspect other components, such as bypass capacitors, when the linearity and sides are pulled in. Check the capacitors for open or leaky conditions. Make sure the supply voltage is correct upon the horizontal output transistor.

TV sound coil

What and where

Distorted or mushy sound found in the TV chassis might be caused by a misaligned sound coil. These sound discriminator coils might drift off frequency and cause hum and distortion within the speaker. The shielded discriminator coil is connected to a preamp transistor or IC component. Drifting of sound can be caused by the moisture affecting the sound coil or small capacitors located in the discriminator circuits. Sometimes alignment of the sound coil is all that is needed to correct the sound problems (Fig. 4-19). If not, replace the small capacitor inside the shielded area.

When

In the latest TV chassis, the sound is taken from the video amp circuits with a 4.5-MHz coil or ceramic filter and fed to a sound IF IC. This audio IC might contain all of the sound circuits. The sound or discriminator coil is tied to pin 4 of IC 201 in Fig. 4-20. When the coil cannot be adjusted to eliminate the hum or distortion, check capacitors in the sound circuits and replace IC 201.

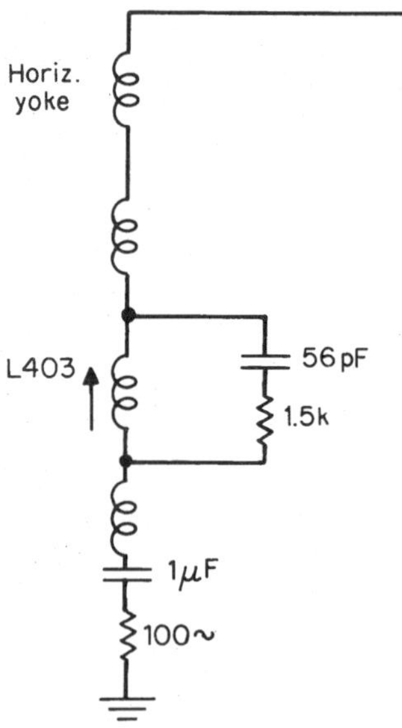

4-18
The horizontal linearity or pincushion coil (L403) is found in the bottom leg of the horizontal yoke winding.

4-19 Adjustment of the sound coil in the sound circuits might cure the hum and distorted audio.

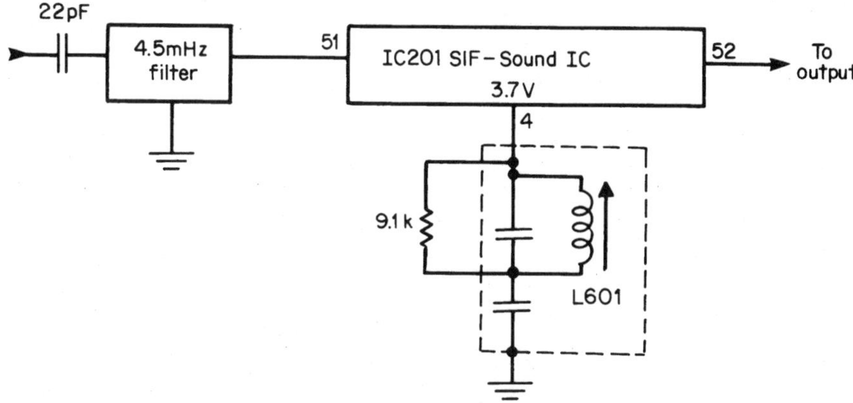

4-20 The discriminator coil is located within the SIF-output sound amp IC circuits.

Testing

Check the coil winding with the low-ohm scale of DMM. Remove the coil shield and test each bypass capacitor in the sound-coil circuit. Test the sound coil with an LCR meter and replace the defective 4.5-MHz discriminator-coil assembly.

Crossover network

What and where

The crossover network blocks and passes certain audio frequencies to the correct loudspeaker. High frequencies are fed to the tweeters and very low frequencies to the woofer speakers. A crossover network is connected to the amplifier output and then fed to each speaker. These networks are found in large amplifier speaker cabinets and in high-powered automobile systems (Fig. 4-21).

The crossover network in the auto sound system might have three basic filters: low-pass, high-pass, and band-pass. A low-pass filter allows low frequencies to pass while blocking higher frequencies. The high-pass filter allows the high frequencies to pass while blocking low frequencies. The midrange driver might be connected to a band-pass filter network that blocks low and high frequencies and allows only a band of frequencies to pass. The crossover network might consist of coils wound with large wire, bipolar capacitors, and large 5- and 10-watt rasters (Fig. 4-22).

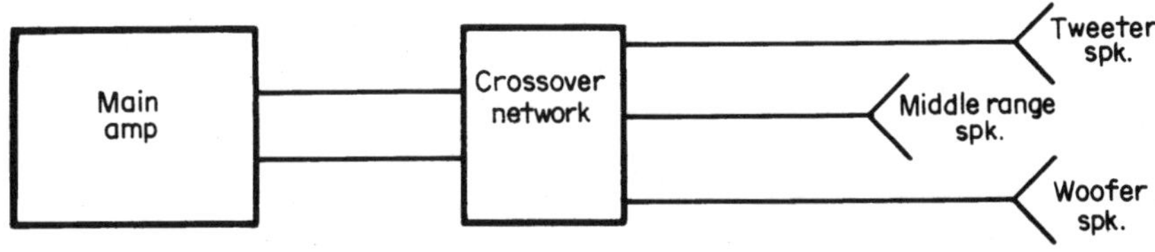

4-21 A crossover network is found between the output power amplifier and speakers.

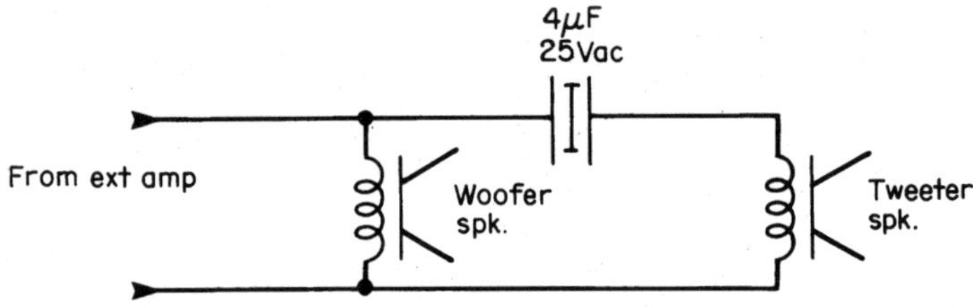

4-22 A simple crossover network with a 4-uF nonpolarized capacitor provides higher frequencies to the tweeter speakers.

Testing

When one channel is dead or distorted, check the crossover network with normal sound output from the amplifier. First sub another speaker. If the speakers and amplifier are normal, suspect a defective crossover network. Check the large coils of wire with the low-ohm range of the DMM. Test each capacitor with a capacitor tester or sub another nonpolarized capacitor. Measure the large-wattage resistors with the DMM. Resolder each coil winding for intermittent audio.

Controls

(See Potentiometers)

CRT (Cathode-ray tube)

The abbreviation for a picture tube is CRT. The cathode-ray tube might be found in the TV receiver, oscilloscope, computer, monitor, and camcorder. A TV CRT tube displays the image received. Sometimes the picture tube in the TV set is called a kinnee or kinescope.

The cathode-ray tube in the oscilloscope might have a gun assembly consisting of a heater element, cathode, anode, and lower- and upper-deflection plates. A stream of electrons from the cathode element is directed towards the phosphor screen with high voltage applied to the anode button. The deflection plates direct the beam to the front of the cathode-ray tube (Fig. 4-23).

TV CRT

What and where

The picture tube within the black-and-white TV chassis consists of a single heater element, cathode, signal grid, screen grid, and high-voltage anode button. While in the color TV set, the color gun assembly has three heater elements, three cathodes, control grid, screen grid and focus grid, and high-voltage anode connections. The front screen has green, red, and blue dots in which each color beam hits each colored dot to form a colored picture. Instead of deflection plates, a deflection yoke with a vertical and horizontal winding provides vertical and horizontal sweep (Fig. 4-24).

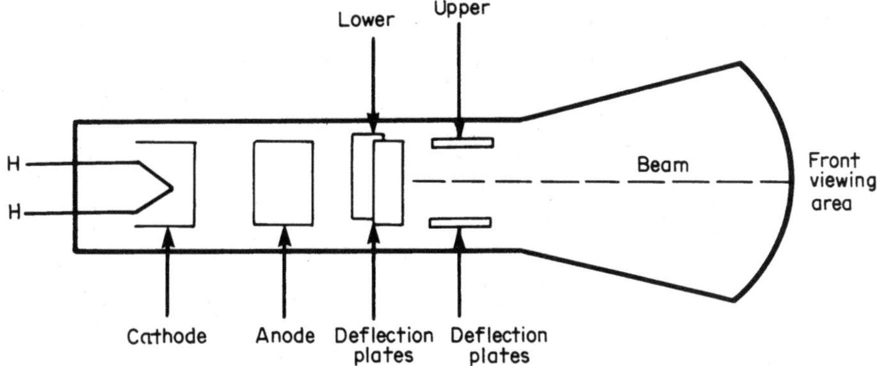

4-23 The oscilloscope CRT consists of a glass tube with a heater, cathode, anode, lower-deflection plates, and upper-deflection plates.

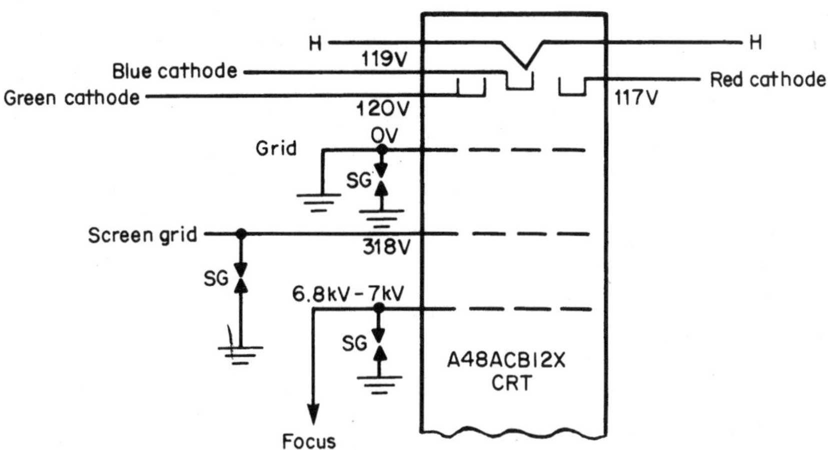

4-24 The CRT found in the TV set consists of heaters, cathodes, grid, screen grid, focus grid, and high-voltage anode elements.

The heater voltage is taken from a separate winding of two or more turns of large wire within the flyback transformer. A color picture tube signal is sent to each color cathode with voltages fairly close, above 100 volts. The control grid is at ground potential. A screen grid and focus elements are taken from a voltage divider network across the lower winding in the secondary of the horizontal output transformer. The focus voltage in early TV sets varied from 3.5 to 5 KV, while in the larger screen sizes the focus voltage may vary from 5.5 to 7.5 KV. The HV anode element is fed 25 KV to 27 KV, and in larger screens should not exceed 32 KV.

When

The defective picture tube might have no raster, weak picture, poor brightness, no control of brightness, intermittent raster, retrace lines, all one color screen, one color missing, or a shorted gun assembly producing chassis shutdown. A washed-out, negative, blotchy, or shiny picture and poor focus can also result from a defective CRT.

A poor heater connection or open heater might prevent the heater element from lighting up in the gun assembly. Weak or poor brightness can result from a weak cathode element. The intermittent raster might be caused by flaking material from the cathode element that has lodged between the grid elements. When one color is missing, suspect a weak color-gun assembly or weak color signal from the color output transistors. Intermittent picture can result from a poor CRT harness or heater connection. Suspect a cracked glass assembly when the picture tube arcs inside the deflection yoke and shuts down the chassis. Chassis shutdown can also result from dust clogging up the spark gaps in the CRT socket assembly.

Testing

Test the picture tube with a CRT tester (Fig. 4-25). Measure the resistance of the two heater pins to determine if the heater elements are open and there is no raster or picture. Check the correct voltage found upon each element and compare with those found on the TV schematic. Measure the high voltage with an HV probe or meter at the anode terminal. Tap the end of the gun assembly and notice if lines appear in the raster. Slightly move the CRT socket and notice if heaters go out and the picture turns black. Replace with a new socket harness assembly. Blow out the

4-25 Check the picture tube with a CRT tester and high voltage with the high-voltage meter probe.

spark gap assembly within the CRT socket when the brightness comes up and the chassis shuts down.

The raster might be dark with an incorrect grid, focus, and anode high voltage. Rotate the brightness control with firing lines in the raster. When the focus control is rotated and the control arcs over, replace both brightness and screen controls. Improper brightness, poor focus, lines at the top, weak picture, insufficient width and height of raster, and a color missing can be caused by defective components in circuits tied to the picture tube (Fig. 4-26).

Camcorder CRT

What and where

The 1-inch CRT found in the early camcorder is used as a display monitor. What you see in the eyepiece is what you get (Fig. 4-27). Often, when the camcorder was dropped, the small cathode-ray tube was cracked or broken. Today, in most camcorders, a fluorescent display serves as a picture monitor.

What and where

The small CRT found in the camcorder might consist of a heater, cathode, grid one (G1) and grid 2 (G2), and high-voltage anode. A cathode element is connected to one of the heater terminals with only a four-terminal tube socket (Fig. 4-28). The small black-and-white CRT might be mounted in a separate eyepiece that is attached to the camcorder.

The electronic viewfinder has video, brightness, vertical, and horizontal sweep circuits feeding the CRT. A flyback transformer supplies heater, screen grid (G2), and high voltage to the picture tube. Focus is obtained with a focus magnet on the long neck of the picture tube. The supply voltage to the cathode-ray circuits might

4-26 Check for defective components in the vertical circuits with bunched-scanning lines at the top of the raster.

4-27 The display monitor is a small 1-inch CRT in the early camcorder and is used to select the viewing area.

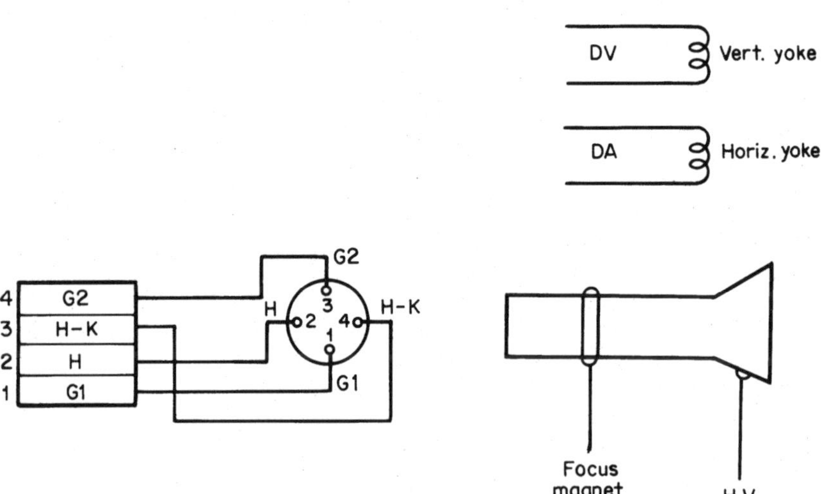

4-28 The four-pin CRT viewfinder has a focus magnet for focusing the picture in the viewfinder of the camcorder.

be only 5 volts. The vertical and horizontal sweep circuits are fed to a very small deflection yoke.

When

Check the heater elements when the CRT has no raster or display. Suspect the video and brightness circuits with poor brightness. Check the vertical section for a white horizontal line and improper vertical sweep. Go directly to the horizontal section with no horizontal sweep or high voltage at the anode terminal. Remember the voltages within the one-inch display tube are very low. Most manufacturers do not place correct voltages upon the CRT schematic.

Testing

Although there are no picture tube testers that will test the condition of the small CRT, a low-ohm test across the heaters will indicate an open tube. Check for light of the heater terminals. Does the tube light up? Suspect a weak gun assembly when the picture is dark and when the video and brightness circuits are normal (Fig. 4-29). Measure the HV applied to the anode terminal. Check the voltages on the grid terminals of G1 and G2. Before ordering a new CRT, make sure the cost will not exceed the price of an electronic viewfinder assembly.

Computer or word processor monitors

The monitor is a device (such as a TV, video, and computer) that displays signals. Check the computer and word processor monitor in the same manner as the TV picture tube circuits (Fig. 4-30). The same symptoms that appear upon the computer and word processor monitors are somewhat like that of a TV picture tube. Measure the voltages upon the CRT and compare to the schematic.

4-29 The small viewfinder CRT has a deflection yoke on the tube to connect to the deflection circuits.

4-30 Check the computer and word processor monitor in the same manner as the TV picture tube.

CRT brighteners

What and where

The picture-tube brightener is an electronic device that plugs over the CRT pins and then plugs into the picture-tube socket. This gadget increases the voltage upon the heater terminals which increases the electron flow from the cathode element, resulting with increased brightness. CRT brighteners are placed on the end of the picture tube when the picture is very dim and out of focus. A small step-up power transformer is located in the brightness assembly (Fig. 4-31). Although the CRT brightener may provide another six months of fair brightness, a picture-tube restorer process will provide longer picture-tube service before actual tube replacement.

Testing

It is cheaper to replace the CRT brightener than to replace the CRT (Fig. 4-32). Make sure the picture-tube pins are clean. Scrape off each pin with a pocketknife. Most CRT brightener problems relate to burned or overheated socket and tube pin terminals. To determine if the transformer is normal, take a continuity test of the primary and secondary leads of the transformer at the socket and plug ends.

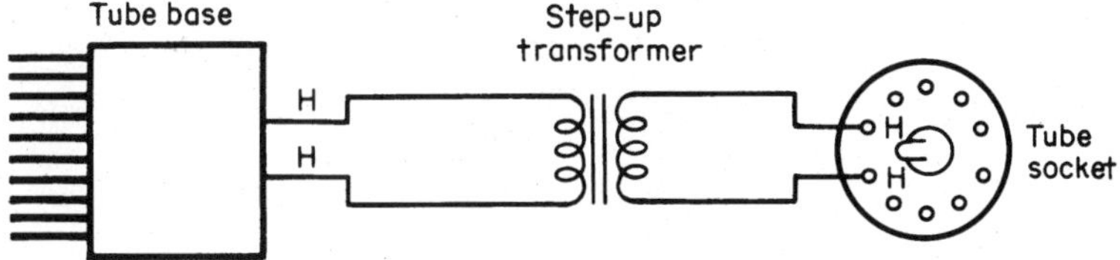

4-31 A small step-up transformer is found in the base of the CRT brightener.

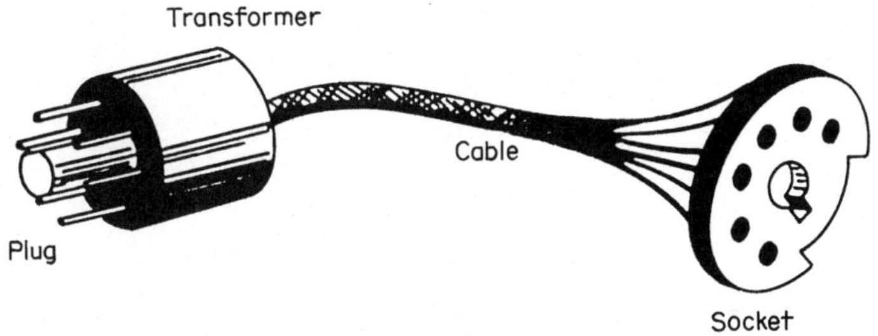

4-32 The brightener plugs into the TV CRT socket and then plugs into the picture tube to increase the heater voltage.

Crystals

A crystal may form a plate or bar cut from a piezoelectric material. The piezoelectric crystal can be made of quartz, Rochelle salt, or other synthetic material that delivers a voltage when a mechanical or voltage force is applied. The crystal detector found in the early crystal set was made of galena, or a homemade crystal detector can be constructed from a sulfur and lead mixture (Fig. 4-33).

The crystal is found in oscillator circuits, amplifiers, calibrators, receivers, earphones, microphones, loudspeakers, meter circuits, phono pickups, probes, sensors, transducers, resonators, piezo tweeter speakers, and many other components. Several crystals might be found in CB or police scanners. The color section of a TV set has a 3.58-MHz crystal.

When a quartz crystal vibrates, ac voltage appears between the pair of plates and is called a piezoelectric crystal. The fundamental frequency of the vibration depends upon the thickness and type of cut of crystal material. The thinner the crystal, the higher the frequency. Both sides of the crystal are placed between a metal surface in a plug-in type holder. Some crystals have wire leads like those found in a crystal-controlled clock.

Ham radio crystals

What and where

Within the ham radio bands, the crystal oscillators are used in transmitting and also in receiving. The oscillator is a circuit that generates a signal for the ham bands.

4-33 A galena or germanium IN34 crystal is found in the early crystal receiver.

The stability of the crystal keeps the transmitter on the right frequency and likewise the receiver, to be able to hear certain tuned ham bands. Crystals are very reliable in oscillator circuits. The local oscillator in the ham transmitter is crystal-controlled for the greatest stability.

The crystal oscillator frequency is determined by a piezoelectric crystal. The early crystal oscillator was built around vacuum tubes. Today, the crystal oscillator can be built with bipolar and field-effect transistors (Fig. 4-34).

Testing

Small crystals can be tested in a crystal test circuit or with a crystal tester. The crystal test circuit might be a simple crystal circuit that will show the crystal will oscillate. Although, it will not show at what frequency the crystal oscillates, the test circuit will cull out the defective crystal. Likewise, a simple crystal tester will indicate if the crystal is good or bad (Fig. 4-35).

Camcorder crystal oscillators

You may find only one crystal oscillator circuit within the camcorder. In the Samsung SCX854 model, a simple 12-MHz crystal is found in the MI-COM process control circuits. There are five different crystal oscillator circuits found in the Canon ES1000 camcorder, X190 (11.895 MHz), X230 (16 MHz), X231 (32 kHz), X1001 (28.64 MHz), and X2101 (3.58 MHz).

Crystal X230 is tied to pins 50 and 51 of the main MI-COM IC231 in the Syscon-Servo section. While crystal X231 is connected to pins 54 (XC IN) and 55 (XC OUT) (Fig. 4-36). X190 controls the ATF IC90 circuits found in the Syscon-Servo section. The crystal oscillator X2101 (3.58 MHz) is found at pin 57 of Y/C process IC2101 controlling the VXO circuits.

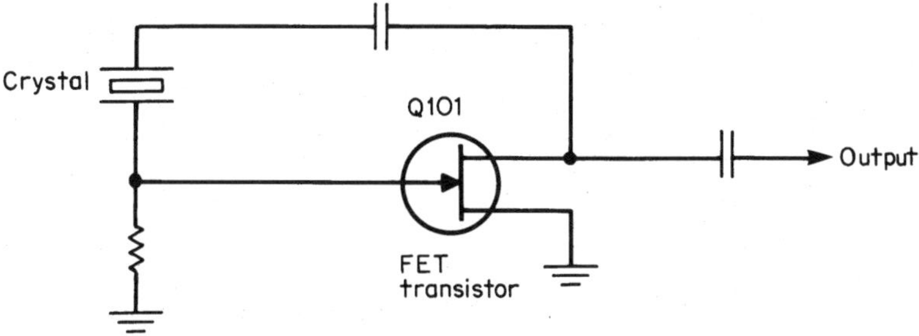

4-34 The frequency of the crystal oscillator circuit is determined by the crystal.

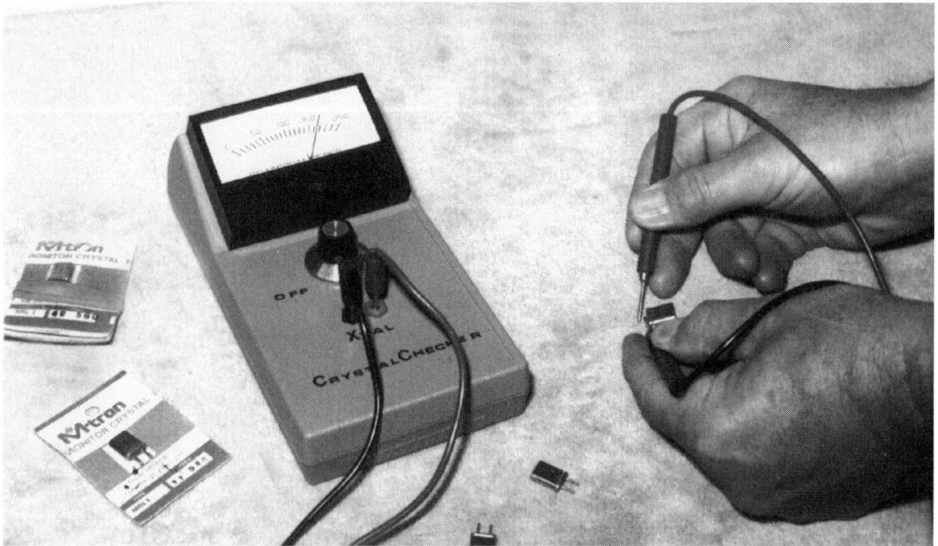

4-35 Check the suspected crystal on a crystal checker to determine if it is good or bad.

Testing

The quickest method to determine if the crystal oscillators are oscillating in the camcorder is to take a scope waveform. Place the scope input terminal on either side of the crystal that ties to a certain pin with the ground lead clipped to a common ground in the camcorder. The amplitude and frequency of the waveform can be seen upon the scope screen. Suspect a defective crystal or IC when no oscillator waveform is measured (Fig. 4-37). Take critical voltage measurements upon the IC to determine if IC is defective.

CD player crystal oscillators

What and where

One or more crystals might be found in compact disc players. These crystal-controlled circuits are found either in the master MI-COM or system control IC. In

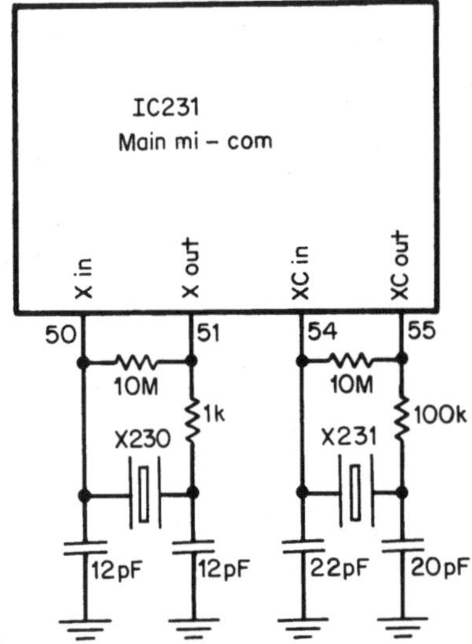

4-36
Crystals X230 and X231 are found in the Canon ES1000 camcorder.

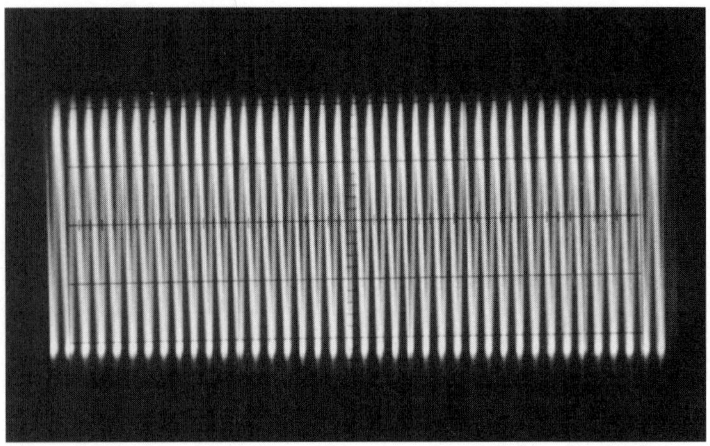

4-37 Check the crystal oscillator waveform on the crystal IC pin terminals to determine if the circuit is oscillating.

a Dennon compact disc player, a 32.768-kHz and 4.193-MHz crystal are found on pins 46 and 45, and 42 and 43 of master MI-COM IC601 (Fig. 4-38).

Testing

Check the waveform upon each crystal terminal to determine if the crystal is oscillating. Spread out the waveform upon the scope so you can see each cycle (Fig. 4-39). If there is no waveform on either crystal, suspect the IC601 or improper supply voltage to the MI-COM IC. Usually, both crystals will not appear bad at the same time. Measure the voltage at each crystal pin and compare to the schematic. Replace the defective IC601 when both crystals indicate no waveforms.

TV 3.58-MHz crystal

What and where

A color crystal at a 3.58-MHz frequency is found in the color circuits of the color TV chassis. Today, the color circuits are found in a large IC that might include the IF, luma, chroma, and deflection processor IC. To locate the color circuits, look for a small crystal mounted close to a large IC (Fig. 4-40).

When

The 3.58-MHz crystal must oscillate before there is any color within the picture. The no-color symptom can be caused by any component in the color circuits. If the 3.58-MHz crystal does not oscillate, no color can be seen in the picture. A leaky color IC can cause a no-color symptom. Bypass and electrolytic capacitors tied to the IC terminals can cause no color in the picture. Improper or no low voltage from the power supply can produce weak or no color (Fig. 4-41). The intermittent color crystal can make intermittent color in the picture and might have a color barber-pole symptom.

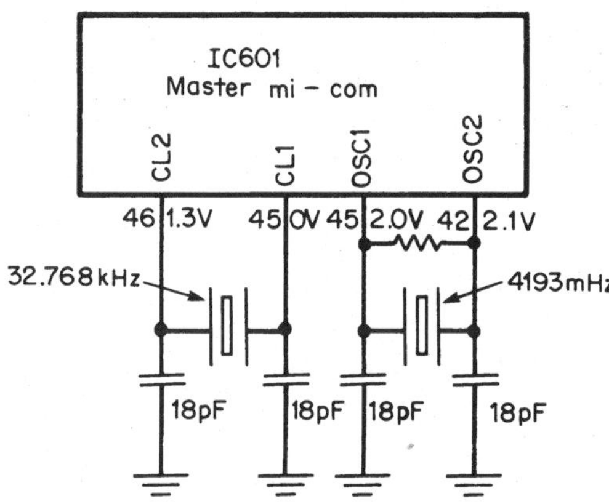

4-38 Two different crystals are found in the master MI-COM IC601 in a Dennon compact disc player.

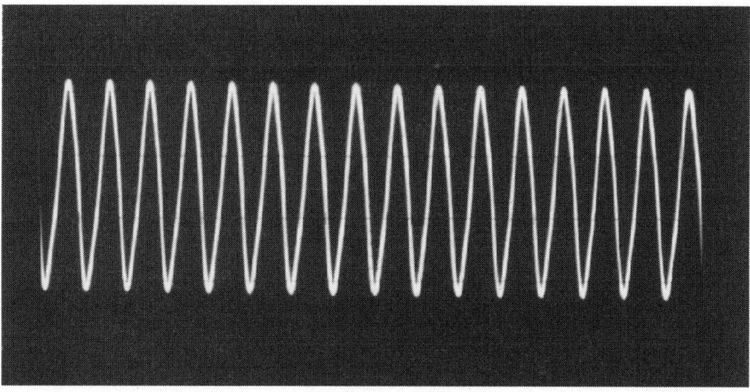

4-39 Check the pin terminals of the IC with a scope waveform to determine if the crystal circuit is functioning.

4-40 Locate the color IC and color circuits by identifying the 3.58-MHz crystal on the TV chassis.

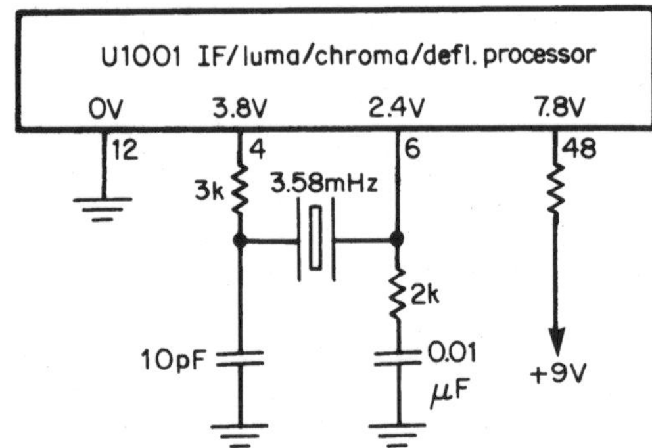

4-41 The color 3.58-MHz crystal is located within one large U1001 IC with the IF/luma/chroma/and deflection circuits.

Testing

Check the color waveform at the 3.58-MHz crystal terminals with a scope (Fig. 4-42). Suspect a defective crystal or color IC with no color in the picture. Measure the supply voltage terminal of the suspected color IC. Check the voltage upon the color terminals of IC. If you do not know what terminals are connected to the color circuits, look the IC part number up in the universal semiconductor manual. Replace the crystal and bypass capacitors in the crystal circuits for no or intermittent color.

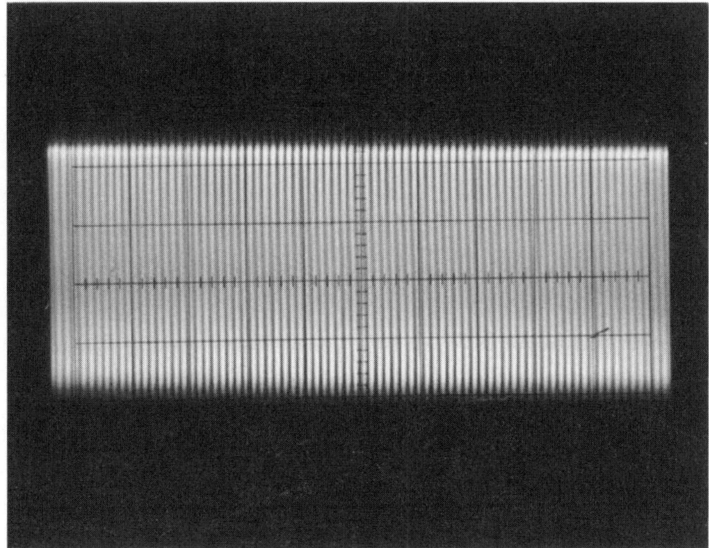

4-42 Check the 3.58-MHz crystal for an oscillating waveform when color is missing in the picture.

5

CHAPTER

Testing components D

(Degaussing coils to diodes)

A degaussing coil comes on automatically when the TV set is turned on. To remove difficult areas that have poor purity, such as the corners of the TV screen, use large outside degaussing coils. These handheld degaussing coils are wound with many turns of copper wire and plug into the ac power line (Fig. 5-1). The delay line, demagnetizer components, and how to test every diode found in the consumer electronics field are also discussed in this chapter.

The degaussing coil

In the early TV chassis, the degaussing coil was turned off and on with a thermistor component. When the chassis was turned on, power line voltage was applied to the degaussing circuit. The thermistor unit has low resistance to current, and after on for awhile, the heat generated offers higher resistance to the degaussing coil and shuts down the degaussing process.

(Fig. 5-2). A degaussing coil magnetic field demagnetizes objects that have been magnetized. The north pole can magnetize the metal hole screen in the color picture tube. Shutting off the floor sweeper in front of the TV set can magnetize the CRT. Large magnetic fields, such as large pm speakers set next to or on top of a TV, can magnetize the screen. When the TV screen becomes magnetized, colored and blotchy areas will appear in the picture, resulting in poor purity. Remember, the degaussing coil is mounted around the outside edge of the picture tube.

In today's TV chassis, you can find a degaussing relay circuit that is controlled by a control microprocessor (Fig. 5-3). Q3000 serves as a degaussing switch, turning on relay K3001, which in turn switches in the degaussing coil circuit. Now the power line voltage is applied across the relay switch points, thermistor, and degaussing coil.

5-1 Use the hand-held degaussing coil to demagnetize the front of the TV picture tube.

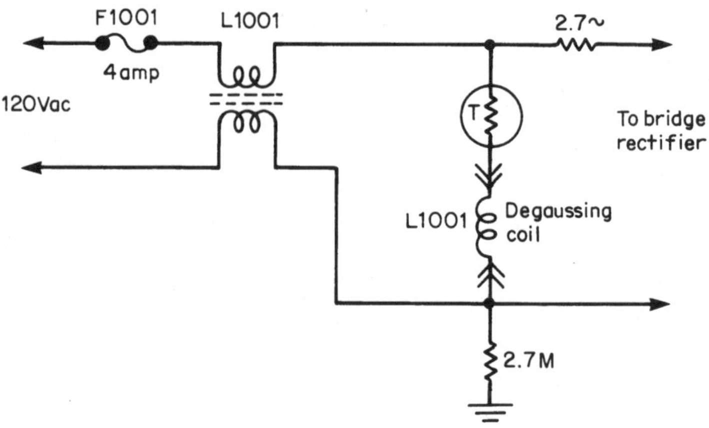

5-2 In the early degaussing circuits, the degaussing coil is in series with a thermistor across the ac power line in the TV chassis.

As the thermistor heats up, the resistance increases and eventually ceases degaussing operation. Each time the TV set is turned on, the degaussing coil removes magnetized areas on the screen.

When

Suspect the degaussing coil circuits when several areas of color are found left on the picture tube. This can happen at any time. To solve this problem, the TV set can be moved to the other side of the room, changing the relative position with the magnetic north pole. By turning the TV off and letting it sit for 20 minutes, the degaussing process should begin when the set is turned on. This area of color can be removed with a hand-held degaussing coil. Check the degaussing circuit when poor purity is noticed on the picture tube.

Testing

One of the biggest problems with the early degaussing circuit was that the terminal wires of the thermistor would burn off and open up the degaussing circuit. Measure the resistance of the thermistor when cold and check the resistance on the schematic. Measure the power line ac voltage across the thermistor and degaussing coil. Shut off the set and check the continuity of the degaussing coil with the low-ohm scale of the ohmmeter.

In degaussing coil relay circuits, check the relay switch contacts. Measure the resistance across the solenoid winding of the relay. It is best to remove the relay from the circuit. This will eliminate a fixed diode wired across the solenoid winding. Check the voltage on the base and collector of Q3000. With very low voltage at the collector terminal, suspect a leaky transistor. Degaussing relays have been noted to cause intermittent degaussing, resulting in intermittent purities on the TV screen.

Delay line

When and where

The delay line in the TV chassis introduces a time lag in a received signal. Usually, the delay line is mounted between video amp and video buffer transistor or between video buffer transistors (Fig. 5-4). You might not find a delay line within the latest TV chassis.

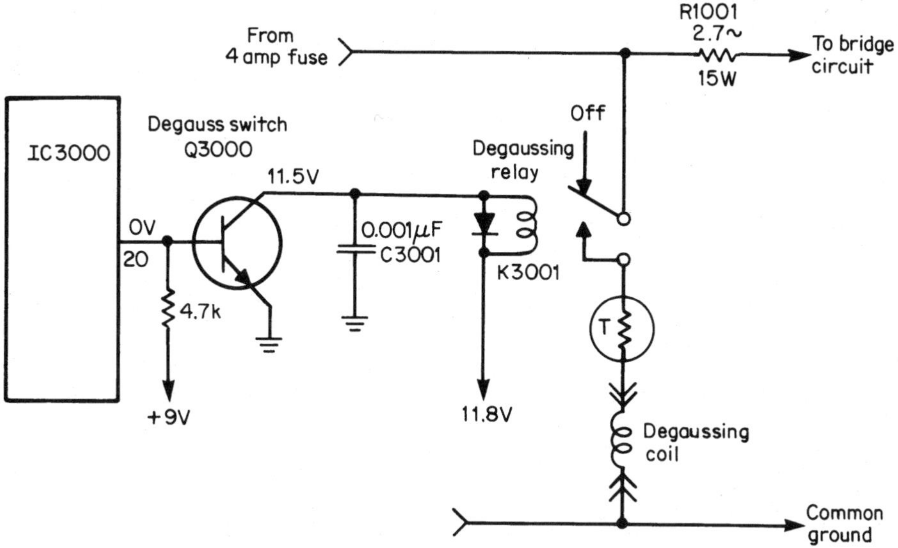

5-3 Today, the degaussing coil might be controlled with a degaussing relay circuit.

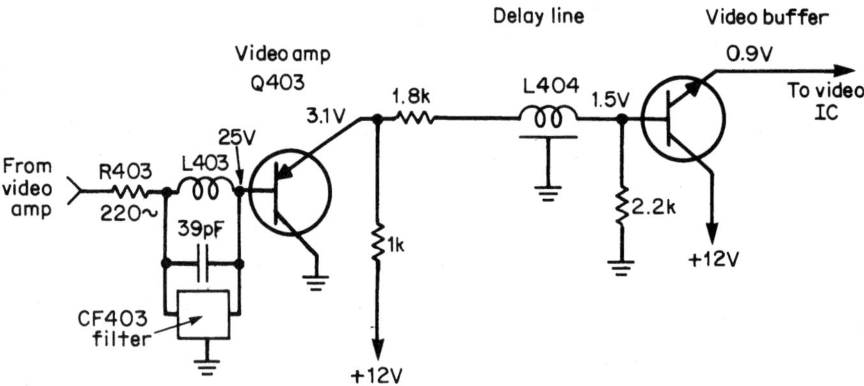

5-4 The delay line is located between the video buffers or amps in the TV chassis.

When

A washed-out picture with retrace lines can be caused by an open delay line. The delay line can short from the coil winding to ground, resulting in no video signal. A normal sound and intermittent video can occur with poor delay-line soldered connections. With a smeary out-of-focus picture, suspect a defective delay line.

Testing

For a no-video and washed-out-picture symptom, take a waveform in and out of the delay line. Scope the in and out video signal. A normal video signal in and weak or no signal out indicates a defective delay line. Double-check the delay line by measuring the resistance of the delay coil with the ohmmeter (Fig. 5-5). Check across the coil and also from each side to chassis ground. It is wise to remove the delay line from the circuit and check from each terminal to the ground terminal, if the resistance is less than 1K ohms while in the circuit. The resistance of the delay line may vary from 50 to 100 ohms. For instance, the delay line in a Sharp 195B60 model is 69.7 ohms. Check the inductance of the delay line with an LCR meter if the inductance is marked on the parts list.

Demagnetizer

What and where

A demagnetizer removes magnetism from a component or object. The demagnetizing force reduces residual induction of material that is magnetized. Often, a demagnetizer tool has many turns of wire that may operate from the ac or dc source. An 8-track or cassette hand-held demagnetizer may operate from the cigarette lighter to demagnetize the tape head of a cassette or VCR player (Fig. 5-6). The large hand-held degaussing coil operates from the power line and demagnetizes CRTs to improve linearity and correct color impurities.

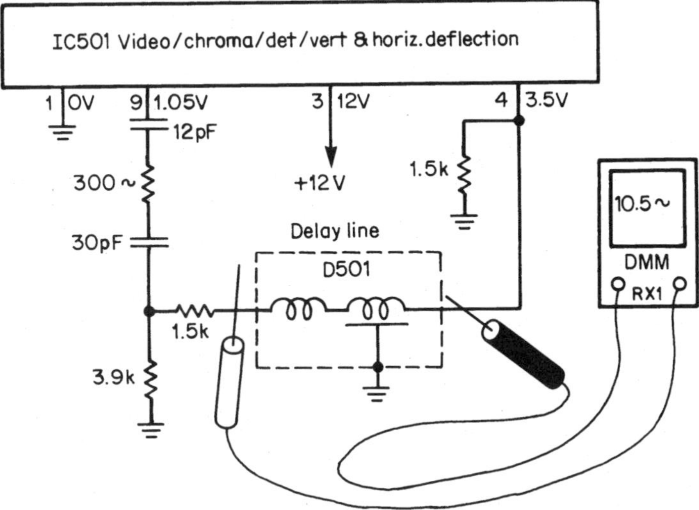

5-5 Check the continuity of the delay line with the low-ohmmeter range of the DMM.

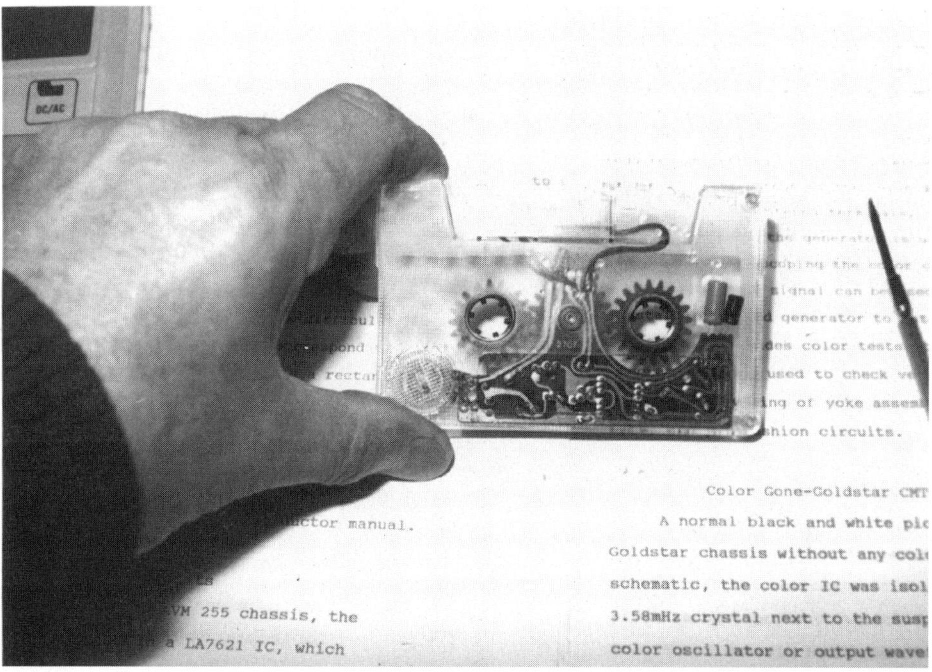

5-6 The outside cassette demagnetizer removes magnetism from the VCR and cassette player tape heads.

A degaussing coil is energized within the TV set and degausses the faceplate inside the picture tube. This prevents spotted or magnetically contaminated areas on the faceplate of the CRT. The cassette recording is erased as the tape passes over the erase head and removes the previous recording. This is another form of magnetic removal. A battery-operated cassette head demagnetizer improves recording and playback quality by inserting the cassette into the player. The head demagnetizer, which operates from the power line, has a curved demagnetizing tip that reaches hard-to-access areas of record and playback heads. The cassette or video cartridge can be placed in a bulk eraser that demagnetizes or erases the whole tape.

Place a screwdriver blade next to the demagnetizer head and see if it vibrates. Measure the resistance across the ac or cigarette plug to determine if the winding is open in the demagnetizer. Remove the cover and check each coil winding connected to the ac cord to see if it is broken or has a bad connection. With the ohmmeter, check for a break in the ac power cord.

Diodes

A diode is a device that contains an anode and cathode element. The semiconductor rectifier with a PN junction is a fixed diode. In the early tube days, a tube with a cathode and plate was used as a rectifier. A rectifier changes ac into dc. The diode comes in many forms and shapes: general-purpose diodes, bridge circuits, fast-recovering diodes, high-speed switching diodes, high-voltage diodes, RF switching diodes, signal diodes, stud-mounted diodes, surface-mounted diodes, varactor diodes, variable-tuning capacitor diodes, and zener diodes. Diodes are found everywhere. The fixed diode is used throughout products designed for the consumer electronics field (Fig. 5-7).

What and where

The fixed rectifier diode may consist of copper sulfide, germanium, galena, selenium, and silicone material. The germanium and galena rectifiers are found as signal rectifiers. A high-voltage stacked diode can also be used in the anode CRT circuit of a black-and-white TV chassis. In the early radio or TV chassis, selenium rectifiers were found to replace the diode-tube rectifier. The silicon fixed rectifier is found in half-wave and full-wave power supplies. The current may range from 1 to 3.5 amps in the TV chassis. A bridge circuit contains four silicon diodes in a rectifier bridge circuit (Fig. 5-8).

The fast-recovery and high-speed switching diodes are found in computer, VCR, and TV circuits. RF switching and signal diodes are found in radio and TV RF circuits. The varactor and variable-capacitor tuning diodes can be located in radios, TVs, and VCR circuits. Surface-mounted diodes (SMDs) are found in practically all small products, cassette players, CD players, TVs, and VCRs. The zener diode can be located in many low-voltage power supplies of consumer electronic products found in the entertainment field.

General-purpose diodes

The general-purpose diode is a small-signal semiconductor that is used in many different circuits. This diode may be used in rectification, signal, and switching-

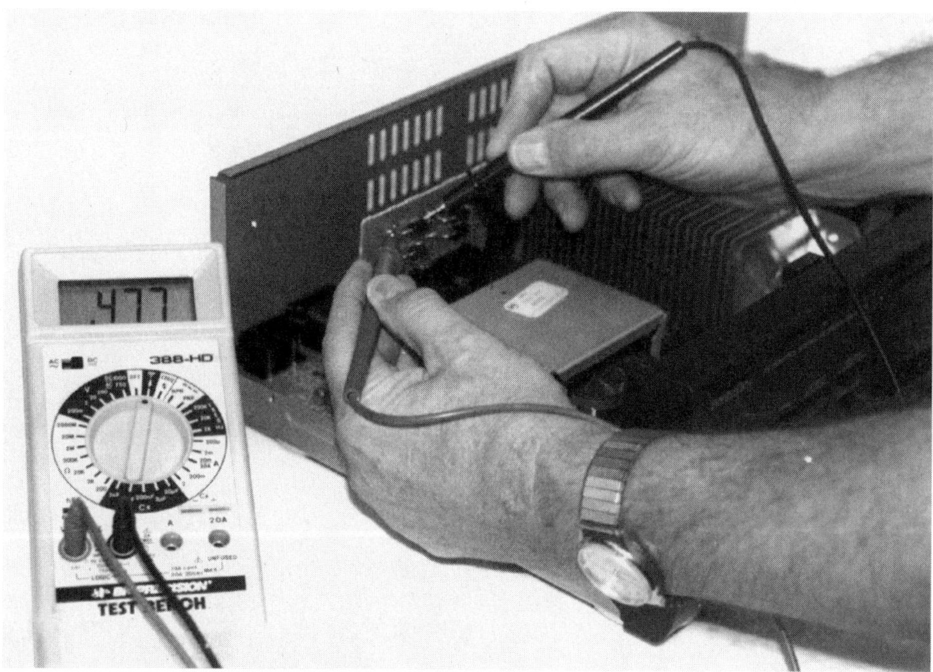

5-7 There are many different types of diodes found in the various electronic products.

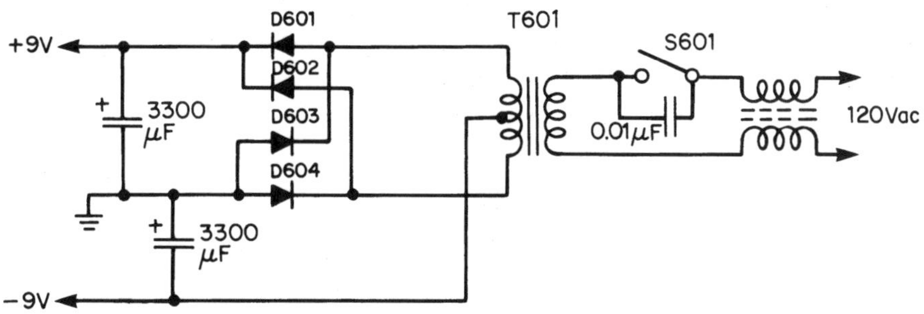

5-8 A bridge rectifier circuit in the compact disc player.

detection circuits. The common IN34A or IN60 and IN914 are general-purpose and switching diodes (Fig. 5-9).

What and where

The general-purpose fixed IN34A or IN60 diode is used as a crystal detector in the early crystal receiver. This same diode is found in the signal-detection circuit of the AM-IF detector radio section (Fig. 5-10). The IN34A and IN60 is a general-purpose signal diode made from germanium material. Most general-purpose diodes operate in circuits that are less than 60PIV (peak inverse voltage) or PRV (peak reverse voltage). The IN34 and IN60 signal diodes have axial-type leads. When soldering into the circuit, use a heat sink or long-nose pliers at the end being soldered, so as not to melt down and destroy the contact point of the germanium diode.

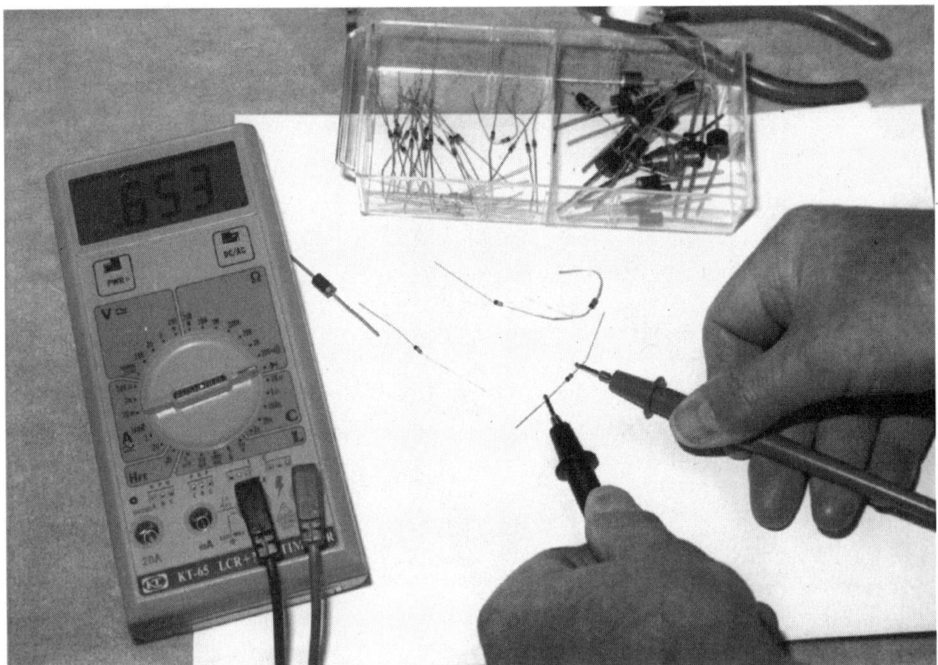

5-9 The germanium diodes are contained in a clear glass enclosure, unlike the regular diode.

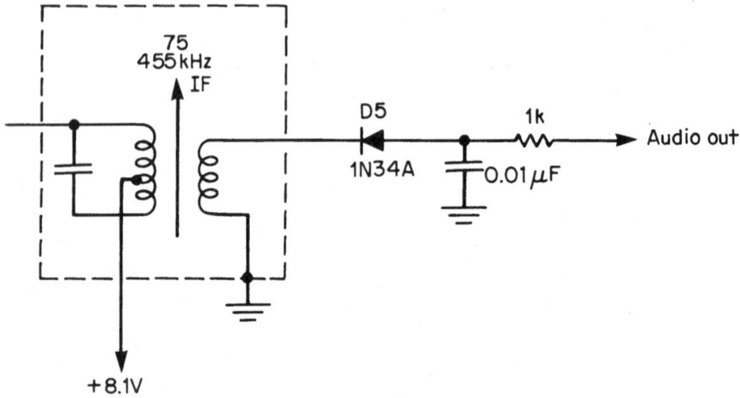

5-10 The IN34A germanium diode rectifies the radio frequency to audio.

The IN914 general-purpose diode is made from silicon material (Fig. 5-11). The IN3600, IN4150, and IN4446 series are also silicon signal diodes with axial terminal leads. The surface-mounted signal diode might have an MMB914 industry number. These SMD diodes are miniature and solder directly on the PC wiring.

When

The defective signal diode might become leaky, have a high resistance, and appear open. The radio receiver might be weak or have no reception. Although the signal diode does not cause too much trouble, an intermittent signal reception can be caused by poor terminal connection. Intermittent radio reception might be caused by an improperly soldered connection.

Testing

Check the signal diode right in the circuit with a diode or resistance measurement. Signal diodes can be tested with an internal audio amp or oscilloscope. With an open signal diode, no or very little radio signal can be seen on the scope. Critical voltage tests can indicate a leaky signal diode.

Check the signal diode with low resistance measurements. A good diode will only show a low reading in one direction. When the test leads are reversed, you should get infinite measurement. The diode measurement in one direction should be around 10 ohms with the pocket VOM and red lead to the collector terminal. The resistance measurement of an IN34A diode with the diode test of the DMM should be around .265 ohms with a red probe at the anode terminal (Fig. 5-12). A FET-VOM resistance test of the signal diode should be about 10 ohms.

Measure the diode resistance measurement with the DMM and the positive (red) probe to the anode and the black probe to the cathode of the signal diode (Fig. 5-13). Reverse these test leads when diode tests are made with the VOM or FET-VOM. A leaky signal diode will show a low resistance measurement in both directions, while the open diode has infinite reading.

The silicon general-purpose diode (IN914 series) will have a very different measurement. The normal silicon diode will have a lower resistance measurement in one direction and no reading with test leads reversed. The silicon signal diode will have a low resistance measurement of around 20 ohms with the VOM, 15 ohms with a FET-VOM, and around .587 ohms with the diode test of the DMM.

General-purpose silicon rectifiers

The general-purpose silicon rectifiers are used in rectifier circuits such as the low voltage power supplies. A silicon diode is a semiconductor processor with silicon material. The abbreviation for silicon is Si. The silicon semiconductor is processed from sand (Fig. 5-14). The silicon rectifier consists of a junction between the N and P silicon wafer or plate. The silicon rectifier has a higher current capacity and surge than the germanium diode. The silicon rectifier charges ac current to dc voltage. The diode passes current in only one direction.

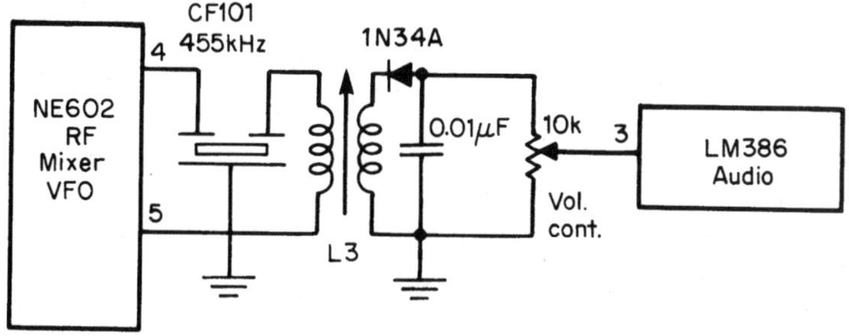

5-11 A IN34A diode found in a shortwave receiver project.

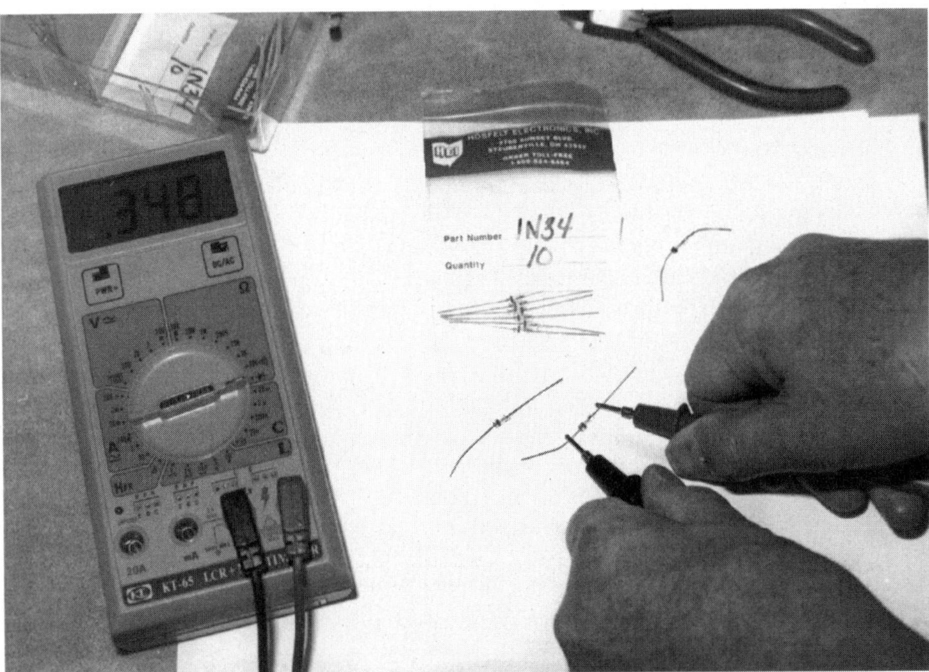

5-12 Taking resistance measurements of a IN34A diode with the diode test of a DMM.

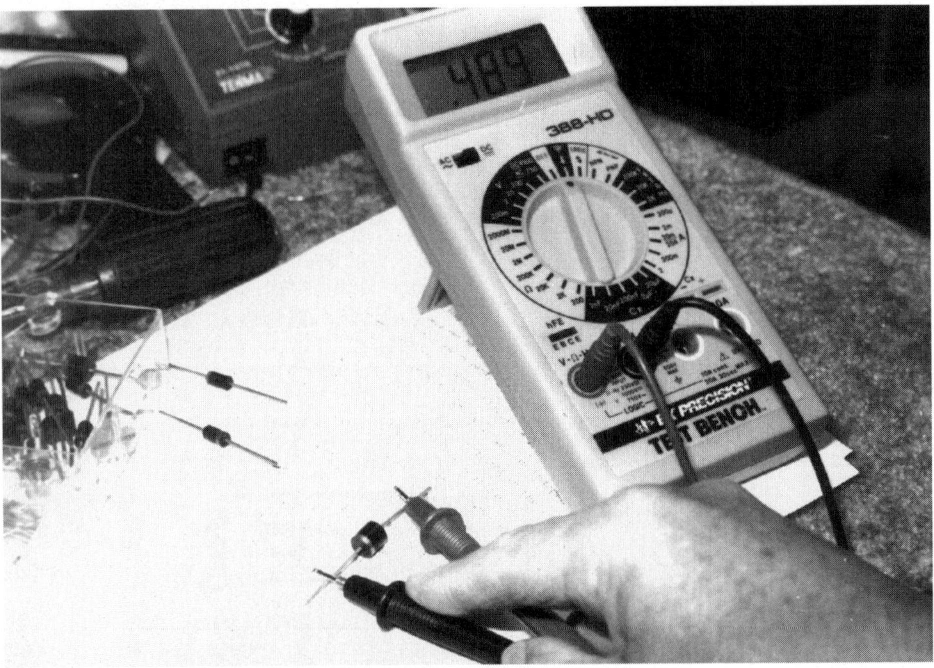

5-13 Place the red probe to the anode and the black probe to the cathode to check the low resistance with the DMM. Reverse the test leads for accurate diode tests with the VOM.

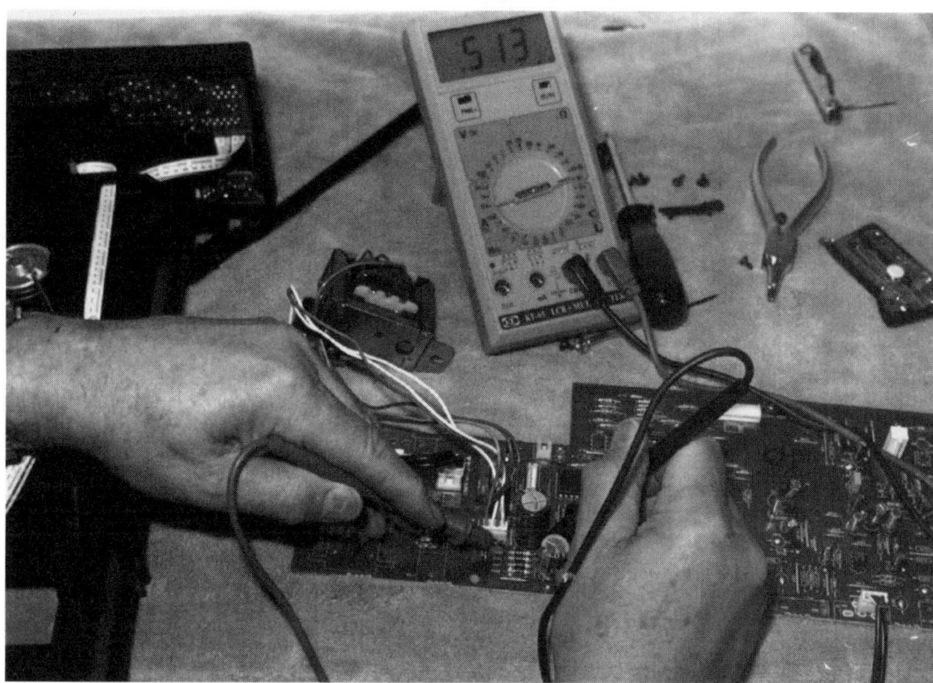

5-14 Four silicon diodes in a bridge circuit of a compact disc player.

What and where

The diode rectifier appears in 1-amp diodes up to 1000 volts, 2.5-amp diodes of 1000 volts, and 3-amp diodes at 1000 volts. A 2.5-amp silicon diode is used extensively as a universal replacement in the TV chassis by the electronics technician (Fig. 5-15). The PTC205 and RL251 series are 2.5 amp silicon diodes. The 1-amp IN4007 is rated at 1,000 volts, while the IN5408 has a 3-amp 1,000-volt rating. Today you will find some 3-amp silicon diodes in the power supplies of the TV chassis.

Current flows when the cathode element of the rectifier diode is negative in respect to the anode terminal. When the cathode is positive to the anode, no current flows or passes through. The half-wave rectifier circuit uses only one diode rectifier, while the full-wave circuit has two or four diodes in a bridge circuit. The cathode is the positive, and the anode is the negative terminal of a diode rectifier.

When

The silicon rectifier can become shorted, leaky, or appear open. A shorted diode might have a resistance lower than 10 ohms, while a leaky diode has a resistance of less than 100 ohms. Often, the shorted diode might open up the primary winding of the power transformer (Fig. 5-16). Today, the step-down power transformer is wound with larger-diameter wire, while the primary winding has many turns of smaller copper wire. A leaky diode can destroy the power fuse and resistor and make the power transformer run quite warm. Of course, the open silicon rectifier has no voltage at the cathode terminal.

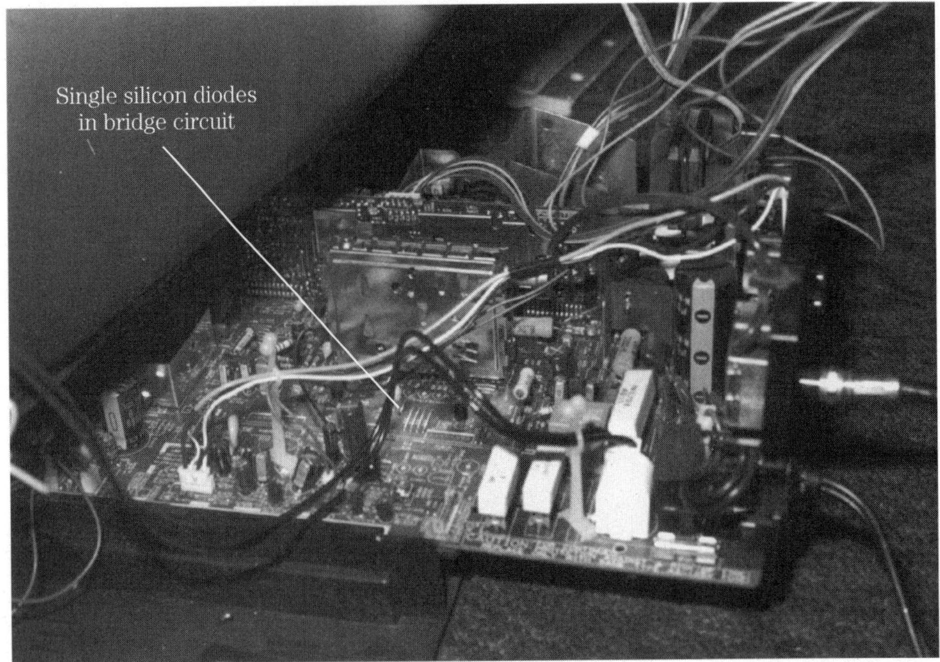

5-15 Replace four silicon rectifiers in the TV power supply with 2.5-amp diodes.

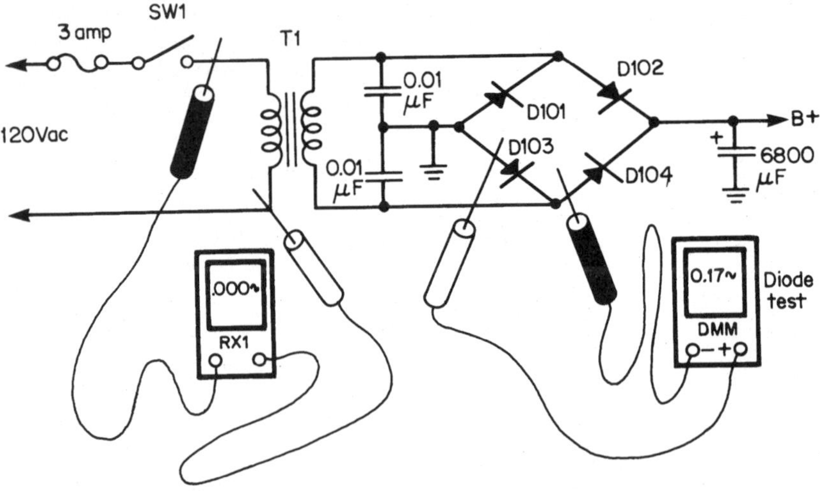

5-16 When a leaky or shorted diode is found within a step-down power transformer circuit, check for an open primary winding at the ac cord.

When installing a new silicon diode, make sure the diode voltage and current ratings are adequate. A 1-amp diode will run warm when it's replaced in a 3-amp circuit. If the voltage and current ratings are low, the diode will overheat and become leaky. Always observe correct polarity when replacing diodes. The cathode (positive) terminal is always connected to the filter capacitor in a low-voltage power supply circuit. If the diode is installed backwards, the diode can be ruined with other damaged components.

Testing

The silicon diode can be tested within a diode tester, DMM diode test, VOM resistance test, FET-VOM resistance measurement, and voltage measurements (Fig. 5-17). Check the voltage output at the cathode terminal of the suspected diode. If the voltage is very low or absent, suspect a leaky diode or defective filter capacitor. Next, take an in-circuit diode test with a DMM or diode tester. Remove one end of the suspected diode for a correct test.

A silicon diode test with the VOM should be around 15 ohms in one direction, .495 ohms with the DMM, and 10 to 15 ohms on the FET-VOM (Fig. 5-18). An infinite measurement should have no resistance measurement on a silicon rectifier with reversed test leads. Notice that the silicon rectifiers used in power supplies have a much lower resistance measurement. Observe correct polarity when installing a silicon rectifier.

5-17 A diode is checked on a diode tester within a bridge rectifier component.

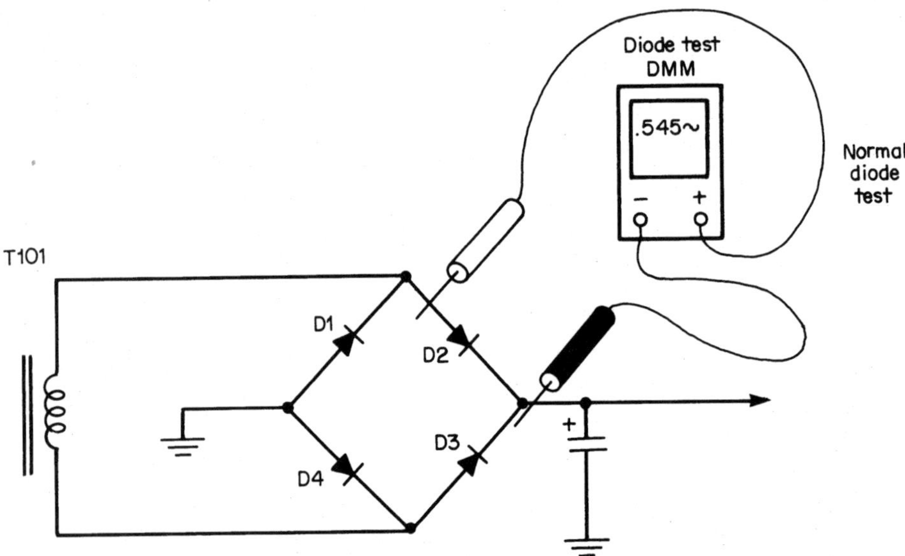

5-18 Test each silicon diode within the bridge rectifier circuits.

Bridge rectifiers

What and where

A bridge rectifier is made up of four silicon diodes in a bridge circuit. Often more than one diode is destroyed in the bridge circuit when one diode shorts out or a faulty capacitor becomes leaky. Excessive lightning or power outage damage can destroy all four diodes in a bridge circuit. The bridge diodes can be enclosed in a full-wave circuit with two or four diodes in one component. You may also find four separate diodes mounted in a bridge circuit (Fig. 5-19).

When four rectifying diodes are connected in a bridge circuit, they are called a full-wave bridge circuit. During each half-cycle, the ac input is rectified by the opposite pair of diodes in series with each other. Ac is applied to the anode terminals while the cathode terminal is connected to the B+ voltage output. A bridge component may have outside markings with a cycle symbol and a positive and negative terminal (Fig. 5-20).

When

When the main fuse blows, suspect one or more diodes within the bridge circuit. A shorted or leaky horizontal output transistor can damage the bridge rectifier. The leaky filter capacitor connected directly to the bridge circuit can destroy one or more diodes. Always check for an overloaded component when one or more diodes are leaky. Suspect a shorted diode when the main fuse is black inside.

A bridge-type symbol might bc found in some of the Japanese models. The diode symbol points to the rectified output voltage (Fig. 5-21). Notice that the anode terminals of the two diodes are connected to common ground, while the other

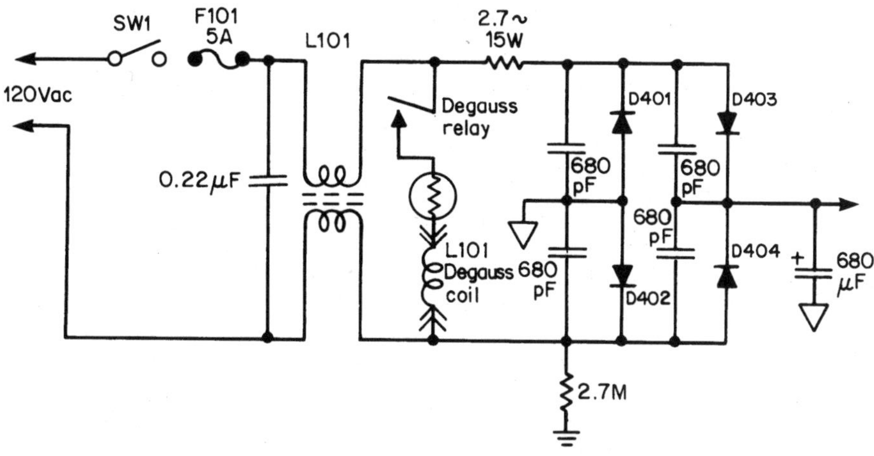

5-19 The bridge rectifiers within a power line ac voltage circuit of the TV chassis.

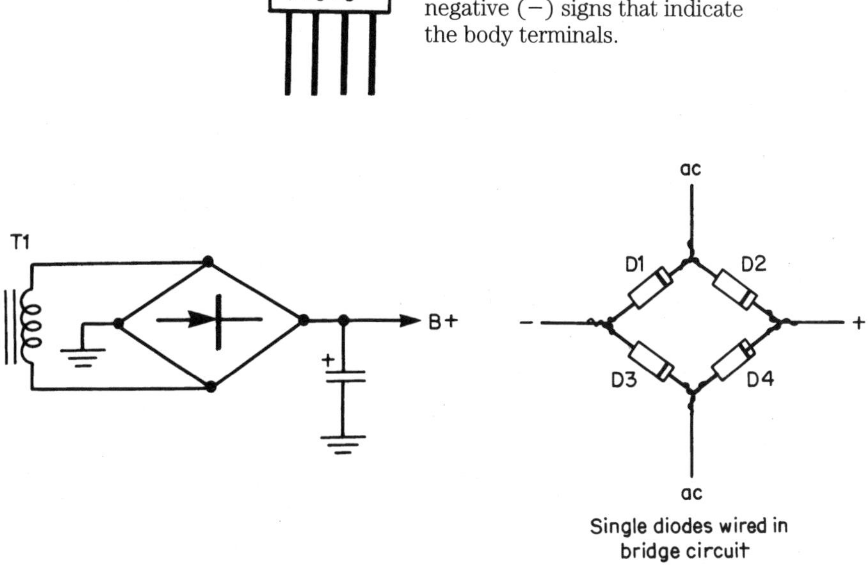

5-20
Some bridge rectifier components have a cycle, positive (+), and negative (−) signs that indicate the body terminals.

5-21 You may find only a bridge-type symbol in a bridge circuit of a Japanese electronic chassis.

two cathode terminals have a B+ output voltage. The two cycle terminals are connected to the power transformer or ac power line.

Testing

Check for an overload resistance across the large filter capacitor. If the resistance is below 500 ohms, check each individual diode within the bridge circuit. Test for open primary winding of the power transformer if one is used in the bridge circuit. Of course, when one diode is found leaky, you must replace the entire bridge component. Replace only the leaky diodes when four different diodes represent a bridge circuit. It's best to remove the end of the silicon diode for correct tests. Check the suspected diode with the diode test of the DMM or diode tester (Fig. 5-22). Remember, four separate diodes can be soldered in a bridge configuration to replace a single bridge component if one is not handy.

Boost rectifiers

What and where

In the early tube circuits, the B+ boost rectifier was sometimes called a damper tube. The black-and-white chassis might have a B+ boost circuit that supplies higher voltage to the picture-tube elements. A B+ boost rectifier in the color TV chassis has an additional positive voltage that might be added to the low-voltage power supply. Usually, the boost rectifier is found in the higher-screen-voltage circuits.

The boost screen voltage might be taken from a separate rectifier off an additional winding of the flyback or horizontal output transformer. Today, the screen and focus voltage is taken from a separate winding of the high-voltage transformer (Fig. 5-23). A leaky boost rectifier symptom is no boost voltage, no raster, no picture, and maybe hum in the speaker. With no raster and a defective boost circuit, high voltage can be measured at the picture tube.

Testing

Check the boost voltage at the screen grid of the picture tube. This voltage is quite high, and in the latest tube circuits might go over 900 volts. With a shorted or leaky diode, suspect a burned isolation resistor in the B+ boost voltage circuit. Most boost diodes become leaky or shorted. Very few appear open. Test the silicon boost rectifier with the diode test of the DMM. Measure the resistance of the diode with a VOM or FET-VOM. A low resistance can be measured in only one direction with a normal boost diode. A low measurement with reverse tests leads, in both directions, indicates a leaky boost diode.

Damper diodes

What and where

The damper diode prevents ringing in the power supply of a TV receiver and oscillations in the horizontal output transistor circuits (Fig. 5-24). In the early tube chassis, the damper diode was called a damper tube. The damper diode was connected from the collector terminal of horizontal output transistor to common ground. Today, the damper diode is found inside the horizontal output transistor.

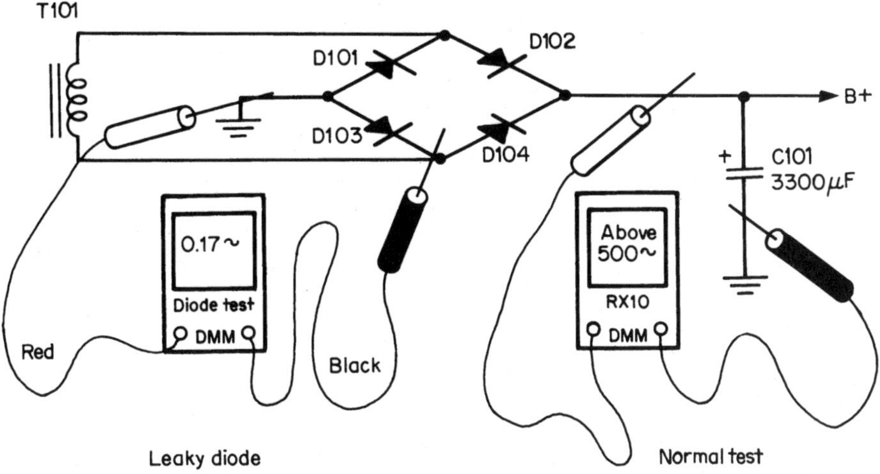

5-22 How to make tests within the bridge rectifier circuits.

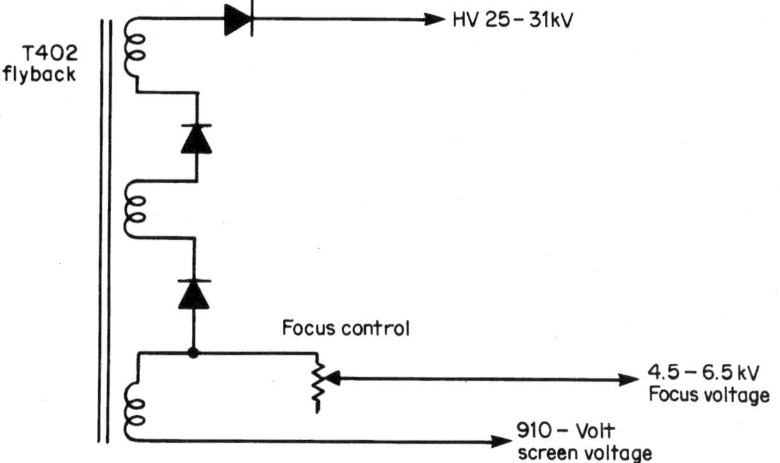

5-23 The screen boost voltage might be taken from the secondary winding of the high-voltage transformer.

A leaky damper diode can blow the fuse or trip the circuit breaker. The open damper diode can cause the horizontal output transistor to run warm and be destroyed. A low leakage measurement between the cathode and emitter terminals of the output transistor indicates a leaky or shorted damper diode. Look for the diode in the collector circuit of the horizontal output transistor in the early TV chassis. Double-check for the damper diode inside the output transistor in the latest circuits. The damper diode should have a resistance of around 400 to 500 ohms on the DMM diode test.

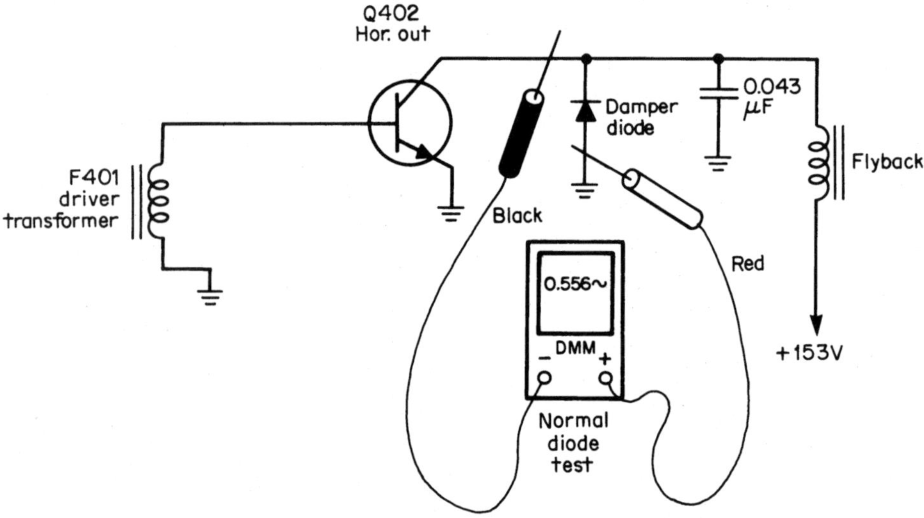

5-24 Test the damper diode with test probes at the collector (body) of the horizontal output transistor and chassis ground.

Testing

Suspect a damper diode when the fuse keeps blowing or the horizontal output transistor runs warm and becomes leaky. Check the damper diode with a resistance measurement (Fig. 5-25). Test the damper diode inside the output transistor with the diode test of the DMM. Low resistance measurements should be found in one direction, while infinite resistance is found with reversed test leads. A leaky damper diode will have a low resistance in both directions, while an open diode will show an infinite reading. Replace the single damper diode with 1200-KV or 1600-KV ratings.

Fast-recovery diodes

What and where

The fast-recovery or high-speed diode is found in the computer having low capacity and fast recovery time. The rapid-switching diode operates well in computer low-frequency and very high-frequency circuits. The high-frequency recovery diodes are found in the HV of the flyback. A defective flyback recovery diode results in replacing the whole flyback transformer. The fast recovery diode conducts with a forward bias and does not conduct with reverse voltage. Fast or superfast recovering diodes can switch at higher rates than mechanical switching.

The fast-recovery diode might be molded glass and surface mounted. The switching diode might be a IN914 or IN4933 series type. The fast-recovery diode might operate at 1.0 to 8 amps and up to 1,000 volts. The superfast-recovery diode must operate up to 30.0 amps at a much lower voltage (200 V). A high-amperage recovery diode might operate at 275 amps at 400 volts. Most surface-mounted (SMD) fast-recovery diodes operate in a 50 to 600-volt circuit. The fast-recovery diode resembles any general-purpose diodes. The high-speed diode or rectifiers are low-voltage types with a high-amperage surge current rating.

Testing

Check the fast- or superfast-recovery diode as any other silicon diode with a resistance measurement. Test the recovery diode within the circuit with the diode test of a digital multimeter (DMM) or diode tester. Remove one end for correct resistance measurement.

High-voltage diodes

What and where

The high-voltage rectifier in the early stack-type rectifiers was found in the black-and-white TV chassis of selenium diodes. When the high-voltage diode becomes overheated or leaky, it emits a sweet smell. The defective black-and-white stack diode becomes leaky, has burn marks, and runs quite warm. With a high-voltage probe, check for HV at the anode button of the picture tube.

In the early days, high-voltage diodes had a tube with heaters, a cathode, and plate elements. Today, the high-voltage diode might be a damper diode or a HV diode found in the microwave oven (Fig. 5-26). The high-voltage diode in the microwave oven operates in a very high-peak voltage range. High-voltage circuits in the oven are dangerous and should be checked with correct HV test instruments. Always discharge the HV capacitor before taking resistance tests or touching any component in the microwave oven.

When

The silicon HV rectifier and magnetron are the most troublesome components in the microwave oven. These diodes come in many sizes and shapes. Often, the HV

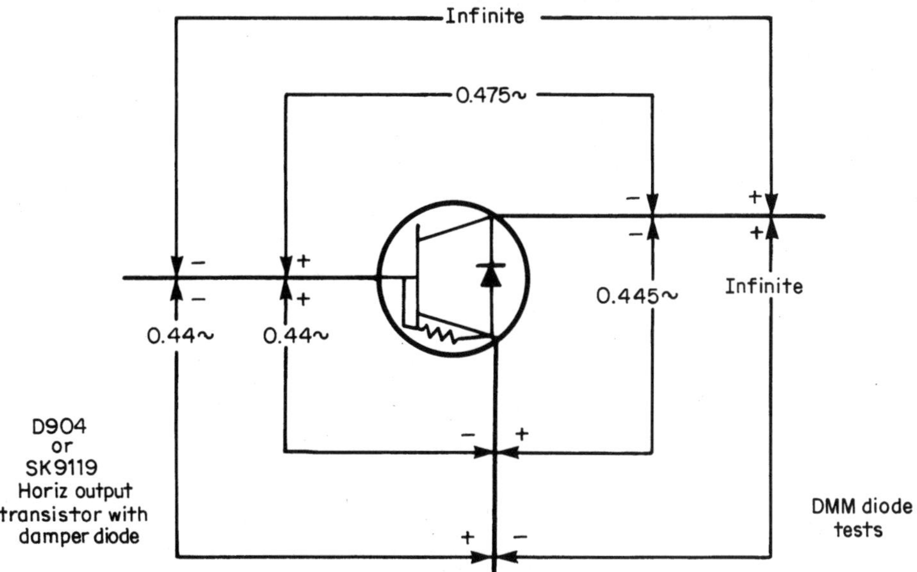

5-25 The different resistance measurements taken on an output transistor with the damper diode inside of the transistor.

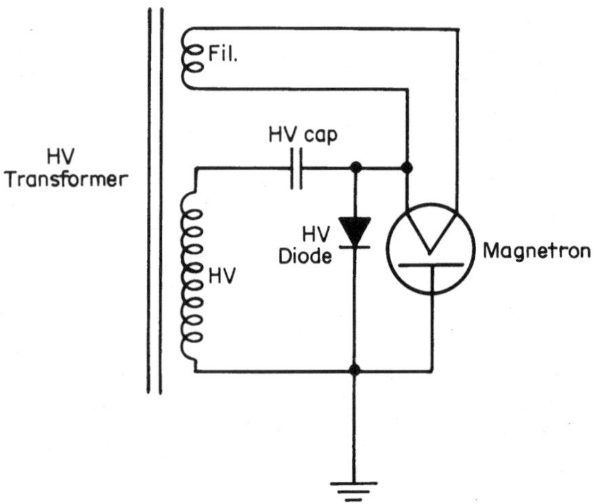

5-26 The microwave high-voltage diode is located in the HV side of the power transformer.

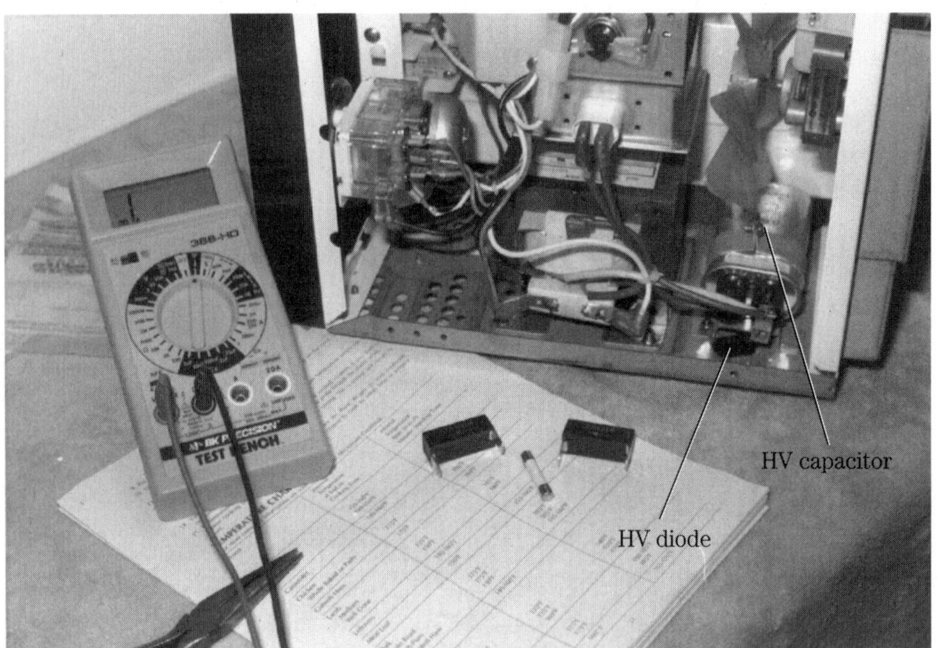

5-27 Locate the HV diode next to the large HV capacitor.

diode is located close to the HV capacitor (Fig. 5-27). When the HV diode cannot be found in the early ovens, check for it inside the magnetron cage. A defective diode might keep blowing the 15-amp fuse or produce a no-heat, no-cook condition. Before attempting to check the diode, discharge the HV capacitor. In most cases, the HV diode shorts or becomes leaky. If the body of the diode appears warm, replace it.

Testing

Discharge the high-voltage capacitor before taking measurements in the oven. The suspected HV diode can be checked with the RX10K ohmmeter range. Preferably, use an ohmmeter with a 6- or 9-volt battery. Do not use an ordinary VOM, or infinite resistance might be read in both directions. Isolate the diode by disconnecting one lead from the circuit, and clip the test leak across the suspected diode. The diode might read several hundred ohms. Now reverse the test leads. No reading should be obtained (infinity).

Replace the diode when a low-resistance measurement or the meter shows continuity in both directions. If the diode measures below 100 kilohms, replace it. A forward continuity will indicate several megohms. Check the leaky diode with the digital multimeter (DMM). The HV diode can be tested safely with a TRI CHECK II and Magnameter tester (Fig. 5-28). These testers are manufactured by:

Electronic Systems, Inc.
624 Cedar Street
Rockford, IL 61102

The TRI CHECK II can test the HV diode like the DMM. A high-voltage test with the Magnameter can also locate a defective HV diode. Do not use a small VOM or DMM to test HV in the microwave circuits. You might destroy them, even if they measure up to 3 KV. Connect the Magnameter to the magnetron tube. Measure the HV and current (Fig. 5-29). Suspect a leaky diode or magnetron tube when no HV or

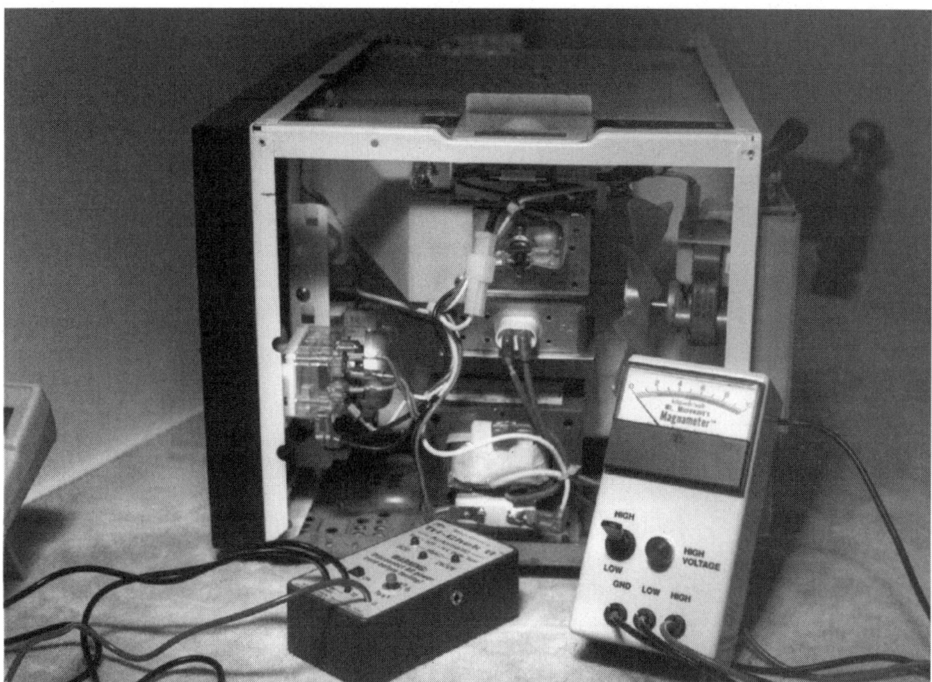

5-28 Check the high voltage and HV diode with a Magnameter and TRI CHECK II diode tester.

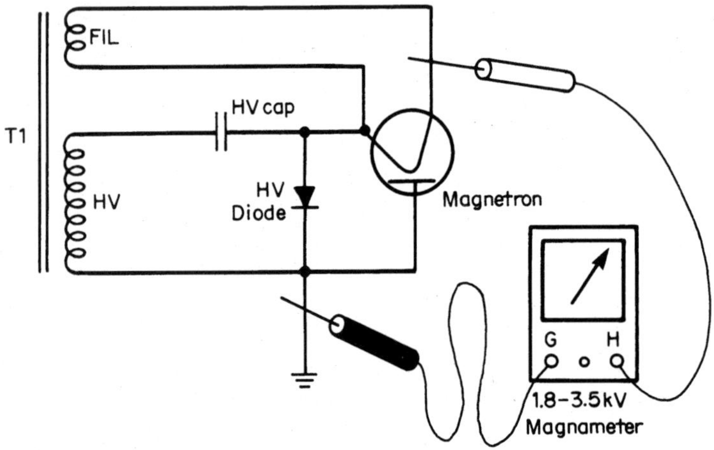

5-29 Measure the high voltage in the microwave oven with a safe Magnameter test.

current is found on the Magnameter. Discharge the HV capacitor. Take a resistance measurement across the HV diode. Remove the ground screw holding the cathode element to common ground. Now take a resistance measurement. Remember to discharge the HV capacitor each time after the oven is fired up.

Pin diodes

What and where

The pin diode can be used as an RF switch and made from a special semiconductor layer between the P-type and N-type material. A pin diode might be found in an RF application, such as tuning in a shortwave receiver (Fig. 5-30). The layer between the P and N material is called an intrinsic semiconductor. The intrinsic layer reduces the capacity of the diode so it can operate at higher frequencies. The diode with intrinsic material is called a pin diode, which has an anode and cathode terminal. Test the pin diodes as a general-purpose diode in a DMM or diode tester.

Stud-mounted diodes

What and where

The silicon stud-mounted diode is bolted to the chassis or heat sink (Fig. 5-31). These stud-mounted diodes are high-amperage rectifiers or high-current recovery diodes. These special mounted diodes are found in high-current power supplies, battery chargers, and welding products. The stud mount is cathode-to-stud unless an "R" suffix is before the part number, then it is an anode-to-stud terminal.

Testing

Check the high-current diode with a diode test of the DMM or diode tester. The normal silicon diode will indicate a low measurement in only one direction. Suspect a leaky stud diode with a measurement in both directions. Remove the anode lead for correct diode tests. These stud-mounted diodes run quite warm and sometimes have burned and poor resistance connections.

Surface-mounted diodes

What and where

The surface-mounted diode mounts directly on the PC wiring. You may find one or two separate diodes in one SMD component (Fig. 5-32). These SMD diodes look just like a three-legged transistor. Do not mistake a transistor, resistor, bypass capacitor, or a feed-through for a fixed diode. Check the schematic and parts layout for a fixed diode. The unidentified SMD can be located with the diode test of the DMM.

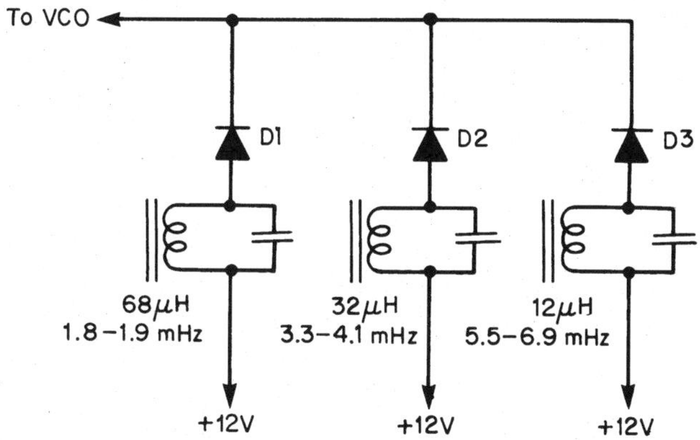

5-30 Pin diodes can be in series with paralleled tuning circuits of a shortwave receiver.

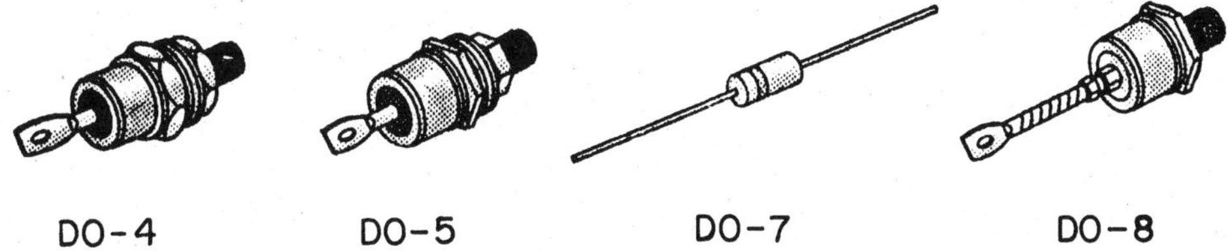

5-31 The high-current stud-mounted diode bolts directly to a heavy metal heat sink.

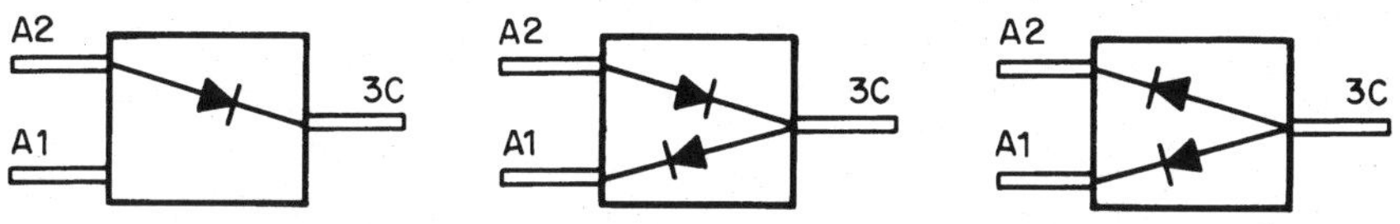

5-32 One or two diodes might be found within the surface-mounted component.

Testing

Check the SMD diode as a regular silicon diode, and test the SMD diode component with the DMM. Make sure the diode tester has sharp probe points to get at the tiny connections. The normal diode will show a low ohm reading in one direction. Suspect a leaky silicon diode with a low resistance and reversed test probes. Remove the SMD diode and test out of the circuit with a diode tester, DMM, or resistance measurements. Observe correct polarity when installing a new SMD diode. You cannot reverse the polarity of the SMD diode when two diodes are found in one package. Test the new SMD diode before soldering into the circuit. Do not apply the soldering iron for more than 4 seconds; you might destroy the chip or diode and PC wiring.

Varactor and tuning diodes

The varactor diode is a special type with a PN junction that has certain internal capacitance. By varying the reverse-bias voltage, the diode acts as a voltage variable capacitor. Varactor tuning is a method of changing the frequency of RF and oscillator circuits in the radio receiver (Fig. 5-33). The varactor diode may also be called a tuning diode. Varactor diodes are found in AM and FM tuning applications.

What and where

Varactor diodes are used in radio circuits to tune a series or paralleled inductance circuit. The early use of varactor diodes was found in the auto receiver. Today, clock radios, autos, AM-FM-MPX receivers, TV tuners, and shortwave radios might have some type of varactor tuning. Replacement tuning diodes for electronic radio and shortwave projects come in many different capacities (Table 5-1). Varactor re-

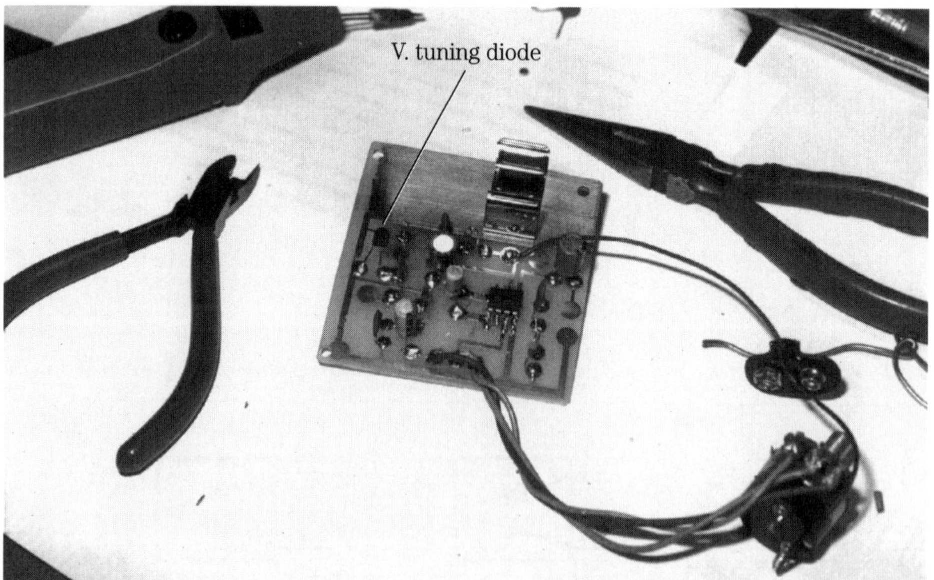

5-33 The varactor tuning diode looks like a transistor with only two terminals.

Table 5-1. Varactor diode and tuning replacement with capacity range

Part #	Capacity
MV2101	6.8 pF
MV2103	10 pF
MV2105	15 pF
MV2109	33 pF
MV2115	100 pF
NTE 618	440 pF

placement diodes can be obtained in matched sets for UHF-VHF TV tuners. Often, the tuning diode does not cover the complete tuning range as a variable capacitor.

The tuning or varactor diode looks like a small general-purpose transistor, except there are only two terminal connections. Like any diode, the varactor diode has a cathode (K) and anode (A) terminal. The anode terminal is always at groundpotential, and the controlled voltage is applied at the cathode terminal.

Testing

The varactor diode can become leaky when excessive voltage is applied. These tuning diodes can operate up to a maximum of 30 volts reverse breakdown voltage. Check the varactor diode like any diode. Place the red probe of the DMM to the anode terminal and the black probe to the cathode terminal. All diodes are tested on the diode test of the DMM. A normal varactor diode will have a low resistance in one direction and infinite measurement with reversed test leads (Fig. 5-34). The normal varactor diode resistance should be around .593 ohms on the diode test of the DMM, with the red probe at the anode and black probe at the collector.

Zener diodes

What and where

The zener diode (usually constructed from silicon) is built to break down with a zener voltage. This special diode responds to reverse-applied voltage. When the source voltage exceeds the voltage rating of the diode, it begins to conduct with reverse-bias voltage. This might be called the avalanched point of the diode. A zener diode might be compared to the avalanche diode. The voltage source will not increase above the zener voltage rating. Usually there is a resistor in series with the diode to increase the series resistance and lower the current in the zener diode circuit.

Zener diode replacement comes in ½-, 1-, and 5-watt varieties. Also there are a variety of different diodes with breakdown voltages (Table 5-2). It is difficult to tell the appearance of a zener diode from a switching or regular diode. Keep them in a separate bin or container. The 500-milliwatt (½) zener diode replacements come in

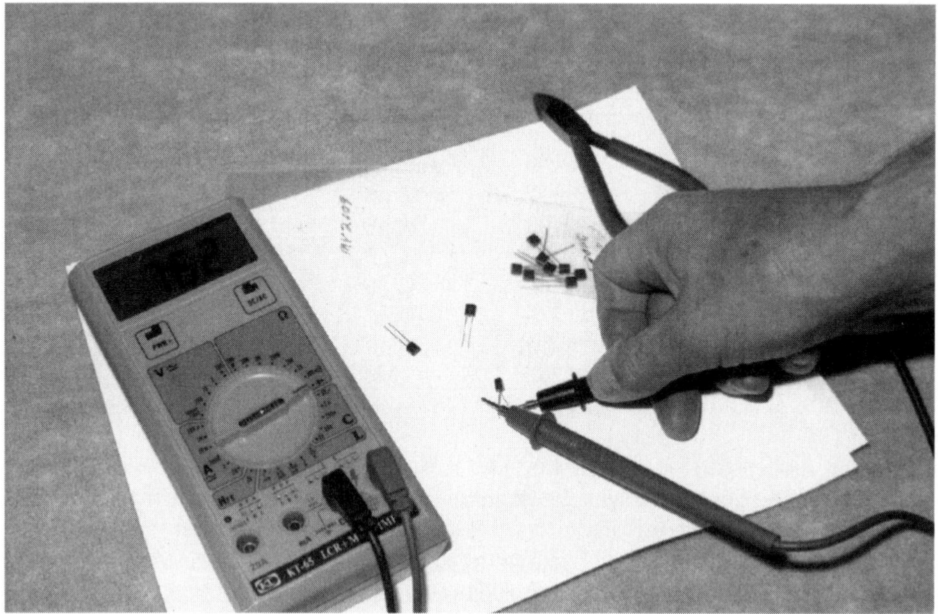

5-34 Testing the varactor diode with the diode test of the digital multimeter.

Table 5-2. Zener diode part numbers and voltage for replacements

Part number	Diode voltage
1N4728A	3.3 volts
1N4729A	3.6
1N4730A	3.9
1N4731A	4.3
1N4732A	4.7
1N4733A	5.1
1N4734A	5.6
1N4735	6.2
1N4736A	6.8
1N4737A	7.5
1N4738A	8.2
1N4739A	9.1
1N4740A	10
1N4741A	11
1N4742A	12
1N4743A	13

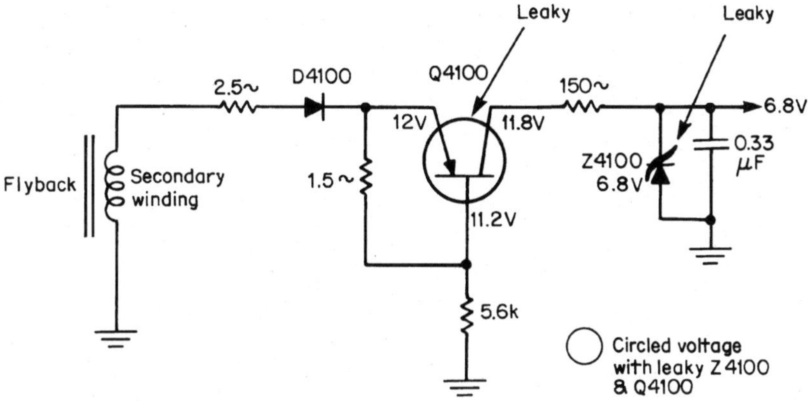

5-35 Notice the low-voltage source with a leaky zener diode.

a IN5226B series, 1-watt zener with an IN4728A series, and the 5-watt variety with part number IN5333 series.

High-voltage zener diodes are found in the TV voltage sources. These diodes might have a voltage breakdown up to 200 volts. Replace special voltage breakdown diodes with the original part number. Most zener diodes are replaced with 1- and 5-watt types. The zener diode is found in most consumer electronic products of various voltage sources.

When

The defective zener diode might become leaky or shorted and sometimes will appear open. Most zener diodes run warm. Often, when a zener diode becomes leaky, there are burned marks on the body area and around the PC board terminals. A shorted zener might be charred. When the zener diode overheats and appears leaky, the voltage source will be lower (Fig. 5-35). If the diode shorts out, the voltage source might be zero. Suspect a higher voltage source when the zener diode opens.

The zener diode is found in regulated low-voltage circuits. The diode might be mounted behind a voltage-dropping resistor in the output voltage source. You can find the zener diode in the emitter circuit of the regulated transistor to common ground. A combination of zener, transistor, and IC voltage regulators might be found in the various voltage sources of the secondary winding of the TV horizontal output transformer (Fig. 5-36).

Testing

Test the zener diode as with any silicon diode. A normal zener diode on the FET-VOM will test around 15 ohms with the red probe at the cathode terminal. This zener diode will test around 1177 ohms with the diode test of the DMM and the red probe at the anode terminal.

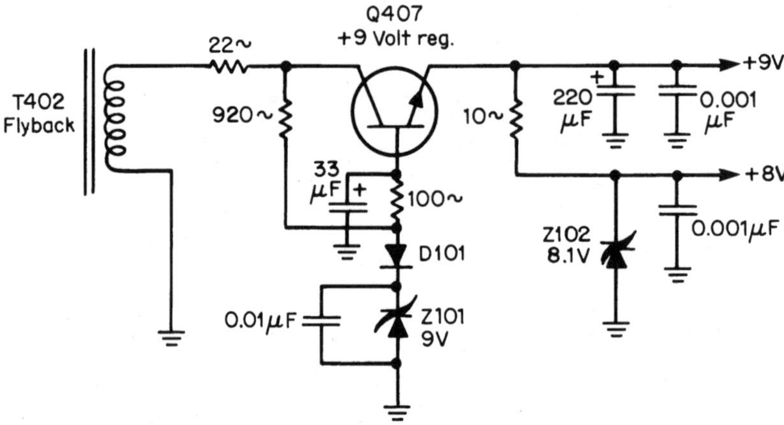

5-36 Transistor and zener diode regulation is found in the scanned-derived secondary voltages from the flyback winding.

6

CHAPTER

Testing electronic components E to H

(Electronic tubes to high-voltage triplers)

Electronic tubes known as vacuum tubes were found in the early radio and TV receivers. The electron tube was used as a rectifier, amplifier, and oscillator in the TV chassis. The only tube found in today's TV chassis is the picture tube. Just about every consumer electronic product found in the entertainment field contains transistors, ICs, microprocessors, and solid-state diodes. Tubes have practically skipped the country, so to speak, except in sound amplifiers, industrial, and communications applications.

Fuses are designed to protect the electronic product from excess voltage, a breakdown of a component, or a shorted and leaky part that can cause fire damage. The fuse may consist of a piece of metal with a low- or safe-level wire that melts down to protect equipment from further damage. Fuses come in many sizes and shapes. They should never be replaced with a larger size than found marked on the fuse holder or within the schematic diagram. Replacement of larger-amp fuses can cause fires and destroy lives.

Today, the portable AM-FM-MPX cassette player may contain a record/playback (R/P) and erase head. The R/P audio head records and also plays back a stereo recording, while the erase head removes the previous recording from the same cassette, so it can be used over and over again. The VCR and camcorder may have video, sound, erase, and flying erase heads. This chapter covers how the electronic components operate, break down, and how quickly you can locate and test each component from E through H.

Electron tubes

The vacuum or electron tube has almost all the air and gases removed from a glass or metal container. Electrons are emitted from a heated cathode element and attracted by a plate element with a high dc applied voltage. A detector tube might consist of a heater, cathode, and plate elements. The heater or filament element heats up the cathode, which emits a stream of electrons that is collected or attracted by a positively charged element called the plate or anode. Usually, the heater is separated from the cathode element, except in low, filament, battery-operated tube radios and shortwave receivers. Here the cathode element is part of the filament element.

The electron tube might have two working elements known as cathode and plate and used as a detector or rectifier. The triode tube consists of a heater, cathode, control grid, and plate elements (Fig. 6-1). The grid within the triode tube is a metallic mesh with holes in it so electrons can pass through it to the plate or anode element. By applying a negative or more positive voltage on the control grid, electrons can be controlled to the plate element. A triode tube can be used as a detector or amplifier within an electronic circuit.

A tetrode tube has four elements with an added screen grid element between the control grid and plate elements. The pentode tube has a suppressor grid found between the plate and screen grid. The suppressor grid is usually tied to the cathode element or at ground potential. The secondary electrons that bounce off the plate element are forced back to the plate element. You may find more than one set of tube elements within one glass envelope, such as dual triodes or triode and pentode.

Magnetron tube

What and where

The magnetron tube is found within the microwave oven. This metal sealed tube has a cathode, heater, and anode or plate elements (Fig. 6-2). The cathode and heater elements are connected to the negative side of the power supply, which has approximately 4000 volts with respect to the anode. The anode or plate terminal is connected directly to ground.

The electrons are negative charges, which means they are strongly repelled by the negative cathode and attracted to the positive anode. Electrons travel straight from the cathode to the anode of the 4000-volt potential were—the only force acting in the magnetron. However, the magnetron is a type of diode with a magnetic

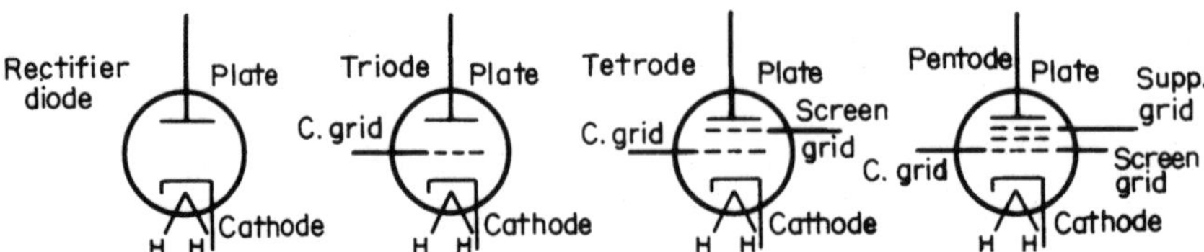

6-1 The most common tubes used today are the triode and pentode tubes in high-priced audio amplifiers.

6-2 The magnetron tube found in the microwave oven has a metal outside case.

field applied axially in the space between the cathode and anode of two permanent magnets.

A circular motion by the electrons induces alternating current in the cavities of the anode. When an electron is approaching one of the segments between two cavities, it induces a positive charge in the segment. As the electron goes past and draws away, the positive charge is reduced while the electron is inducing a positive charge in the next segment. This inducing of alternate currents in the anode cavities can be assumed as lumping together of the resonant circuit, producing oscillations. The high-frequency energy, produced in the resonant circuit (cavities), is then taken out by the antenna and fed into the oven cavity through the waveguide (Fig. 6-3).

Before making any tests within the microwave oven, discharge the HV capacitor. The charge from this high-voltage capacitor can cause extreme shock hazard and damage. Discharge the HV capacitor's two terminals with two long insulated screwdrivers. Discharge this capacitor before attempting to make voltage or resistance tests.

When

Most defective magnetrons become shorted between the heater and anode terminals. A leaky magnetron between the heater and anode terminals may result in a few thousand ohms, while a shorted magnetron tube has only a few ohms of resistance. Remember, a leaky or shorted diode can cause the same symptoms. An open magnetron can have an open filament or heater connection. Usually the high voltage is higher than normal without any current measurement with an open magnetron tube.

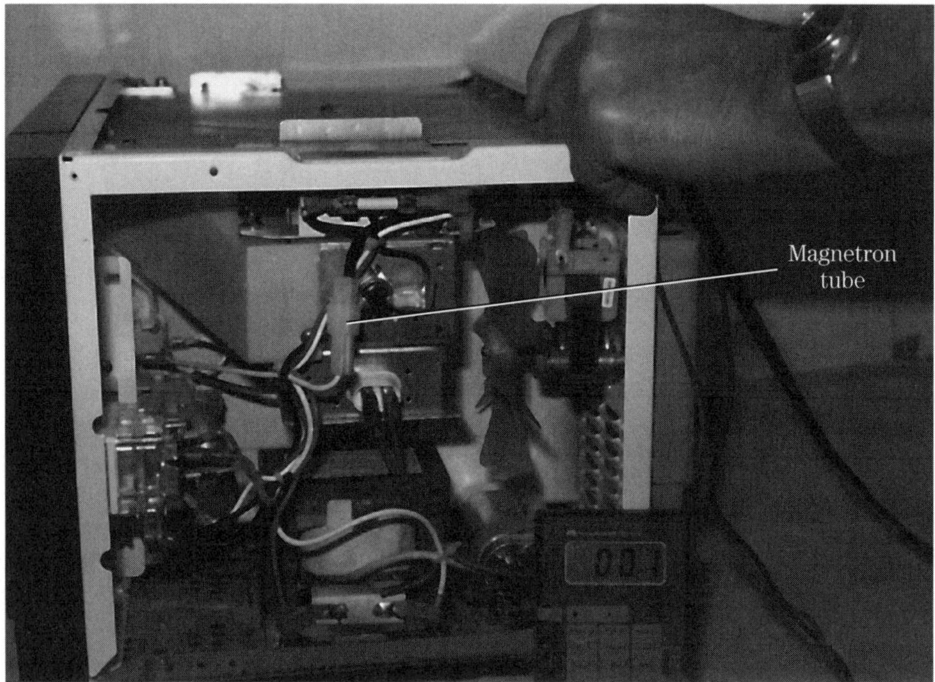

6-3 The small antenna of the magnetron feeds RF signals through a waveguide assembly to the oven cavity.

Testing

The defective magnetron can be located with resistance, voltage, and current measurements. Take a resistance measurement across the HV diode after discharging the HV capacitor. This resistance measurement should be above 10 megohms. If the resistance is below 10 ohms, remove the ground lug of the HV diode and take another measurement (Fig. 6-4). Suspect a leaky magnetron tube with any type of resistance measurement. Sometimes the HV diode will short out and place a heavy load on the HV circuits, providing the same service symptoms, unless removed from the circuit.

To determine if the high-voltage circuits or magnetron are normal, take a critical voltage measurement. Do not use a DMM or VOM portable meter for this measurement. Use only a Magnameter tester. You can quickly destroy the portable voltmeter within seconds. The Magnameter is designed to take safe voltage and current measurements. No voltage measurement might indicate a defective HV circuit. Low dc voltage indicates a leaky magnetron or HV diode, while higher than normal voltage indicates an open magnetron with very little current drain. The Magnameter can be switched to current measurement, indicating if the magnetron tube is operating.

Remove the heater terminals to take a very low resistance measurement of the heating element. This measurement will be less than 1 ohm. Replace the magnetron

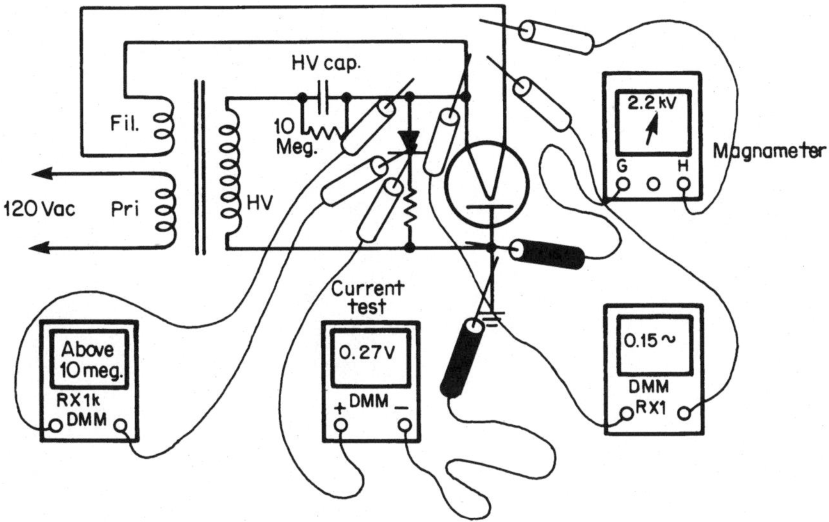

6-4 Take critical resistance and Magnameter tests to determine if the magnetron is leaky or shorted.

if the heater element is open or shows no resistance measurement. Sometimes the heater connectors will overheat and become burned or oxidized, leaving a very poor connection. Either replace the connectors or solder the connector directly to the heater terminals.

If a Magnameter test instrument is not handy, you can make a quick-safe current test with a DMM or VOM. Remove one end of the HV rectifier or diode and clip in a 10-ohm, 10-watt power resistor in series with the HV diode and ground terminal. Clip the voltmeter leads across the 10-ohm resistor. Set the dc voltmeter at 10 volts. Fire up the microwave oven. Do not touch the portable voltmeter at any time. If the magnetron is operating, a small voltage measurement is indicated on the dc meter. Discharge the HV capacitor after turning the oven off.

Sound amplifiers

What and where

Today the electronics technician might find expensive sound amplifiers using the vacuum tubes. Tube amplifiers are found in many European countries such as England, Russia, and France. The most common signal amplifier and rectifier tubes are a 5AR4, 6SN7, 12AT7 (ECC81, 6201, 7728), 12AV7 (ECC82, 6067, 7730), 12AX7 (ECC83, 7025), 6922 (6D58, 7308, ECG88, 7199). The power amplifier tubes consist of a 6L6WGB (5881), 300B, 5881WX7 (6L6), 655OC (KT88), EL34 (6AC7), and EL84M (6BQ5, 7189) (Fig. 6-5). Although there are no longer tubes manufactured in the United States, replacements are shipped in from Russia and China. Test the suspected tube within a tube tester or sub another one.

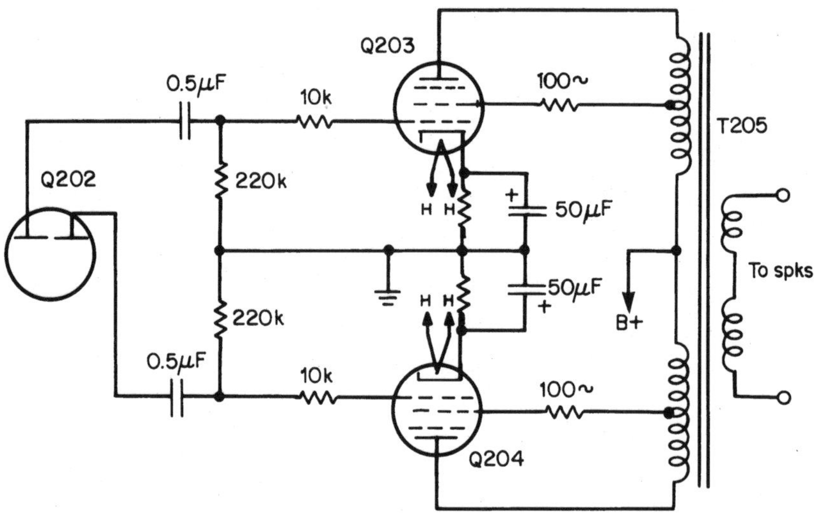

6-5 Two tubes in push-pull operation might be found in low-powered tube amplifiers.

Fans

Blower fans

What and where

The small blower motor moves air around overheated components within the ham transmitter and large-power receivers. Several fan motors can be found in the microwave oven to move cool air over the magnetron tube to remove moisture and odors out the back air vents (Fig. 6-6). Most fan blowers are ac operated, although dc fan motors are found in portable electronic equipment.

The defective fan blower might not rotate, rotates slowly, becomes intermittent, and has a dead condition. Most of the time, the fan motor might be clogged with dirt and dust or have dry bearings. Sometimes the motor bearings become dry and let out a screeching noise. Most motors can be repaired by removing them from the equipment, cleaning them with cleaning fluid, and properly lubricating them. Wash out each bearing thoroughly. Intermittent rotation is mostly caused by a poor terminal connection or worn brushes.

Testing

Check the ac voltage across the motor terminals with an ac meter. If voltage is normal and there is no motor rotation, place the blade of a screwdriver near the end bell of the motor and see if it vibrates. Ac voltage is being applied to the motor with a slight vibration of the screwdriver. Remove the motor, clean it up, and lubricate it. If there is no vibration with normal input voltage, suspect an open motor winding. Remove one lead to the motor and take a low-resistance continuity measurement. Usually the small fan blower motor resistance is under 10 ohms. Replace the defective motor with an exact replacement part.

6-6 In the early ovens, the fan blower was attached to the bottom of the magnetron tube for cooling.

Microwave fan blower

What and where

The fan blower within the microwave oven can be mounted on or near the magnetron tube. Since the magnetron runs quite warm during the cooking process, the fan blower moves cool air over the tube and out the back vented area. In the early microwave oven, the fan motor comes on when ac power is applied to the power transformer (Fig. 6-7).

Testing

Measure the ac power line voltage at the motor terminals, and take a quick resistance continuity test to determine if the motor winding is open. Inspect the ac motor connections for poor or broken wiring (Fig. 6-8). If the motor rotates slowly, remove the motor, clean it up, and wash out the bearings. Relubricate.

Microwave convection blower

What and where

The fan blower in the present-day microwave oven operates when the oven is turned on. Besides circulating cool air over the magnetron, the fan blower might also move air across the heating coils and direct the air through the oven cavity. You might find a separate convection blower motor besides the regular fan blower in the larger ovens. The fan and convection motors are classified as fast-operating blower motors.

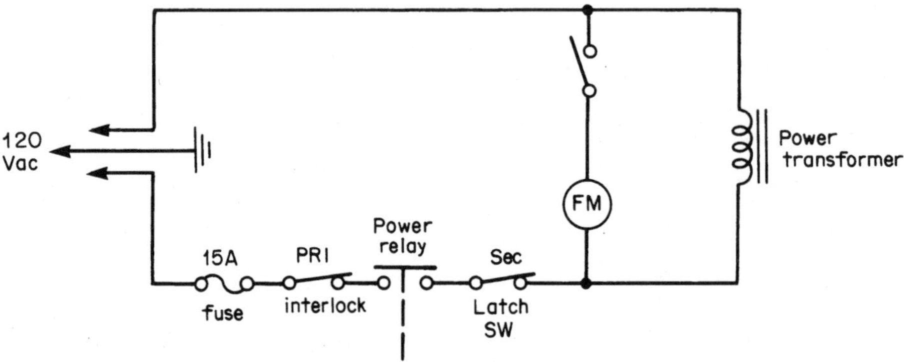

6-7 The fan motor is turned on when ac power is applied to the HV transformer.

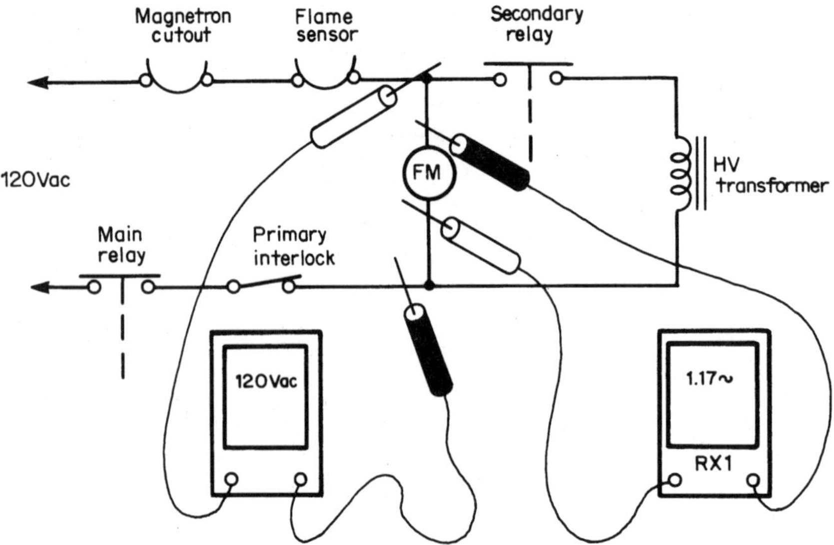

6-8 Check for power line voltage across the motor terminals and take a quick resistance measurement to see if the winding of the fan blower is open.

Testing

Notice that the blower or convection fan in this oven connects across the power line (Fig. 6-9). Ac power is applied to the 1000-watt heating element with the browner relay. The browner relay can be energized for convection operation without microwave oven cooking. Notice that the power relay controls the ac voltage applied to the primary winding of the high-voltage transformer for microwave cooking.

Check the ac voltage applied to the fan motor. If there is no voltage, suspect an open interlock, fuse, or thermal cutout. Simply clip a 120-volt ac pigtail light across the suspected interlock or thermal cutout. If the light comes on, the interlock or thermal cutout is open and defective. Most fan motors slow down or will not rotate

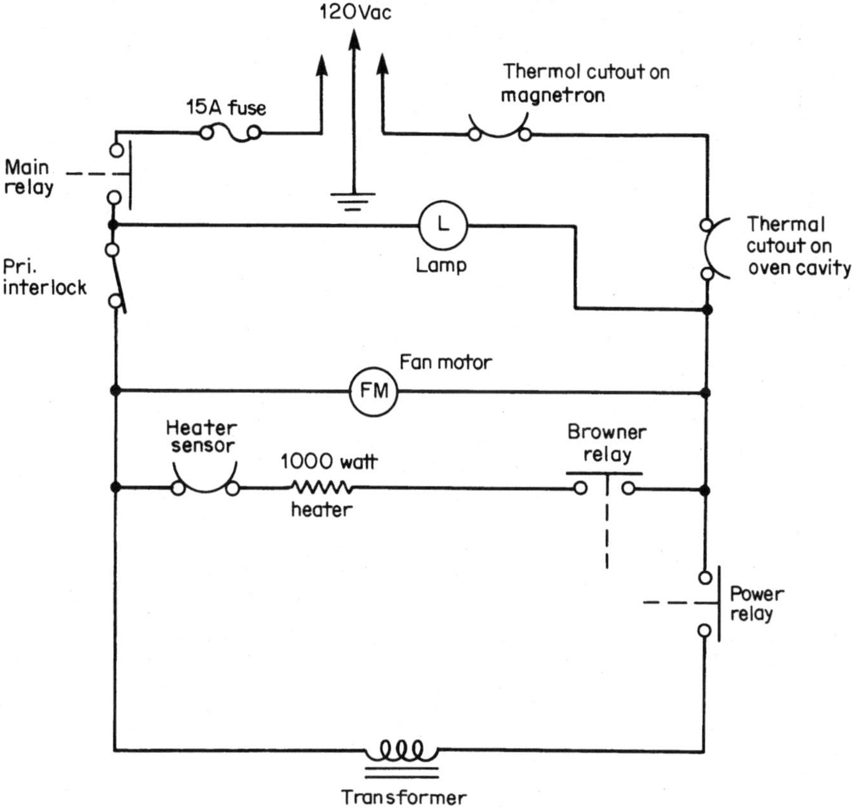

6-9 The convection fan blower circulates the hot air within the oven cavity.

with clogged bearings. Suspect dry or clogged bearings when the fan motor starts to operate after the oven is turned on for several minutes.

Microwave damper door blower

What and where

The damper motor operates at a much slower speed than a regular microwave fan blower. This motor opens a damper door in larger microwave convection ovens. The damper door remains open during microwave cooking and is closed for convection and combination cooking. The damper door is closed during this cooking function because the magnetron fan motor is always running when the oven is on. This will allow more air (cooler room air) to be blown into the oven cavity, keeping down moisture buildup.

Testing

Check for power line voltage across the damper motor. No ac voltage might be caused by an open damper relay, controlled by the front control panel. Take a quick resistance measurement of the motor winding. Like the stirrer motor, the damper motor has a very high resistance. The damper motor in a GE JEBC200 oven has a resistance of 3.21K ohms (Fig. 6-10).

Microwave stirrer motor

What and where

The stirrer motor rotates or circulates the RF energy emitted from the magnetron waveguide assembly. The stirrer motor provides even cooking and eliminates dead spots in the oven. Some ovens rotate or turn the food with a turntable motor instead of using a stirrer motor. Usually, the stirrer motor is located at the top of the oven (Fig. 6-11). The stirrer blades might be rotated from a regular fan motor driven by a large belt.

Testing

When the oven takes longer to cook food and has uneven cooking, suspect that the stirrer motor is not rotating or has a broken belt. Check for power line voltage across the stirrer motor terminals. Shut the oven down and take a continuity resistance measurement of the motor windings. Like the damper motor, the stirrer motor has a high resistance measurement (Fig. 6-12).

Filters

Filter networks are used to eliminate hash, spikes, and excessive noise from entering a TV set or computer. Electronic sign switching, neon signs, large industrial motors, and manufacturing processes are found on the electric power lines. Elec-

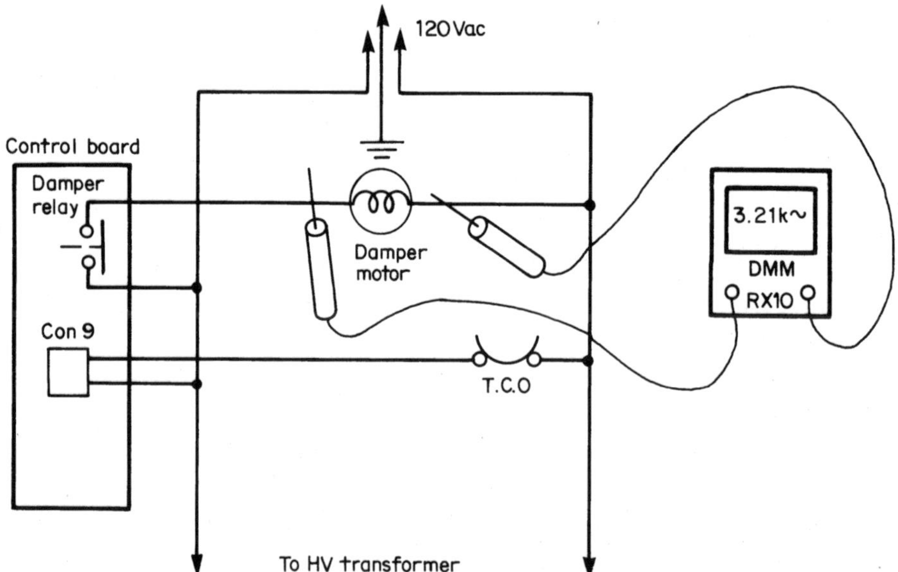

6-10 Notice that the damper motor has a higher resistance measurement than most motors in the microwave oven.

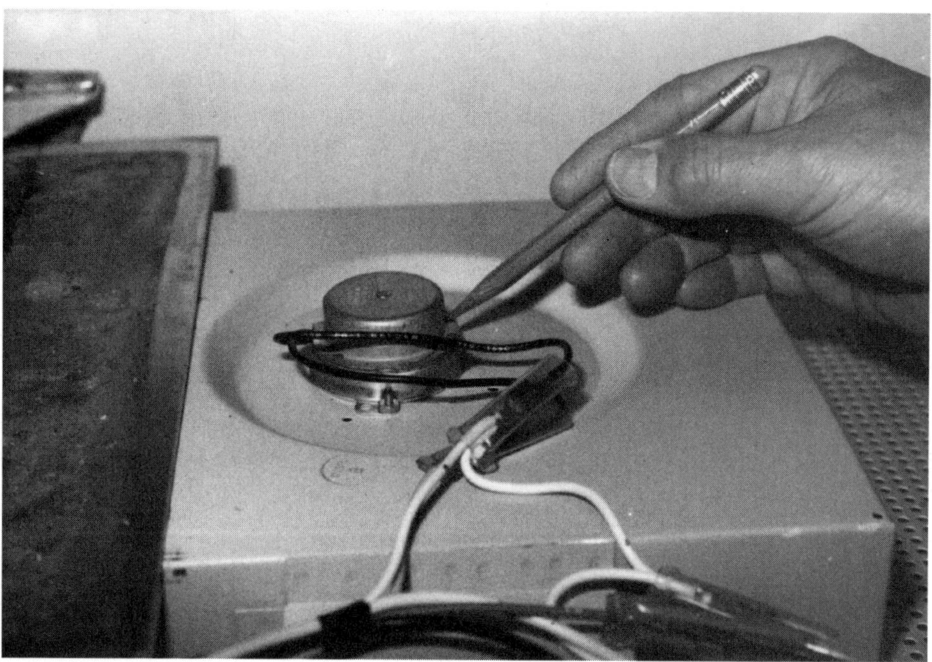

6-11 The stirrer motor rotates a plastic fan blade to help circulate the RF energy for even cooking.

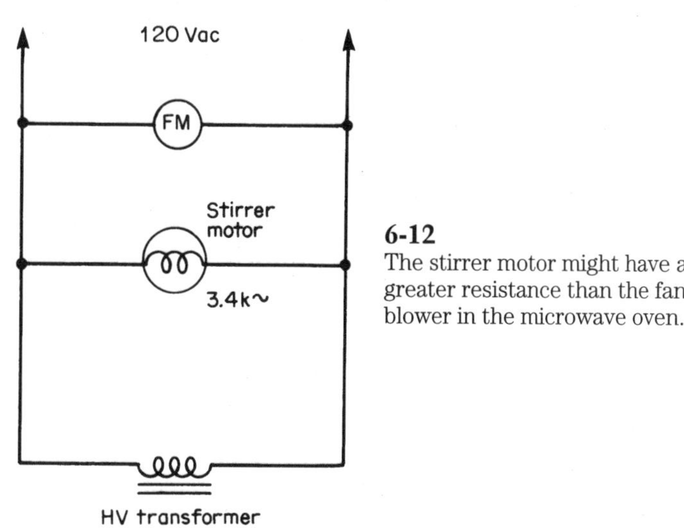

6-12
The stirrer motor might have a greater resistance than the fan blower in the microwave oven.

tronic filtering is used to eliminate in-line noise from car stereo systems. This type of filtering eliminates alternator and generator whine, as well as some RF noises. You might find electronic filter networks in large amplifiers. Larger units use passive toroid coils as well as power transistor electronic filtering. The electronic filtering circuits screen a broad band of radio frequency noise and shunt the high-frequency noise to ground.

Ac line filter

What and where

A typical general-purpose ac line filter might have choke input with an inductance of 1.1 mH to 3 mH at 2 to 10 amps. These EMI filters provide suppression of line-to-ground and low-frequency, line-to-line noise filtering. Some of these line filters have a separate fuse, and lower-priced models do not have any fusible protection. A medical line filter might have capacitor input instead of choke input, and the medical filter provides very low-leakage current, meeting medical and dental equipment requirements.

Testing

Very little problems are related to a line filter network. Check the power line voltage in and out with an ac voltage meter. Check for an open fuse or poorly soldered connection of the choke coil for no ac output voltage. Suspect lightning damage when both the in-line filter and electronic product are damaged (Fig. 6-13).

Auto ignition hash filters

What and where

A filter network eliminates certain frequencies from a given signal. Basically, there are four different types of filters: low-pass, high-pass, band-pass, and band-reject. The in-line filter for car stereo systems helps to remove whine produced by alternators and generators. A simple auto hash filter might have choke input and output (Fig. 6-14). The hash filter might have a 100-pF capacitor providing suppression at 350 to 550 MHz, 1000 pF at 100 to 250 MHz, and 10,000 pF at 25 to 120 MHz.

The toroid noise chokes might be wound with number-12-gauge enameled wire for hash and RF frequencies. The heavy-duty noise suppression network suppresses engine noise interference for receivers, equalizers, and audio amplifiers up to 35 amps. They may also contain CB noise suppression. Connect the noise suppressor within the power lead of the receiver and amplifier. The auto noise filters might have heavy-duty wound coils for high-powered amplifiers. A deluxe in-line noise filter goes between amplifier and radio to help eliminate noise due to signal path ground loops.

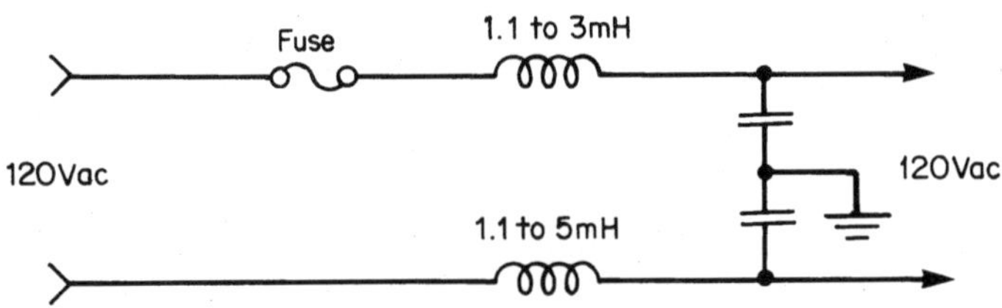

6-13 The line filter might consist of inductance and capacitors connected ahead of the consumer electronic product.

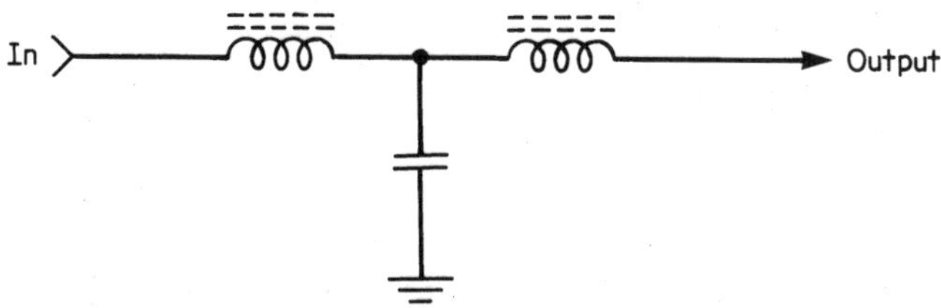

6-14 A simple auto-hash filter consists of two coils and one bypass capacitor.

Testing

Check the continuity or coil resistance at in and out terminals. Test for correct capacity with a capacitance meter, and check the battery voltage at the input and output terminals.

EMI/RFI choke filter networks

What and where

Electromagnetic interference (EMI) and radio frequency interference (RFI) can be created from power lines, arcing, transformers, insulator arcover, motors, medical equipment, neon lights, and ham and police operations outside the home. Inside the home, vacuum cleaners, hair dryers, telephone lines, fluorescent lights, and CB and ham operation can produce EMI/RFI interference. By installing a surge or noise filter ahead of an electronic product, such as a computer or TV set, spike and hash-type interference can be eliminated. EMI filters can be installed for communication equipment, cordless phones, auto phones, and digital audio-video equipment. EMI/RFI line filters plug into the ac power line, and the TV or computer plugs into the line filter unit (Fig. 6-15).

Testing

When the computer or TV set does not come on, suspect an on/off switch or open fuse inside the line filter. Inspect the ac cord and output receptacles. Check for ac voltage in and out. Take a continuity resistance measurement of each side of the power line to see what component is open. It's best to replace the entire

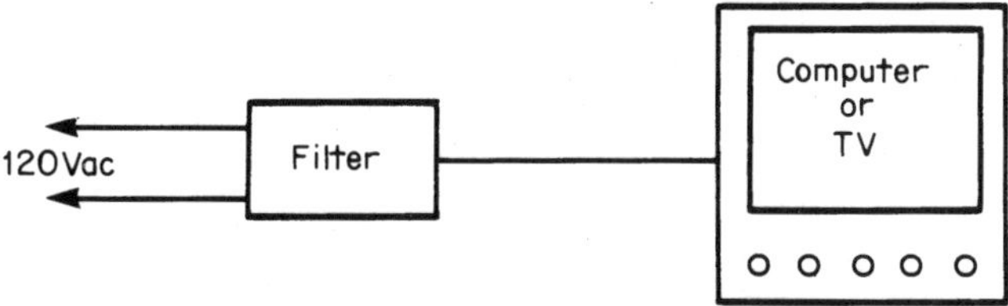

6-15 Plug the TV or computer into the EM1/RF1 line filter and then insert the filter ac cord into the receptacle for protection.

line filter when lightning has struck the outside power lines and electronic products inside the house.

TV interference

What and where

Noisy lines running across the TV screen might be identified as picked up interference. Usually, noisy lines might appear on channels 2 through 6 (Fig. 6-16). Even in extremely noisy areas, very little noise interference is found from channels 8 through 13. To determine if the noisy interference is inside the TV set or picked up, remove the antenna from the TV. If all noisy lines disappear, then the noise was picked up by the TV antenna.

Testing

Most TV interference is caused by power lines in front of the antenna, manufacturing plants, motors, and neon signs. Auto motor noise against snowy streets can reflect up to the outside antenna. To help reduce the amount of interference picked up by the antenna, raise it above the noisy area. Replace the flat TV lead-in with shielded coaxial cable and matching transformers at each end. The 300-ohm antenna terminals should match a 300-ohm input and 72-ohm output for coaxial cable. The 72-ohm lead-in is again matched at the TV receiver with a 72- to 300-ohm matching transformer. A 72-ohm to 300-ohm booster system might help TV reception in fringe areas.

Flyback transformer

The horizontal output transformer is also noted as the flyback transformer. The flyback supplies horizontal scanning and kickback voltage, which in turn is rectified

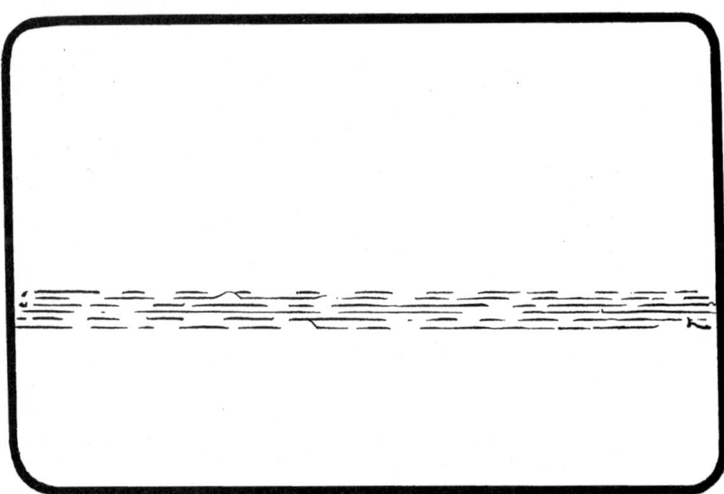

6-16 Noisy lines across the TV screen might be picked up by the outside TV antenna.

to produce high voltage for the picture tube (Fig. 6-17). Besides high voltage, the transformer supplies focus and screen voltages for the CRT.

What and where

The flyback is located in the output circuit of the horizontal output transistor. Dc voltage is fed through the primary winding to the collector terminal of the output transistor (Fig. 6-18). In the early TV sets, a damper diode is found on the collector terminal of the horizontal output transistor (H.O.T.) to common ground. The high-voltage winding supplies a very high-anode voltage to the picture tube. The bottom half of the HV winding contains the focus and screen voltages. Instead of an HV rectifier, several HV silicon diodes are found molded inside the HV winding.

When

Today's horizontal output transformer is rather small in size compared to those found in the tube chassis. Then came the tripler unit, which was attached to the small flyback transformer. The tripler unit consisted of high-voltage diodes and capacitors. Now these same components are found built right inside the transformer-molded windings. When a tripler unit was used, the tripler caused more service problems than the flyback.

The integrated transformer with several windings and molded high-voltage diodes can cause a lot of service problems. Suspect a shorted flyback transformer when the horizontal output transformer is arcing or begins to run warm. The integrated flyback

6-17 The flyback or horizontal output transformer is quite small in the latest TV chassis.

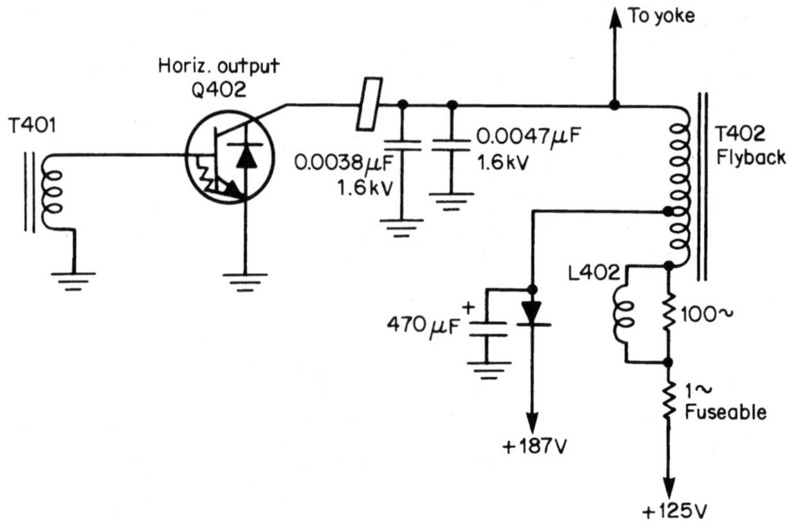

6-18 The primary winding of the flyback connects a B+ voltage to the collector terminal of the horizontal output transistor.

can run hot, pop and crack, and cause chassis shutdown. Like the diodes and capacitors in the tripler unit, the high-voltage diodes break down inside the transformer. Excessive arcover can cause overheating or arcing of the plastic material. Sometimes the flyback arcs over to the metal core area. The defective transformer can become leaky, shorted, and intermittent and have open windings. A flyback with shorted turns is very difficult to locate with only resistance measurements.

The normal flyback should operate fairly cool, unless overloaded by a leaky output transistor or internal high-voltage diodes. An overload in one of the secondary voltages can cause the transformer to shut down the chassis. Often, a shorted horizontal output transistor might point to a defective flyback.

Testing

Often the horizontal output transistor is damaged by a defective flyback transformer. Check the output transistor for leakage between collector and emitter terminals. Replace the transistor with a new replacement. Do not fire up the chassis after replacement. Insert the TV ac power lead into a variac or step-up and step-down isolation transformer. If the chassis is turned on at once, the new transistor replacement might be destroyed.

Slowly raise the power line voltage and notice if the output transistor is operating warm. Monitor the dc voltage applied to the output horizontal transformer. A leaky flyback or improper drive voltage might indicate a hot transistor with lower dc voltage. If the transistor runs cool and the dc voltage to the output transistor raises as the line voltage is raised, the chassis comes up and everything appears normal. Scope the input (base) and output (collector) terminal of the output transistor.

To quickly test the flyback circuits, place the scope probe alongside the flyback for a sharp horizontal pulse. If there is no waveform, suspect a horizontal output transistor, flyback, or no horizontal drive voltage. Take a quick resistance test of the primary winding of the flyback when no voltage is found on the collector terminal of the output transistor (Fig. 6-19). Sometimes you might find an isolation resistor open in series with B+ power supply and flyback winding. Check the flyback with a ringing flyback tester.

TV IHVT transformer

What and where

The integrated high-voltage transformer (IHVT) can have several different voltage windings for low-voltage sources and high-voltage diodes molded inside the same plastic housing (Fig. 6-20). Instead of a separate tripler unit with the low-voltage flyback, the integrated transformer has diodes built right inside the windings.

The integrated flyback can run warm, pop and crack, and cause chassis shutdown. Like the diodes and capacitors in the tripler unit, the high-voltage diodes break down inside the transformer. Excessive arcover can cause overheating or arcing of the plastic material. Sometimes the flyback arcs over to the metal core area. The integrated transformer should operate fairly cool, unless overloaded by a leaky horizontal output transistor or internal high-voltage diodes. An overload in one of the secondary-derived voltages can cause the transformer to shut down the chassis.

Often the horizontal output transistor is damaged by a defective flyback transformer. Use the variac or universal isolation transformer to locate a defective flyback and prevent damaging another output transistor. Before replacing the

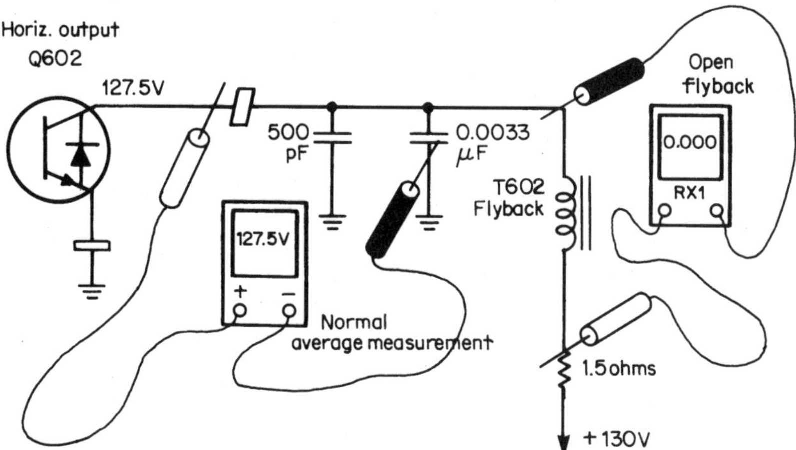

6-19 Test the flyback with a scope waveform and voltage and resistance measurements.

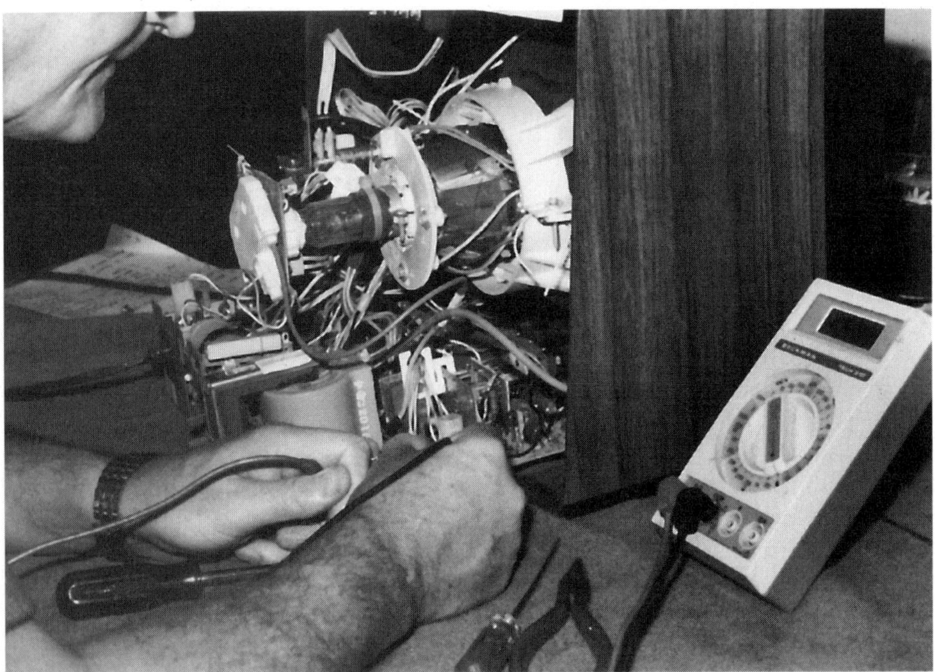

6-20 The integrated flyback (IHVT), output transistor, and damper diode are checked with a diode measurement of the DMM.

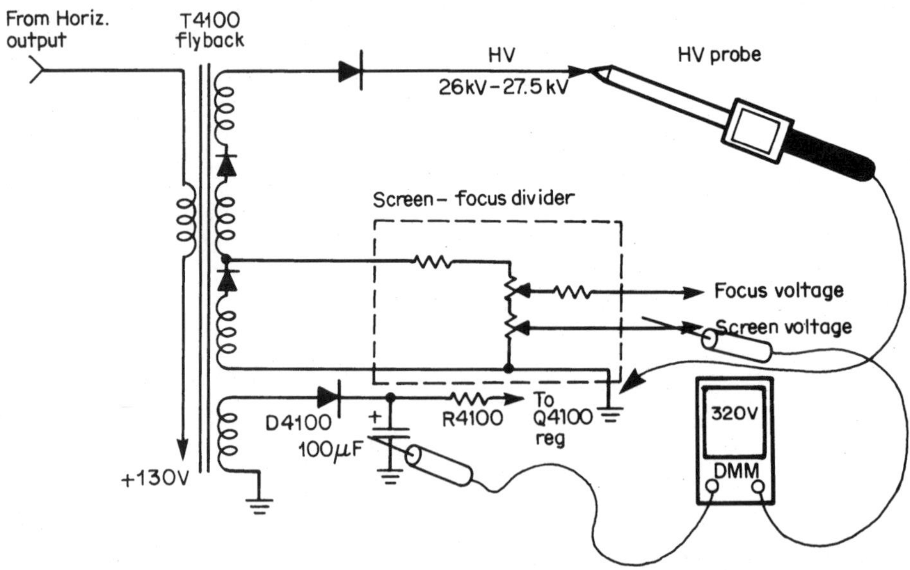

6-21 A quick high-voltage measurement with the HV probe indicates if correct HV is present.

flyback, check the output transistor and the damper diode. Sufficient drive and regulated low voltage should be found at the base and collector terminals of the output transistor.

Take a high-voltage measurement at the anode terminal of the picture tube (Fig. 6-21). Quickly scope the flyback waveform. Take another drive waveform at the base terminal of the horizontal output transistor. Remove the red yoke lead to see if it is loading down the flyback circuits. Check all secondary voltage circuits for overloading the horizontal output circuits. Replace the defective flyback when the high voltage remains low or quickly destroys another output transistor.

Camcorder flyback

What and where

A very small flyback transformer is found in the camcorder with an electronic viewfinder. The small black-and-white CRT anode voltage is derived from a horizontal output or flyback transformer (Fig. 6-22). The horizontal output circuits operate with horizontal and vertical sweep circuits, just like the regular TV chassis. The flyback transformer provides high voltage to the anode of the CRT, heater voltage, and screen and focuses voltage to the picture tube. These voltages found in the camcorder viewfinder operate at 5 or 9 volts dc.

Testing

Take a quick HV test at the anode lead. Check the primary lead of the flyback on the RX10 scale. Double-check all silicon diodes within the scan-derived secondary voltages. Take a quick drive waveform at the base of the horizontal output transistor. Check the sweep circuit IC for a lack of drive waveform, and determine if the horizontal and vertical sweep waveforms are present at sweep IC. Check the horizontal output transistor for leakage.

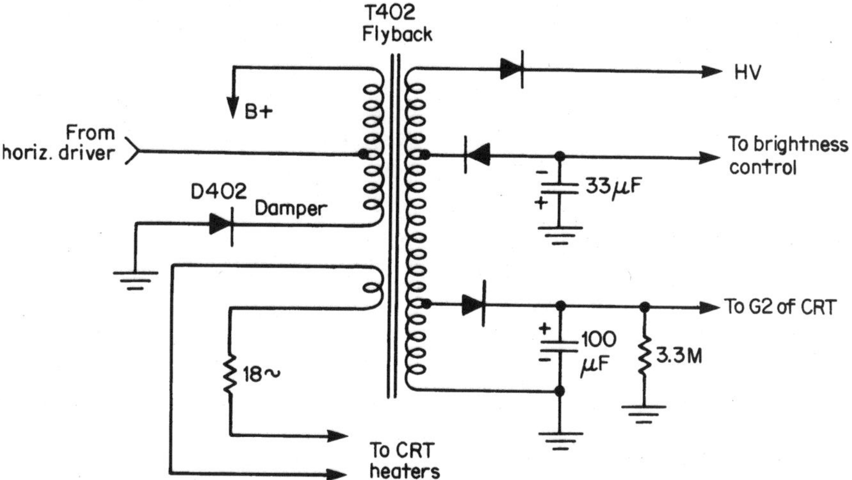

6-22 The flyback circuits in the electronic viewfinder (EVF) of the camcorder are tiny compared to the TV chassis.

Focus dividers (TV)

What and where

Most TV chassis contain a focus and screen voltage-dividing network to obtain correct brightness and focus of picture (Fig. 6-23). A defective tripler, dividing network, or components in the high-voltage circuit can produce poor focus problems. The leaky flyback transformer with enclosed high-voltage capacitors can have low, high voltage, resulting in poor focus voltage. A defective focus control or bad high-ohm dropping resistors can cause poor focusing. Check the focus pin at the tube socket for corroded connections. Arcing inside the focus control can cause tiny white lines firing across the picture. Simply rotate the focus control and see if the lines come and go. Replace the focus and screen-dividing network.

Testing

First measure the high voltage at the anode terminal of the CRT. Poor focus will result if the high voltage is low. Normal high voltage with poor focus can be caused by a poor focus assembly or picture tube. Check the focus voltage at the focus pin of the CRT socket. Place the end of a paper clip inside the focus socket and measure the focus voltage from it. Low focus voltage can be caused by a poor socket or spark gap inside the CRT socket. The average focus voltage should vary between 3.5 KV and 6.5 KV. With the latest and larger picture tubes, the focus voltage can vary from 6 KV to 9 KV. Check the focus voltage with the high-voltage

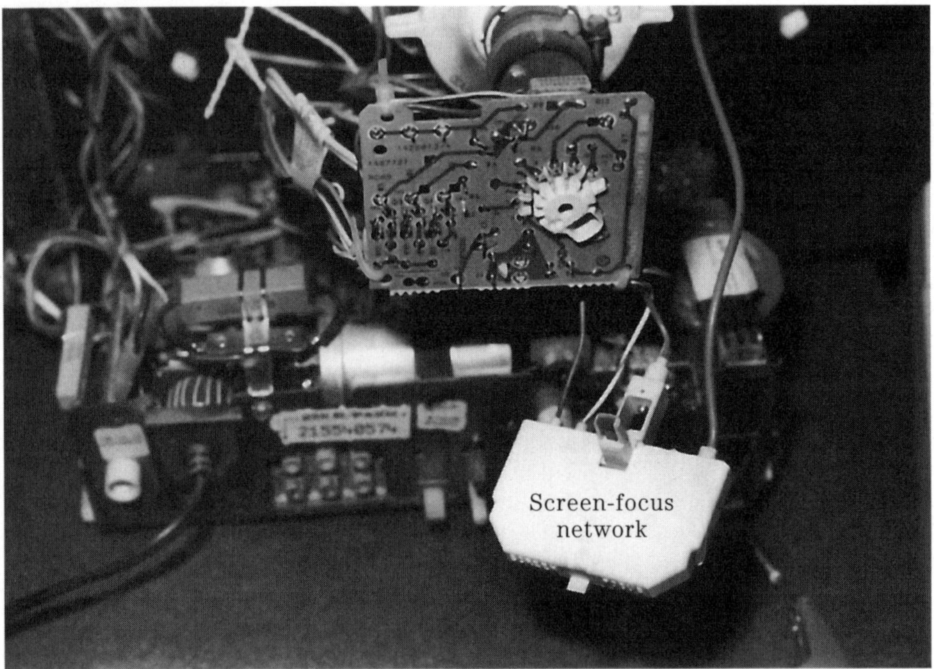

6-23 The screen and focus network plugs into the CRT tube socket of an RCA CTC107 chassis.

probe. The normal focus control should vary the voltage at least 1 KV at the CRT focus terminal (Fig. 6-24).

When the focus voltage compares to that found in the schematic, suspect a defective picture tube. Check the picture tube for correct emission in the CRT tester. Measure critical voltages on the picture tube terminals. A weak or gassy picture tube can cause poor focus. Suspect a focus spark gap or focus control when the focus voltage at the CRT socket will not change or focus. Blow out the socket and the focus spark gap assembly. Do not overlook poor heater terminals.

Fuses

Fuses come in many sizes and shapes. There are ceramic, fast-acting, slow-blow, pico, thermal, and digital-type fuses. A fuse is a safety device to protect a consumer electronic product from further damage when a component breaks down, protecting equipment from damage. Always replace the fuse with correct amperage and voltage. Do not insert a larger-amp fuse to keep the unit operating. You are inviting a fire and possible house damage when replacing a larger fuse.

Automotive

What and where

A 10- or 15-amp fuse was installed in the early car radios that operated with a vibrator power supply. Later, a 3- to 5-amp fuse was installed for protection in the solid-state receiver. Today, in high-wattage amplifiers, you may find a 30- to 200-amp fuse.

Little fuses (C)—fast-acting type of 3 to 30 amps at 32 volts—are found in the latest auto receivers (Fig. 6-25). The mini auto fuse (R) are found in the newer

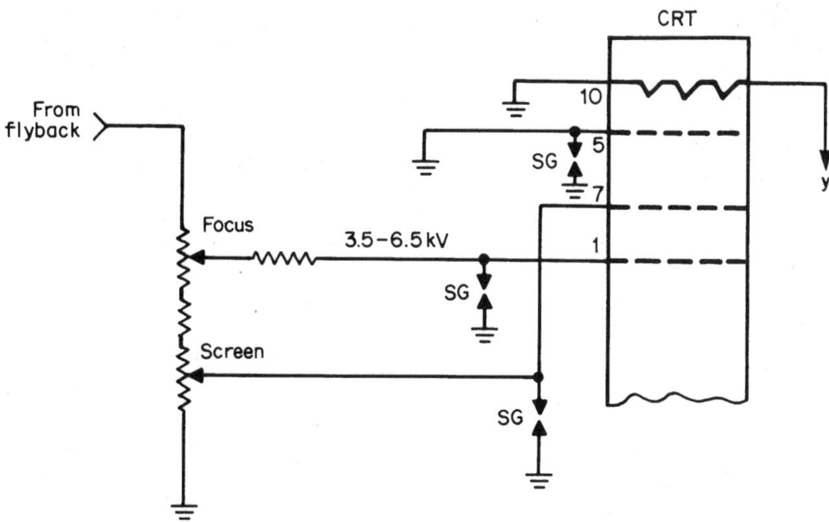

6-24 The focus voltage within the early TV chassis might vary between 3.5 and 6.5 KV.

automobile electronic systems. You may find auto circuit breakers from 70, 100, and 140 amps for use with larger in-car hi-fi installations.

Testing

When the glass envelope of a fuse is very black, suspect a dead short in that line. Sometimes the middle area of a fuse can see that it is blown apart. When in doubt, check the suspected fuse on the RX1 scale of the DMM.

Ceramic fuses

What and where

The microwave oven is protected by a 15- or 20-amp ceramic fuse. Suspect a defective HV diode, HV capacitor, or magnetron when the fuse keeps blowing (Fig. 6-26). Install a Circuit Saver test instrument instead of a fuse. This prevents damage to costly ceramic fuses until the trouble is located (Fig. 6-27). After repair, replace the 15-amp ceramic fuse. Check the suspected fuse with a continuity measurement of the DMM, since you cannot see inside a ceramic fuse.

Fast-acting fuses

What and where

The very fast acting fuse might come in pico, thin film, micro fuse, glass, pigtail, cartridge, and surface-mounted types. The operating temperature range of a very fast acting fuse might be 55 degrees C to 129 degrees C. The tin/lead-plated fuses are designed to open in seconds. Check the suspected fuse with an ohmmeter continuity measurement, or simply replace with a new fuse.

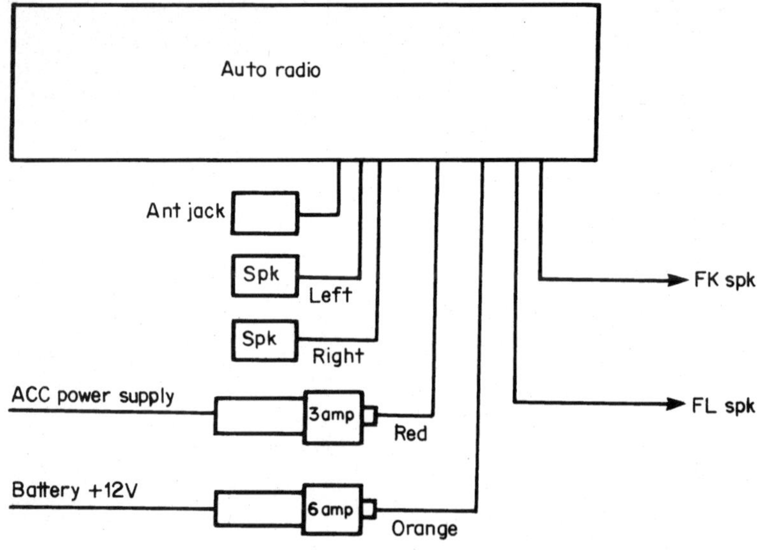

6-25 The auto receiver line fuse may vary between 3 and 6 amps.

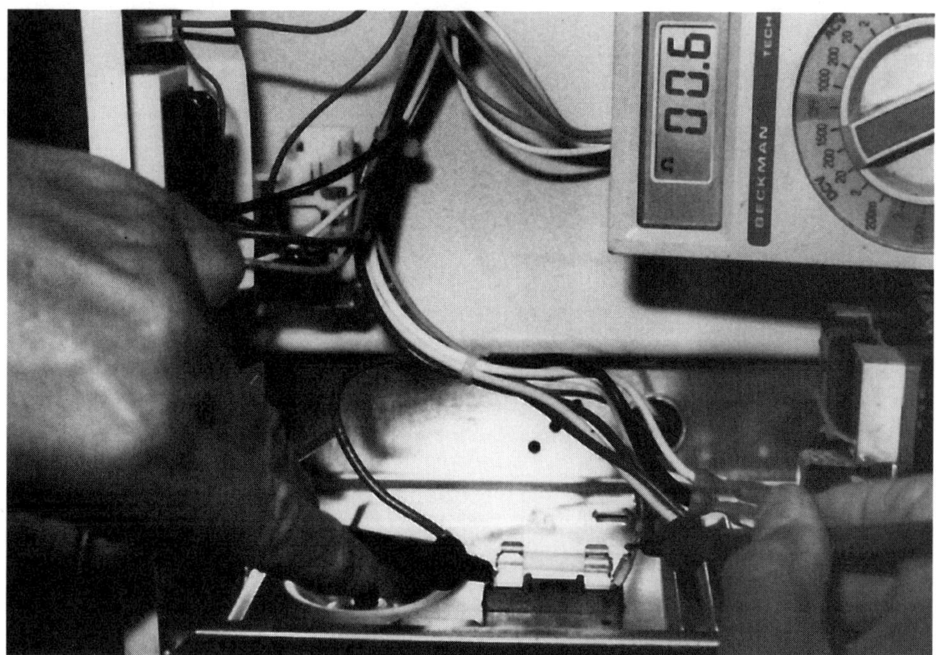

6-26 Check the ceramic microwave oven fuse with the low RX1 range of the DMM.

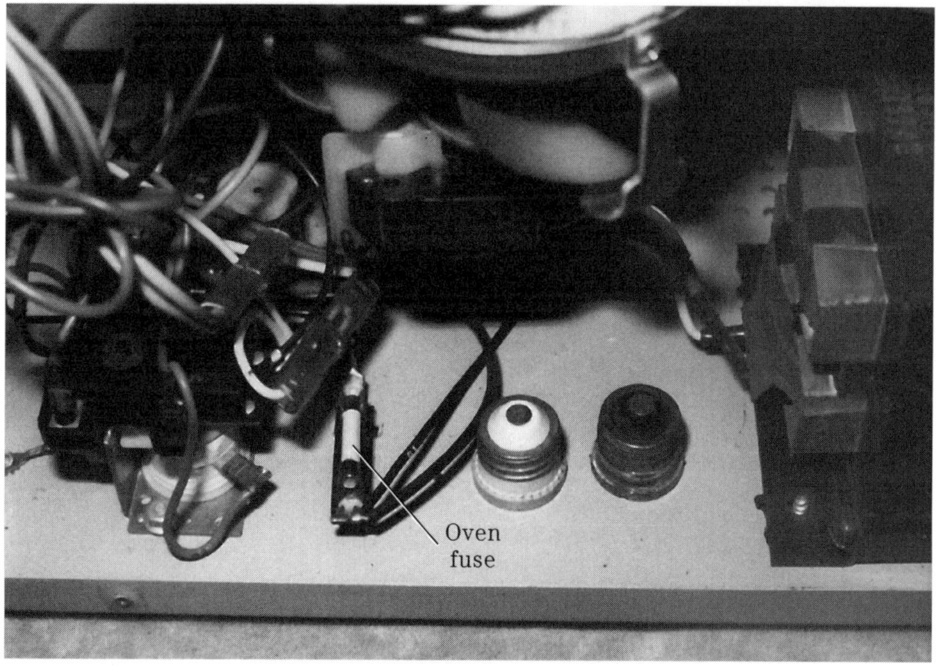

6-27 A 15- or 20-amp ceramic fuse is found in the power line of a microwave oven.

Slow-blow fuses

What and where

The slow-blow fuse is designed for circuits that might have a heavy load at turn-on. The fuse might have a copper-appearing spring or beaded construction (Fig. 6-28). The slow-blow fuse might be contained in a glass enclosure, fuse holder, surface-mounted, and pico fuse. Slow-blow fuses are found in equipment where heavy turn-on or current is found. The slow-blow ceramic fuse replacement might be rated from 1 to 3 amps, pigtail types from 1/4 to 7 amps, and regular slow-blow fuses 1/10 to 30 amps.

Testing

Check for open fuse with the RX1 ohmmeter range. Open power line fuses in the TV set can be caused by leaky silicon diodes, horizontal output transistors, shorted yokes, and leaky flyback and safety capacitors. A repeated blowing of a fuse in the large, high-powered amp and receiver might result from shorted silicon diodes, power transformers, and electrolytic capacitors within the low-voltage power supply (Fig. 6-29). Do not overlook leaky power output transistors or ICs that keep blowing the ac line fuse.

Pico fuses

What and where

A pico fuse is located where space is limited. The pico fuse might be color-coded. The first three bands indicate current rating in milliamperes. The fourth and wider bands designate the time-current characteristics of the fuse. For instance, red is fast acting. The subminiature fuses are enclosed with an epoxy coating or surface-mounted types. They are found from 1/16 to 15 amps at 125 volts.

Thermal protector fuses or cutout types

What and where

The thermal cutout or resettable fuse might be called a circuit breaker. A circuit breaker was found instead of a line fuse in the early TV set. Thermal reset fuses

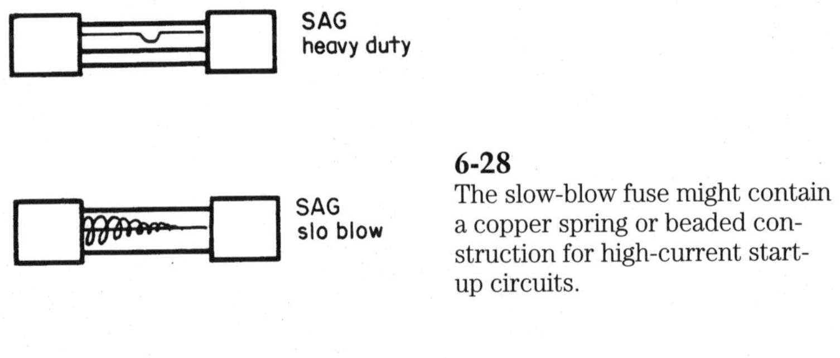

6-28
The slow-blow fuse might contain a copper spring or beaded construction for high-current start-up circuits.

might be found inside the small transformer ac-dc plug-in type power supply. There are several cutout or thermal fuses found inside the microwave oven. The convection microwave oven might have a heater thermal cutout assembly attached to the wall of the oven (Fig. 6-30). When the magnetron overheats, a thermal protector cutout disconnects one side of the power line from the HV transformer. The magnetron thermal switch is located directly on the magnetron tube.

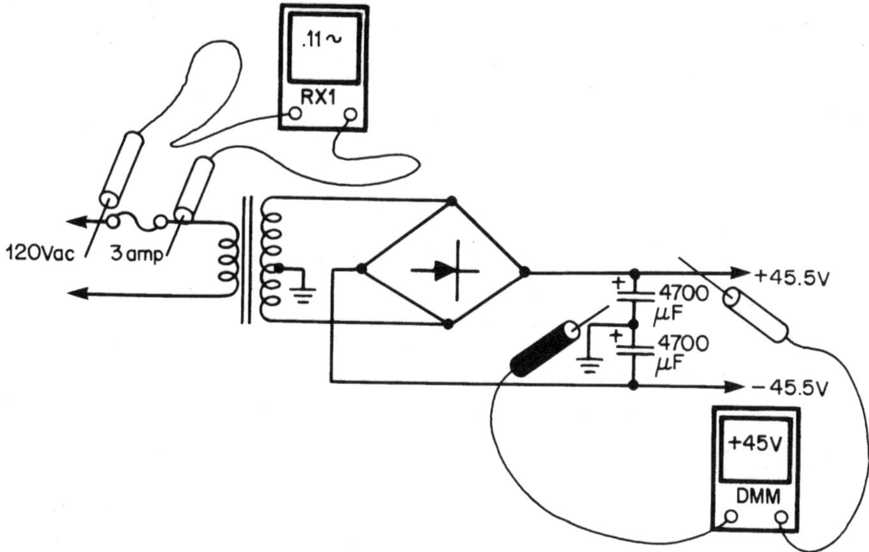

6-29 Check the fuse with the low-ohm resistance test and take a quick voltage test across the filter capacitor to determine if low-voltage power supply is normal.

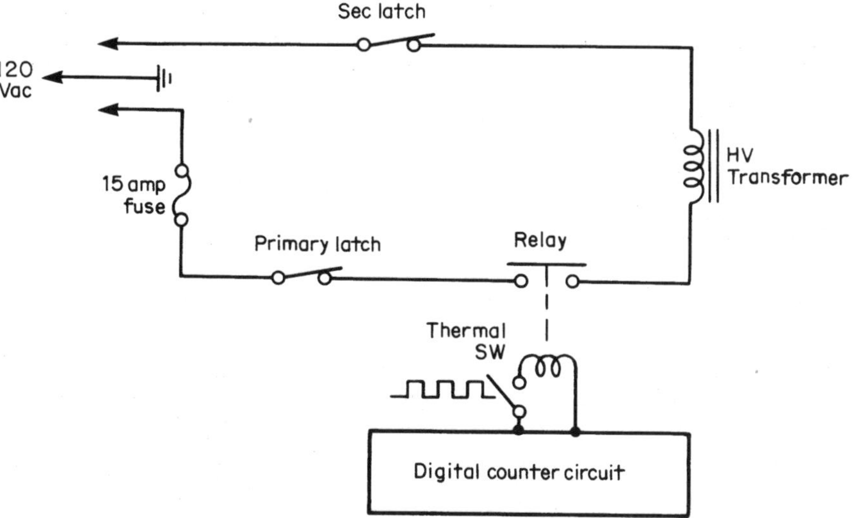

6-30 The thermal cutout might be attached to the outside of the magnetron to protect it from overheating.

Testing

Clip a 120-volt pigtail light across the thermal cutout terminals after discharging the HV capacitor. If the light comes on and there is no oven operation, the cutout assembly is open and must be replaced. A quick low-ohmmeter measurement across the thermal switch will indicate whether the switch terminals are open or normally closed. Replace any thermal cutout that might appear defective or intermittent. No resistance should be measured across the thermal terminals. Replace the thermal cutout when a few ohms of resistance are found between the terminal connections.

Headphones

High impedance

What and where

A pair of magnetic headphones was used in crystal sets and tube radios for private listening (Fig. 6-31). The magnetic earphones have a high impedance, from 1 K to 2000 ohms. Each magnetic earphone consisted of many turns of wire wrapped around a magnetic core with a diaphragm constructed of flexible metal for reproduction of sound. The old telephone receiver can be called a high-impedance listening device. Today, the high-impedance headphones are quite expensive and difficult to obtain (Fig. 6-32).

6-31 High-impedance headphones were used with early tube shortwave receivers.

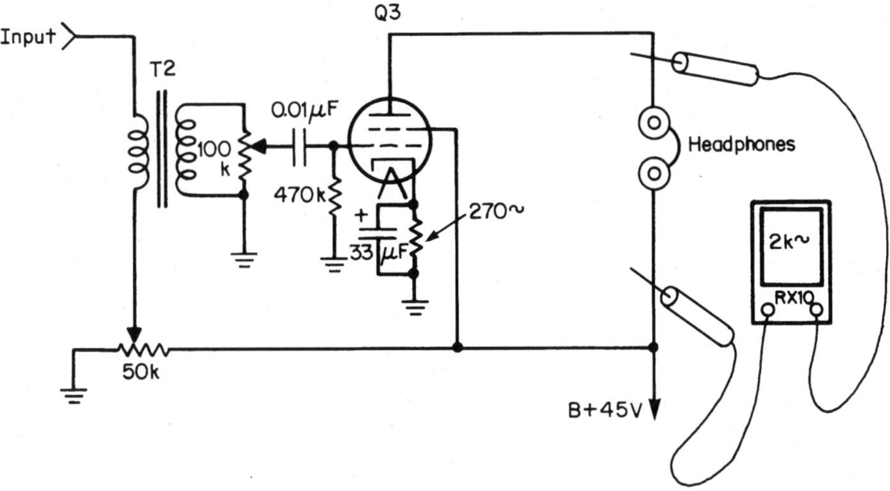

6-32 For private listening, the high-impedance earphones (2000 ohm) were found in tube output circuits.

Testing

Take a continuity test with the RX10 scale of the ohmmeter at the male phone plug. Often, the actual resistance measured of the headphone is much lower than the impedance. Check both phones for possible breakage where the cord enters the phone or at the male plug (Fig. 6-33). Intermittent headphone reception must be caused by a break in the flexible cord or inside the headphone.

Low impedance (8 to 70 ohms)

What and where

A pair of low-impedance headphones can be found with the walkie-talkie, portable radio, cassette, and CD player for personal listening. These earphones

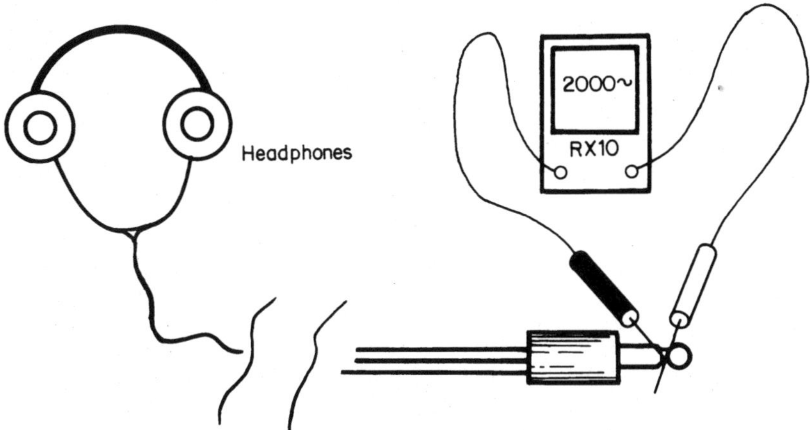

6-33 Check the headphones with a resistance measurement at the male plug.

consist of small pm speakers housed inside an earphone container. Some Walkman stereo earphones have 23mm driver speakers. You may find that studio-quality headphones are made of a neodymium iron boron magnet that deliver a high magnetic field strength. A pair of featherweight digital earphones is for personal stereo and CD player reception (Fig. 6-34). Most low-impedance earphones have a 20-Hz to 20-kHz frequency response, while an expensive pair might have a frequency response of 15 Hz to 25,000 Hz.

The low-impedance earphones have a resistance from 8 to 59 ohms. The actual resistance measurement of a low-impedance earphone might equal the impedance reading, when taken with a DMM. Do not try to use a pair of low-impedance earphones where high-impedance headphones are required. Although no damage results to the earphones, the sound level might be very low in tube output circuits.

Testing

Check the total resistance of each earphone with the RX1 range of the DMM. Sometimes one earphone might be open or have a broken cable wire. Suspect the flexible headphone cable where it enters the headphone or molded stereo plug. Cut off the male plug and install a new one with a broken connection. Push a sharp hairpin right through the cable wire to test for possible cable breakage. Then take another measurement. Clean up around the male plug and female jack with cleaning fluid when intermittent reception is noted.

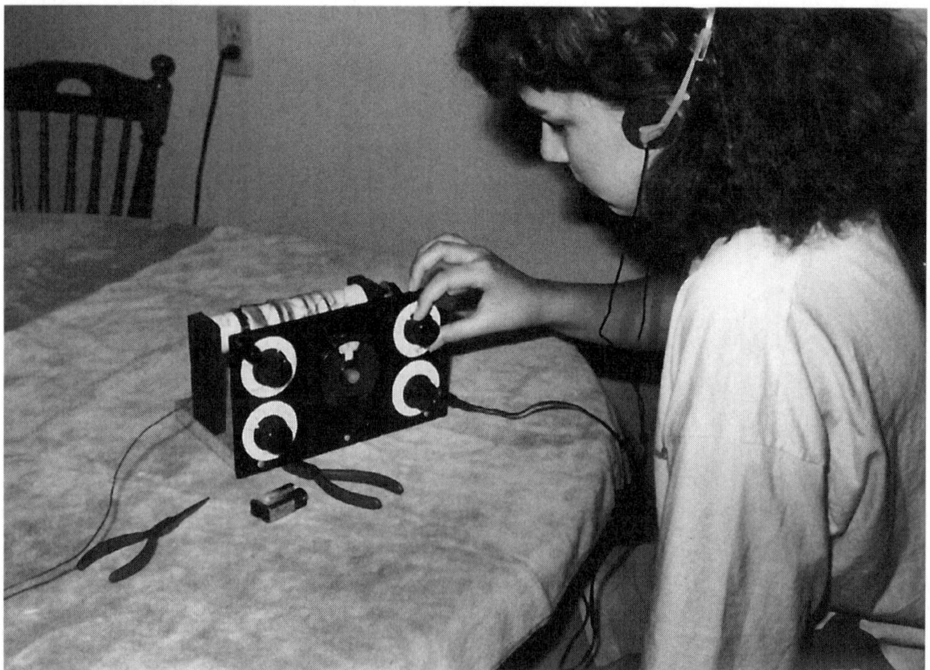

6-34 Low-impedance earphones (8 to 70 ohms) are found in solid-state output receiver circuits.

Heads

The magnetic tape head is found in the tape recorder, cassette player, camcorder, and VCR. A mono winding consists of many rounds of wire over a metal head with one audio output. A stereo tape head has two separate windings to record and play both left and right stereo channels. The erase head is mounted ahead of the R/P head to erase the previous recording (Fig. 6-35).

Testing

A defective tape head might be open, intermittent, worn, clogged, or contain a broken lead at the tape head terminal. Check the resistance of the tape head with the RX10 range of the DMM. Both windings should have a similar measurement with a stereo cassette tape head. Clean up the gap area with alcohol and a cleaning stick. A clogged or packed oxide in the gap area can produce weak reception. Move the metal blade of a screwdriver back and forth in front of the tape head with volume wide open to determine if the head and amp are working. The worn head produces a tinny or high-frequency sound of music in the speakers. A loud rushing sound can be heard with the volume wide open and an open head winding or broken cable wire terminal. The tape head can be tested with an external amplifier connected to tape head connections. The same recording head is used as a record or playback (R/P) head (Fig. 6-36).

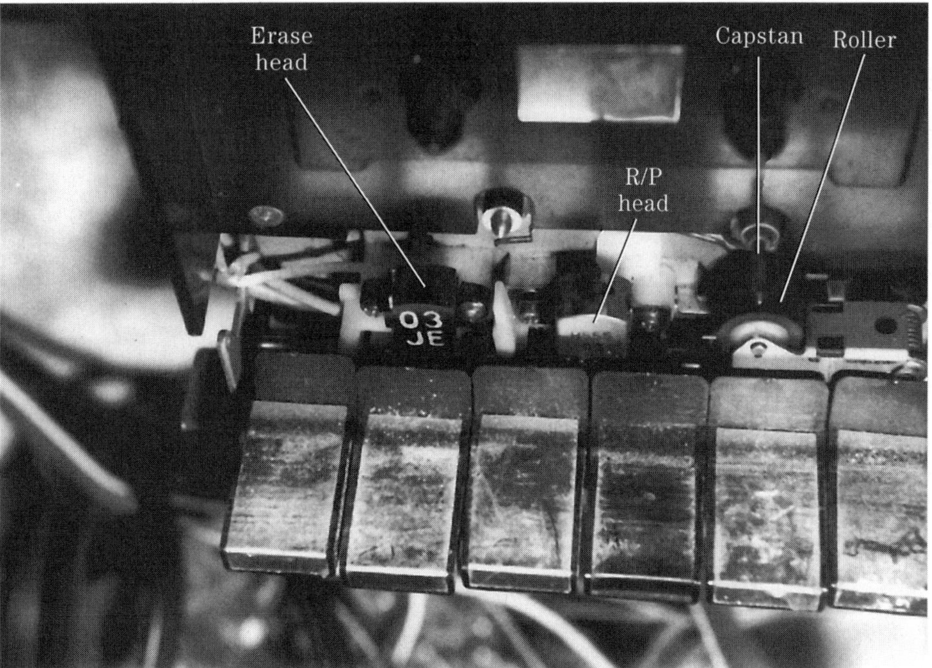

6-35 The erase head is mounted ahead of the R/P head in the cassette player for erasing a previous recording.

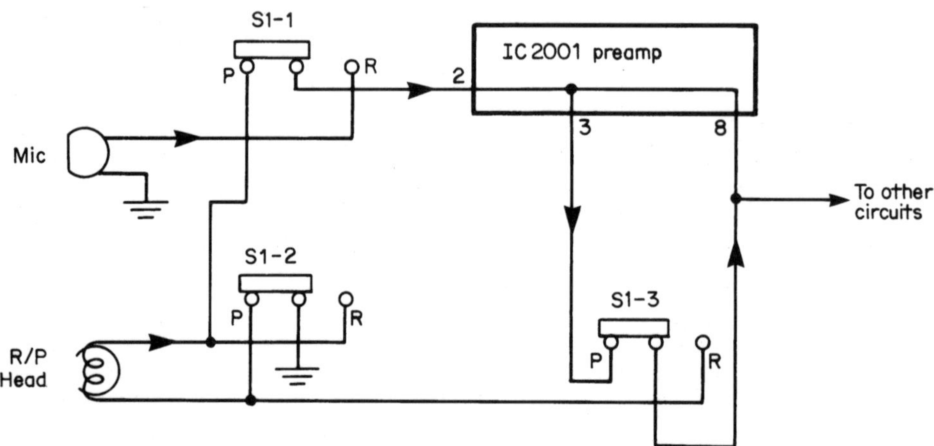

6-36 Often the record/play circuits are switched into an audio preamp IC for cassette recordings.

Sometimes the tape head comes loose from the mounting, or one mounting screw is missing and the head is out of line with the moving tape. Replace the tape head if it is broken loose from the mounting plate. Some heads are just spot-welded to the mounting plate. Before replacing the stereo head, check the color coding of the connecting wires and write them down for easy replacement. Demagnetize tape heads after repairs.

Cassette erase head

The cassette erase head is much smaller than the stereo R/P head and is mounted ahead of the regular tape head. The erase head might be excited by a simple dc voltage or with a bias oscillator (Fig. 6-37). The low-priced cassette player may have a dc-operated erase head. Suspect a defective erase head or circuit when more than one recording is heard as it is played back. A jumbled sound of several recordings indicates the erase head is not functioning. Clean up the erase head with alcohol and a cleaning stick.

Testing

Check the resistance of the erase head. Of course, the erase head will not operate with an open winding. Measure the dc voltage switched from a dc source with the cassette player in record mode. Suspect a dirty erase head switch. Clean up by spraying cleaning fluid down into the R/P switch. Scope the erase head to see if the bias oscillator is operating. Check the bias oscillator when no waveform is found at the erase head.

Camcorder and VCR video heads

What and where

The upper cylinder or drum in the VHS and VHS-C video recorder has a 41mm diameter with four heads. The tape wrap for this system is 270 degrees, and it ro-

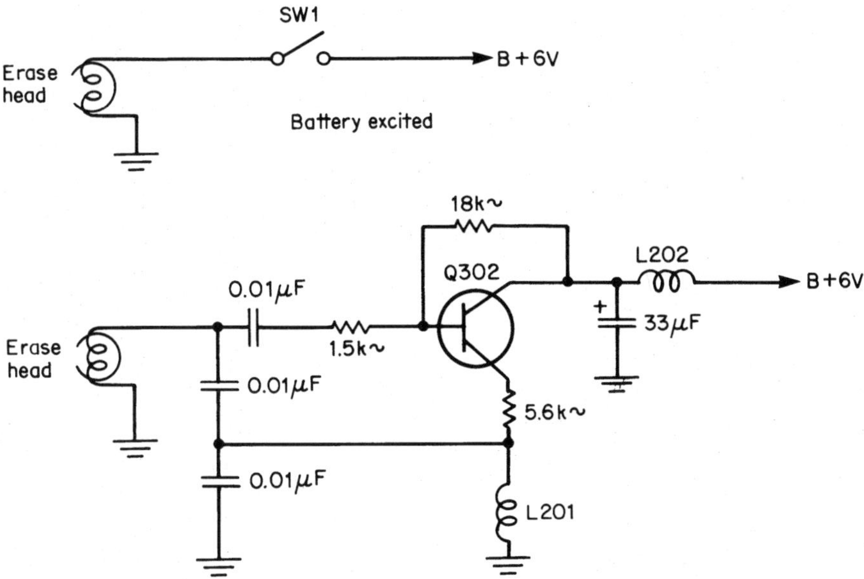

6-37 The erase head might be excited by a dc voltage source or a bias oscillator circuit.

tates at a speed of 2700 rpm. This system produces a more compact camcorder or VCR, providing compatibility with previous VHS recorders (Fig. 6-38). The 8mm drum is smaller in diameter and contains two channels and a flying erase (FE) circuit. The FM audio signal is recorded along with the video, rather than at the edge of the VHS tape.

First use a cleaning tape to keep the heads and tape guides free of oxide. If the dirt on the video head is too stubborn to be removed by the tape-cleaning cartridge, use a cleaning stick and solvent to clean the audio head. Some manufacturers recommend using a chamois leather stick with solvent (Fig. 6-39).

Do not try to clean the head down through the cassette cover vent. Remove the cassette cover to get at the video head to clean it. Some manufacturers recommend that you hold the stick at right angles. While holding the top part of the head only, gently turn the rotating cylinder to the right and left. Do not move the stick vertically or you might damage the head. Thoroughly dry the head after cleaning. If cleaning fluid remains on the video head, the tape could be damaged when it comes in contact with the head surface. The cleaning cassette will clean up some of the rollers and guide pins, but they should be cleaned with alcohol and the cleaning stick.

Testing

With the top half of the picture missing and the bottom okay, suspect a defective video head. When thin lines run through the picture, sub another cartridge or cassette. Check the tape head continuity with the RX1 range of the DMM. Scope the video and switching circuits, and inspect the video heads for worn areas. Take the waveform on the head-amp IC, and check the supply voltage fed to the head-amp IC.

6-38 The VCR or camcorder video head contains two channels of video within the video head.

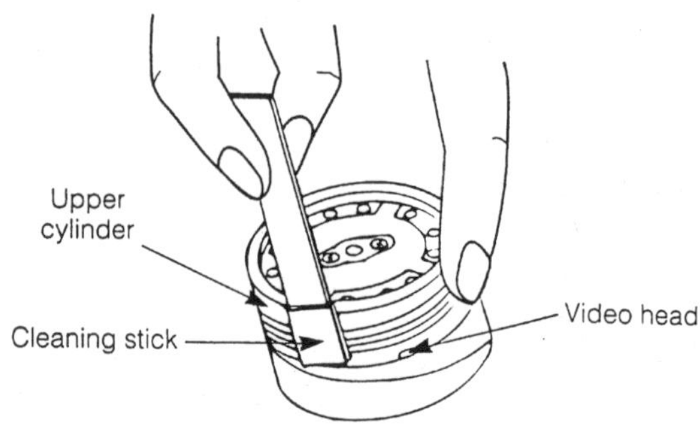

6-39 How to correctly clean up the video heads in the camcorder or VCR.

Camcorder and VCR sound head

What and where

The audio sound head is found along the tape path within the VHS and VHS-C camcorder or VCR (Fig. 6-40). The audio signal in the 8mm camcorder and VCR is processed to FM and mixed with the video heads. The audio record/playback (R/P) tape head might have a bias signal applied from the bias oscillator, which also sup-

6-40 The audio control head is found within the tape track area of a VHS and VHS-C camcorder or VCR.

plies a bias oscillator frequency to the audio-erase and full-erase heads. The audio A/C control head is located after the upper cylinder or drum assembly. Often, the audio head is excited from the bias oscillator circuit and is used to erase audio portion from the tape in VHS and VHS-C units.

The video and audio signals are applied to the tape with the REC/PB drum and A/C tape head in the helical scan format. Of course, the video signal is recorded in the center of the tape with the drum or cylinder head and the audio track at the edge of the tape in the VHS and VHS-C tape format (Fig. 6-41).

Testing

Scope the sound input to the head-sound-amp IC. Check the audio head continuity with the RX10 ohm scale of the DMM. The sound stages within the 8mm audio circuits can be checked with an external audio amp or scope from the stereo microphone through a high-pass filter (HPF) and audio amp. Measure the supply voltage to both the audio FM IC and head-amp IC. Check the tape head with a tape-head tester.

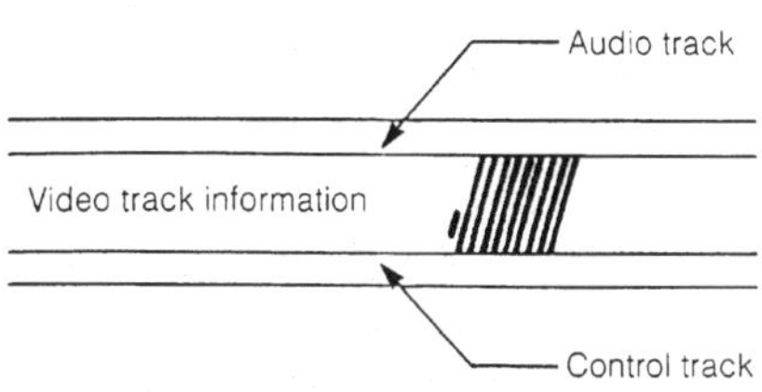

6-41
The sound track of a VHS and VHS-C camcorder or VCR is found at the edge of the tape.

Camcorder flying erase head

What and where

The flying erase head is located between the video heads, CH1 and CH2 on the drum or upper cylinder. The fly erase head (FE) prevents the color "rainbow" effect when recording. The FE on signal starts at pin 38 of the main MI-COM (IC251) and feeds into the electronic switch and to the flying-erase-head oscillator-head circuit (Fig. 6-42). A FE oscillator circuit operates somewhat like the bias oscillator in a record-cassette erase head. This FE signal is amplified and connects to pin 1 of a connector. Connector pin 1 ties directly to the flying-erase-head winding of the 8mm drum. The flying erase head erases the previous recording on the 8mm tape.

Testing

Check the erase head oscillator waveform at the FE OSC head output IC or transistor. Scope the FE signal up to the erase head. Take a quick continuity test of the erase head winding. Check the supply voltages of the FE head oscillator, FE amp, and electronic switch circuits. Clean up the FE and video heads on the 8mm drum or cylinder.

High-voltage divider

(See Focus divider in this chapter.)

High-voltage tripler

What and where

In the early solid-state chassis, a tripler unit was found after the flyback transformer to build up the high voltage applied to the TV picture tube. Very few service problems were found with the small flyback transformer that fed the tripler unit.

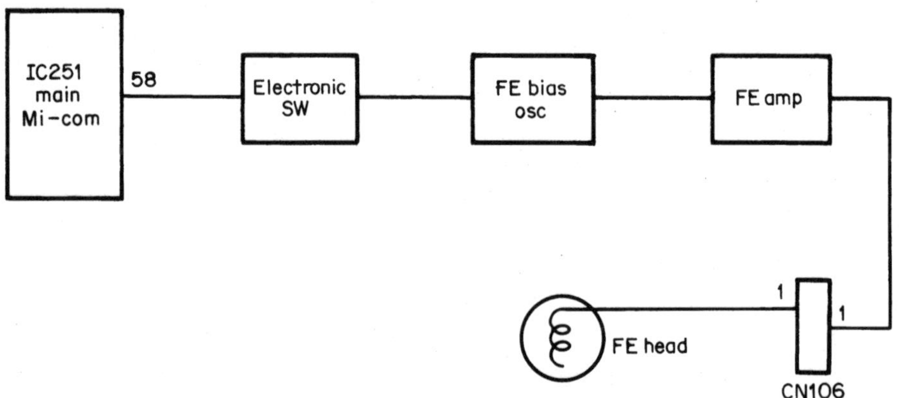

6-42 The flying erase (FE) head erases the previous recording and also prevents the color "rainbow" effect when recording.

These transformers operate at a very low voltage compared to the integrated flyback, and this reduced transformer breakdown causes problems. Often, the RF output voltage was under 10 KV, while the integrated flyback could have over 32-KV output. Actually, the tripler unit produces more breakdown problems than the horizontal output transformer.

The tripler unit consists of an input, focus, ground, and HV anode lead (Fig. 6-43). The tripler unit is made up of HV capacitors and diodes. This tripler unit might arc internally or draw an arc between tripler and metal chassis. Some tripler units have a spark gap placed between ground and input terminals. The insulation between windings can break down, producing high-voltage arcover in the integrated flyback. A breakdown of high-voltage diodes molded inside can cause arcing or a warm flyback.

A shorted tripler unit might destroy the main power fuse or cause the circuit breaker to trip. The leaky tripler can load down the flyback and horizontal output transistor and make it run very warm. A leaky tripler can keep tripping the circuit breaker, and the defective tripler can damage horizontal output transistors. Do not try to repair a tripler unit that is arcing inside or to the metal chassis. Replace it at once.

Testing

Remove the input terminal from the flyback to the tripler unit to check for a leaky tripler. The output lead of the tripler flyback can be checked by creating

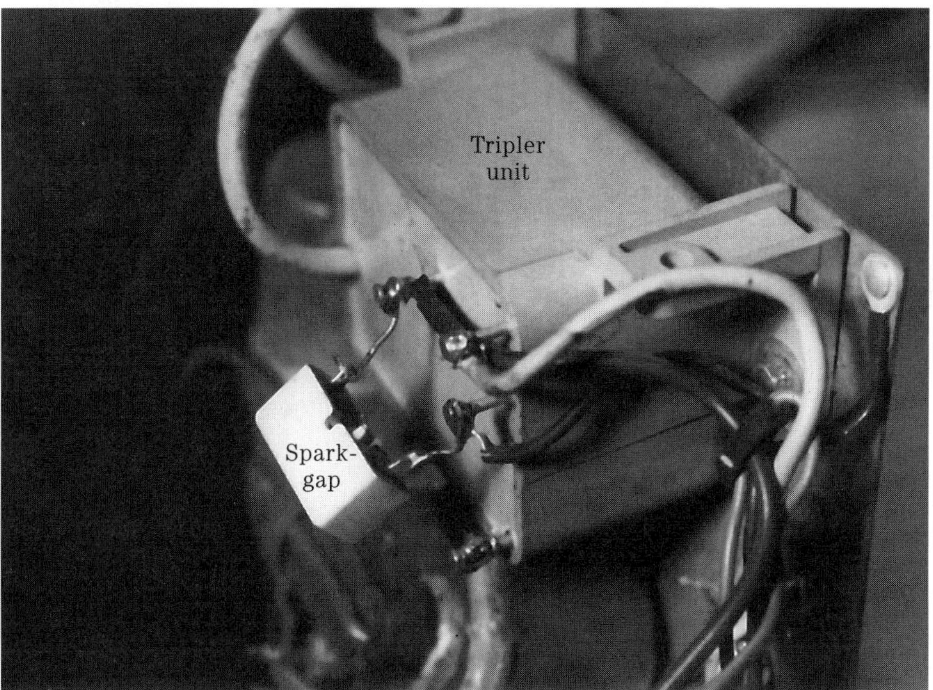

6-43 The tripler unit in the early solid-state TV chassis provides HV, screen, and focus voltage for the picture tube.

a 0.25-inch arc between a screwdriver blade and the integrated transformer. A high-voltage measurement at the anode CRT lead can indicate high voltage of the tripler unit (Fig. 6-44).

The tripler flyback transformer secondary winding can be checked with the low-ohm scale of the VOM or DMM. The resistance of the high-voltage winding can vary from 55 to 350 ohms. This depends on the output voltage and manufacturer. Although the resistance measurement does not indicate a shorted turn, the measurement does tell us if the winding is open. In the integrated transformer (IHVT), winding continuity cannot be measured because of the high-voltage diodes in the circuit.

Feel the outside case of the suspected tripler with chassis shutdown. If red hot or very warm, replace it. All tripler units should run cool. Intermittent or constant arcover can occur inside the tripler. Suspect a defective tripler when the flyback has a tic-tic noise with no high-voltage output. Firing lines can appear on the raster with an internally arcing tripler unit. A defective tripler can produce poor focus of the picture and raster.

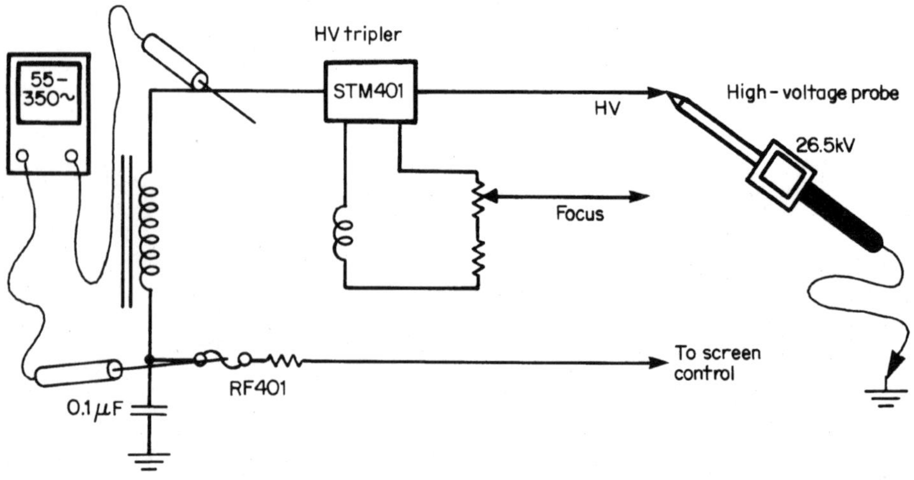

6-44 Check the HV at the anode of the CRT with a high-voltage probe.

7 CHAPTER

Testing electronic components I to P

(Inductors to noise filters)

The TV, VCR, compact disc, and receiver operations are made easy to operate from the armchair or sofa with infrared detectors and emitters from the remote transmitter. The wireless intercom can be placed in any room in the house and correspond with each individual unit. Replacing the lamp in that car radio might be more difficult if you do not know how to retrieve it. Today, LEDs are used in many consumer electronic products as indicators and monitors. A discussion of testing and repairing consumer electronic components from intercoms to motors is found in this chapter.

Inductors

(See Coils)

Infrared detectors and emitters

What and where

Infrared detectors and emitters are found in the TV set, VCR, camcorder, compact disc player, phone, tape player, and the high-priced receiver (Fig. 7-1). Also, the infrared devices are found in burglar alarms, touch screens, conveyor beams, short-distance links, scanners, and other infrared equipment. The infrared diode is a semiconductor diode that emits infrared rays. The infrared frequency is just a little lower than visible red light. The infrared diode found in the TV remote contains an

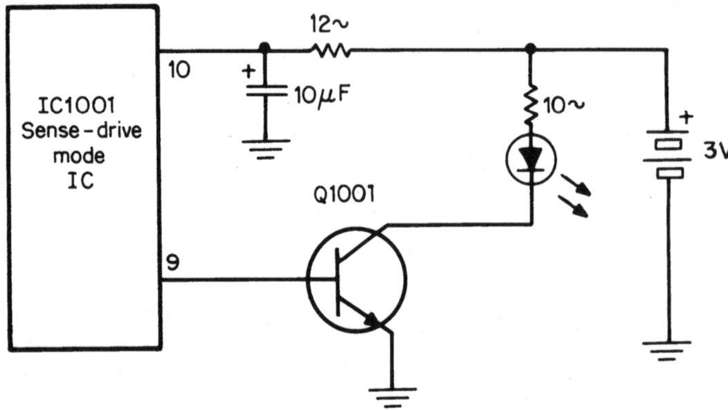

7-1 The infrared emitting diode found in the TV remote transmitter.

emitter made up of gallium arsenide LED (AlGaAs), while the infrared detector is made of a silicon phototransistor (NPN). The photo diode is called a light sensor, pin diode, or remote sensor.

Testing

The infrared remote operates in direct line of sight. If a body or object is found between remote and TV receiver, no infrared action is noted. An infrared remote will not interfere with any other TV receiver in the house. The infrared diode and phototransistor can be tested with the diode test of the DMM, infrared detector card, infrared detector probe, infrared tester or a compact disc laser tester. You can even check the remote with a portable radio. The infrared detector card is held in front of the emitting diode and will show a fluorescent light when the remote button is pressed. If no light is visible on the card indicator, the remote is not functioning (Fig. 7-2).

Check the emitting infrared diode on the diode test of the digital multimeter (DMM). Place the positive terminal to the anode lead and the negative to the cathode terminal and you should get a resistance measurement. This measurement is much higher than a regular silicon diode. Now reverse the test leads and a normal infrared diode should have an infinite measurement (Fig. 7-3). Suspect a leaky infrared diode with a measurement in both directions. Test the phototransistor for a leakage test between collector and emitter.

The infrared diode or TV-VCR remote can be tested for emitter and infrared signal with a CD infrared laser tester. Set the indicator of the laser tester in front of the emitting diode and push down on any remote button. Notice the measurement upon the laser meter. Start with the lowest measurement (0 to 0.3 mW). If the results are low, move up to the next scale. The laser power meter might have a 0.3-mW, 1-mW, and 3-mW range with switchable wavelengths of 633 nM and 750 to 820 nM.

Place the remote 3 inches away from the pick-up probe, and register the measurement on the power meter. Likewise, check a powerful remote control unit, and write down the measurement. By making comparison tests with new remotes, the defective or weak remote measurements can be compared to the new infrared measurements. Move the remote back and forth to acquire the best reading. A weak or dead remote will have a low or no measurement compared to a normal remote transmitter.

Check the infrared remote with the table or portable AM radio. Place the remote a few inches from the radio and press any button on the remote. You should hear a chirping, gurgling, or audio tone when any button on the remote is pressed. Move the remote back and forth until the loudest sound is heard. If any button appears erratic or intermittent, suspect a dirty contact. When you press hard down on each remote button or it takes a little time for the remote to act, suspect weak batteries. No sound indicates a defective control unit or dead batteries.

7-2 Place the infrared indicator card in front of the remote transmitter and notice the different color on the indicator when a remote button is pressed.

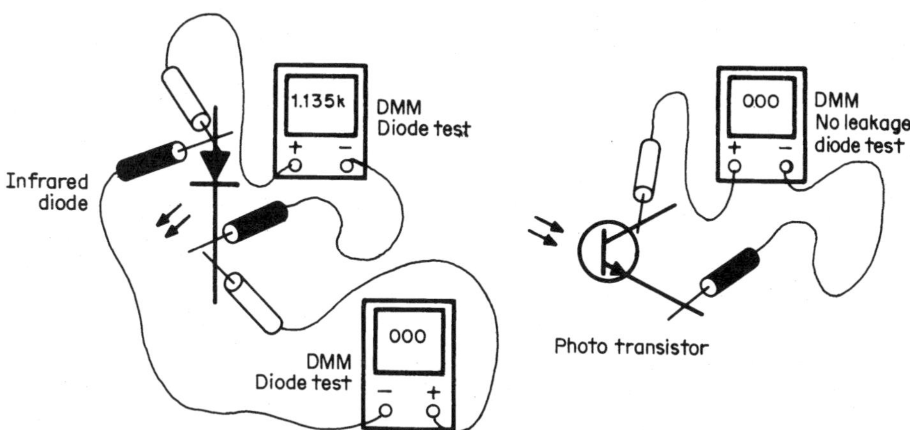

7-3 Check the infrared diode and phototransistor with the diode test of the DMM.

Intercom

What and where

You might find a master unit with several slave stations in the larger intercom units. A two-station intercom has a master and one remote unit. Today, the wireless intercom can be placed in any room for transmission and reception over the ac power lines which it operates from (Fig. 7-4). The wireless intercom can be placed or moved to the basement, garage, or kitchen for messages.

Testing

The early intercoms consisted of one or two dual-purpose tubes, while today, solid state transistors and IC components are used. A simple IC intercom might contain one audio IC, with interconnecting cable between speakers, and operates from a battery or wall power supply adapter (Fig. 7-5). First check the battery or power supply source with the voltmeter. Inspect the interconnecting cable for breaks. Double-check the two remotes by connecting a few feet of cord between the units. When the talk button is pressed you should hear a loud feedback sound if both units are working. A distorted message might be caused by a defective speaker located outdoors. By taking a voltage test on the transistors or ICs, you can quickly locate a defective component. Loud oscillations and hum can be caused by a dried-up electrolytic capacitor upon pin 6 of IC1.

Integrated circuits (ICs)

(See Semiconductors)

Inverters dc to ac

What and where

The inverter power supply converts battery operation to ac power operations. An inverter unit changes direct current to alternating current. The power inverter is ideal in running computers, FAXes, stereos, TVs, VCRs, microwave ovens, power tools, coffee makers, and many other small ac appliances. Some of the latest inverters use modern high-frequency switching power supply technology with the latest power MOSFETs. A digital circuit with quartz crystal control correctly regulates the output voltage and frequency. These commercial power inverters may have a continuous power output of 225 watt to 1500 watts. Commercial inverters are found in today's solar systems. The inverter operates in 12-V and 24-V battery systems.

Testing

A simple dc to ac converter might consist of two power transistors, transformer, resistors, and a 9- or 12-volt battery (Fig. 7-6). The step-up transformer converts the low ac voltage to a HV output. Check the fuse with no voltage output. Make sure the battery voltage is normal. Test all diodes with the diode test of a DMM. Check each transistor with a transistor tester. Inspect overheated connections. Clean up the connections and resolder all corroded contacts.

7-4 A wireless intercom can operate in any room in the house, even to the garage or workshop.

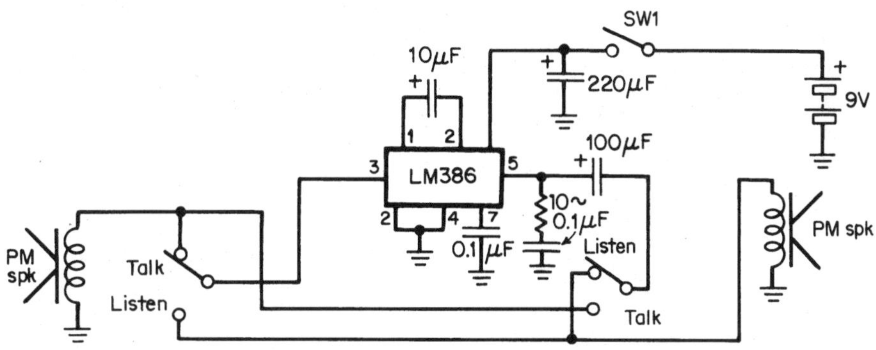

7-5 A simple intercom unit built around a popular LM386 OP-Amp IC.

Kits

What and where

The electronic project kit might consist of several different electronic components, with hookup wire, PC chassis, and instructions for building a piece of electronic equipment. You can purchase an electronic kit from many different manufacturers for just about any type of an electronic product (Fig. 7-7). The electronic kit might consist of electronic components to construct a light display, monitor kit, audio amplifier,

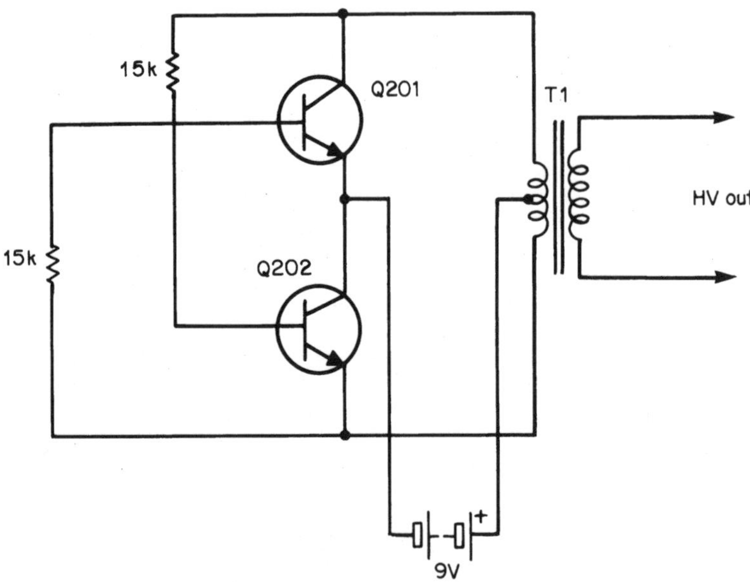

7-6 A simple dc-to-ac inverter circuit with two transistors and resistors and a power step-up transformer.

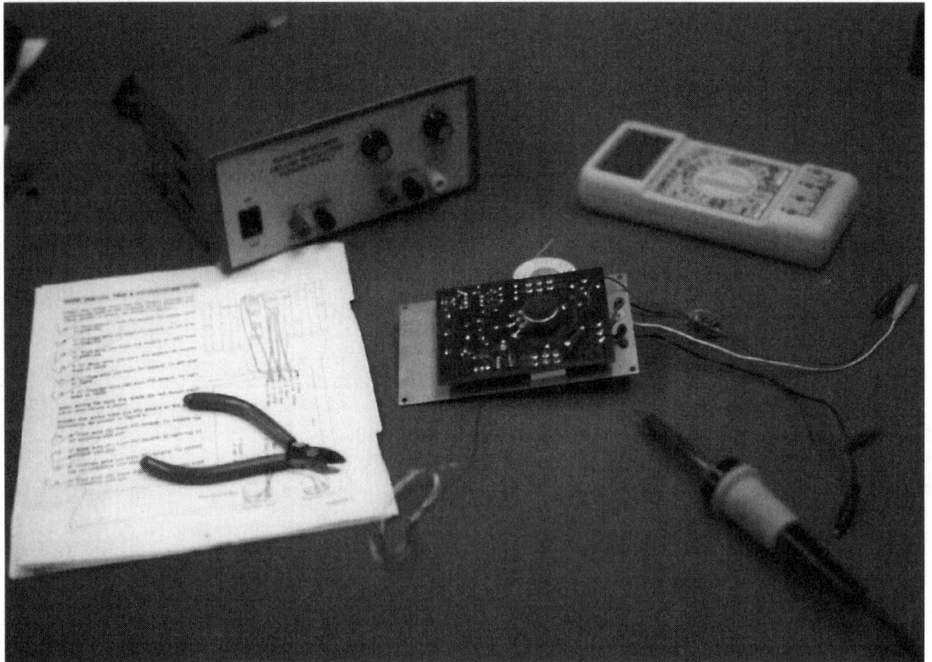

7-7 With kit building, you can quickly learn how to build, solder, and put together electronic projects.

telephone transmitter, blinking LED circuit, voice- activated switch, code oscillator, RF amplifier, electronic tuner, power supply, and on it goes. Here is a list of electronic distributors that sell electronic kit projects:

624 Kits
171 Springdale Dr.
Spartansburg, SC 29302

All Electronic Corp.
P.O. Box 567
Van Nuys, CA 91408

Alltronics Check Kit List
Mfg.(Book):
2300 Zander Road
San Jose, CA 95131-1114

Antique Electronic Supply
622 S. Maple Ave.
Tempe, AZ 85283

C.& S. Electronics
P.O. Box 2142
Norwalk, CT 06852-2142

Electronic Goldmine
P.O. Box 5408
Scottsdale, AZ 85261

Electronic Kits
International, Inc.
1663 Noyes Ave.
Irvine, CA 92714

Electronic Rainbow, Inc.
6227 Coffman Rd.
Indianapolis, IN 46268

The Electronic Rainbow, Inc.
6254 La Pas Trail
Indianapolis, IN 46268

Elenco Electronic, Inc.
150 West Carpenter Ave.
Wheeling, IL 60090-6062

Gateway Electronics, Inc.
8123 Page Blvd.
St. Louis, MO 63130

Graymerk International, Inc.
P.O. Box 2015
Tustin, CA 92681

C. & S. Sales
150 W. Carpenter Ave.
Wheeling, IL 60090

Cal West Supply, Inc.
3120 Via Colinas #105
Westlake Village, CA 91362

Rutenlier Engineering
38045 10th St. E. #1145
Palmdale, CA 93550

D. C. Electronics
P.O. Box 3203
Scottsdale, AZ 85271

Herback & Rademan
P.O. Box 122
Bristol, PA 19007-0122

Hosfelt Electronics
2700 Sunset Blvd.
Steubenville, OH 43952-1152

Interactive Electronics
P.O. Box 913 Dept. K
Eilers, TX 70039

Kevin Electronics
10 Hub Dr.
Melville, NY 11747

Marcraft International
Corp.
1620 E. Hillsboro St.
Tri-Cities, WA 99302

Mark V Electronics
8019 E. Slawson Ave.
Montebello, CA 90640

MCM Electronics
650 Congress Park Dr.
Centerville, OH 45459-4072

Matco A-3
P.O. Box 509
Roseville, MN 48066-059

Oak Hill Research
20879 Madison St.
Big Springs, MI 49307

Ocean State Electronics
P.O. Box 1458
Westly, RI 02891

PAM CON International
P.O. Box 130
Paradise, CA 95267-0130

Ramsey Electronics, Inc.
793 Canning Pkwy.
Victor, NY 14564

Xandi Electronics, Inc.
1270 Broadway Rd. #113
Tempe, AZ 85282

L pads (speaker)

What and where

The L pad is an inverted letter L attenuator consisting of one arm in series and one as a shunt arm (output). The L pad has two separate sources of variable resistance. These early L pads were used in sound systems with the resistance pad component mounted inside the speaker cabinet and can be adjusted from the outside with a control knob.

The L pad is adjusted for the correct amount of audio applied to the speaker. The L pad is installed with one resistance element connected in parallel to the voice coil of the speaker and the other resistance in series (Fig. 7-8). This L pad is connected directly to the speaker voice coil and is similar to a car fader speaker control.

Testing

Today instead of an L pad, a rheostat is used on the primary or 70 volt line instead of at the speaker. This offers greater audio control and frequency response. Sometimes the L pad collects dirt and dust, producing noisy audio. Check the L pad with the low ohm scale of the DMM. Replace the intermittent or noisy L pad with a 5K rheostat that is connected to a 70-volt audio line output transformer (Fig. 7-9).

Lamps

Bayonet and screw types

What and where

The bayonet and screw-type lamps or bulbs are found in table and auto radios (Fig. 7-10). Pilot lights are found in just about any electronic product. Colored subminiature bulbs with long wire leads were used later with a red, green, blue, and yellow 12-volt lamp. A lamp is an electronic device that converts electrical energy into usable light. The lamp bulb might consist of an arc lamp, halogen, fluorescent, incandescent lamp, mercury-vapor, neon bulbs, and LEDs. A lamp bulb might be found in a flashlight, microwave oven, as pilot lights, sensing, subminiature, strobe tubes, tubular, etc (Fig. 7-11). The light bulb might screw into a socket, while a bayonet-type lamp locks into position with two small tabs. These bulbs might operate at 1.70 to 37.5 volts (Fig. 7-12).

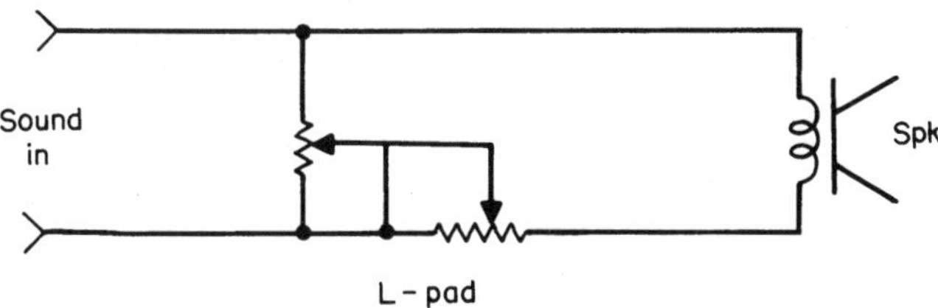

7-8 The L pad varies the audio at the voice coil of a PA speaker.

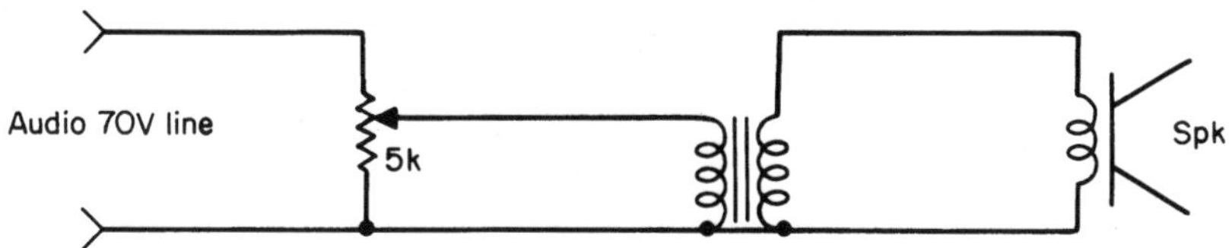

7-9 Today, the L pad is replaced with a simple rheostat placed within the 70-volt line of the amplifier.

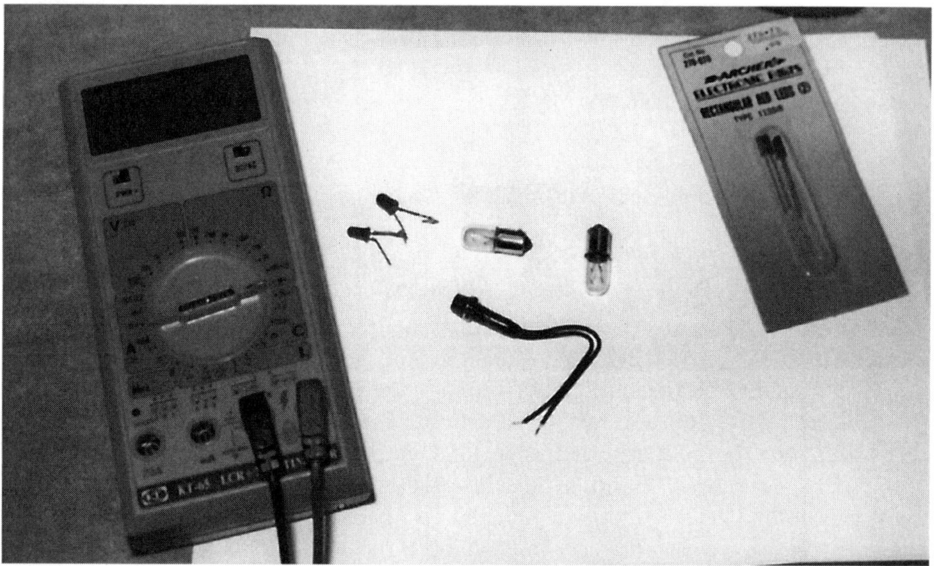

7-10 Pilot lights are found in most consumer electronic products as indicators.

Halogen lamp

What and where

The halogen lamp might be used in projectors, lighting effects and high intensity applications, microfiche, auto headlights, and now in home floor lamps. The halogen nonmetallic elements might consist of chemical properties such as astatine, bromine, chlorine, fluorine, and iodine. The halogen projector bulb 12-volt types

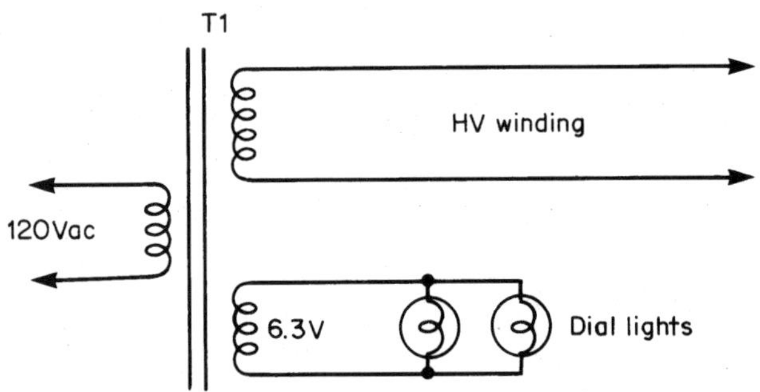

7-11 In the early radio circuits, the pilot lights operated from a separate transformer winding.

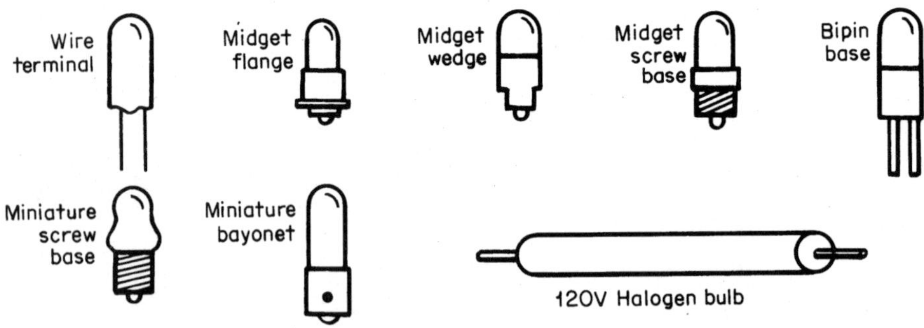

7-12 Pilot light indicators come in many different mounting sockets.

come in 20 W, 35 W, 50 W, 75 W, and 100 W. The halogen microfiche lamp might be replaced with a 13.8 V at 24 W, or a 82 V at 360 W.

Testing

Check the ac voltage applied across the halogen light bulb terminals. Remove the bulb and take another resistance measurement. The lamp is open with no measurement. The new type halogen floor lamp might be replaced with a 100-, 250-, or 500-watt lamp. Handle the bulb with a cloth. Do not leave finger marks upon the bulb. These lamps run red hot and can cause serious burns. Double-check the end contacts of a high-wattage bulb for burn marks. File off or sandpaper the burned contact areas for better bulb contacts.

Fuse-type bulb

What and where

The fuse-type bulb looks just like a small cartridge fuse and clips into a pilot light socket at each end of the bulb (Fig. 7-13). A regular fuse-type bulb comes in 6 V at 250 mA, 8 V at 250 mA, and 12 V at 150 mA. The subminiature fuse bulb might appear in 6.3 V, 8 V, and 12 V at 70 mAs, while a submini stereo fuse bulb comes in a 8-V and 12-V 35-mA and 150-mA fuse lamp. Check for continuity across the fuse type terminals with the low RX1 scale of the DMM or ohmmeter. No reading indicates an open bulb. Inspect the fuse clips for poor connections.

Krypton lamps

What and where

A Krypton lamp consists of a gaseous element and is found in some flashlights. The krypton bulb has wire contacts with a 4.8 V, 0.75 A for the professional torch. Check for bulb continuity with the RX1 scale of the DMM.

LED lamps

(See LEDs)

Neon bulbs

What and where

A neon bulb might also be called a neon glow lamp or neon tube. The neon bulb is a neon-filled gas diode and has a pink glow when a certain voltage is applied. The neon lamp might have short or long pigtail wires and mounts in a bayonet socket (Fig. 7-14). These neon bulbs can be used as dial lights, circuit control applications, or indicators. They might come in red, amber, white, green, and clear glass colors.

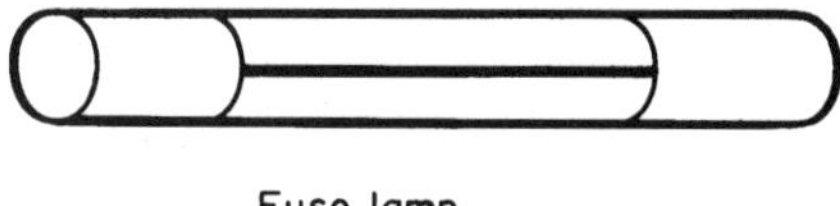

7-13 The fuse bulb resembles a cartridge-type fuse and appears in 6-, 8-, and 12-volt units.

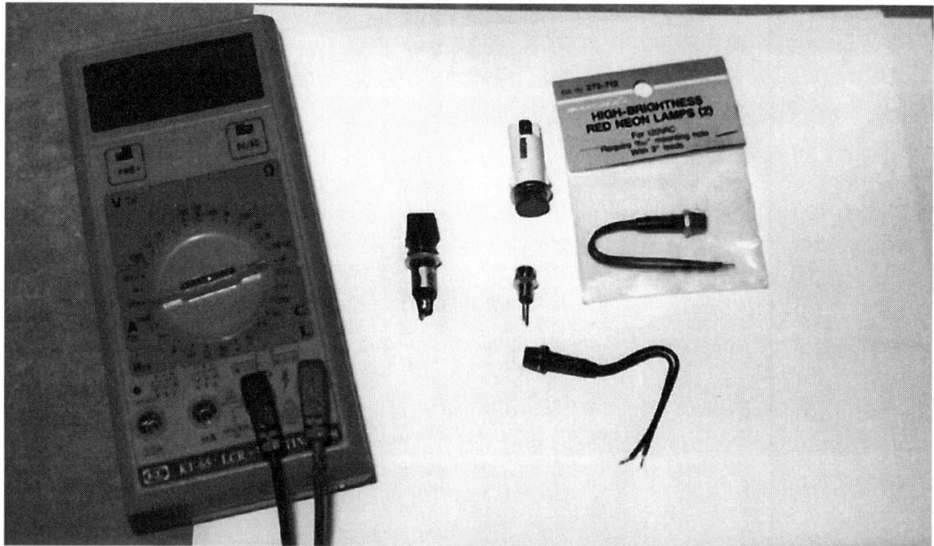

7-14 The neon pilot light might be used as an indicator or within a small ac-dc tester.

A panel mount neon indicator might operate between 55–60 volts with a 100-K resistor in series with one lead. A neon indicator tester might be made up of a neon bulb, voltage-dropping resistor, and pigtail probes. The NE-2 neon bulb is very popular in electronic applications.

Testing

The neon bulb can only be tested when a voltage is applied to it. The neon bulb has no continuity measurement. If the bulb does not light up with a dc or ac voltage from 55 to 100 voltages across its terminals, replace it with another. Make sure the voltage-dropping resistor has not increased in value (Fig. 7-15). A power line voltage tester can be constructed of three different neon bulbs on a three-prong plug (Fig. 7-16).

Panel lamps

What and where

Front-panel lamp bulbs are found in just about every piece of electronic products within the commercial entertainment field (Fig. 7-17). Besides front-panel indicators, the panel light might be found mounted behind a panel meter. The standard incandescent panel lamps might have a screw, bayonet, solder lugs, plug in tips and come in red, amber, yellow, and clear colors. A panel lamp with a lens cap and long pigtail leads might operate from 2 to 28 volts in green, white, red, blue, yellow, and orange colors. Today, you might find surface-mounted (SMD) bulbs and snap-in LED lamps as indicators.

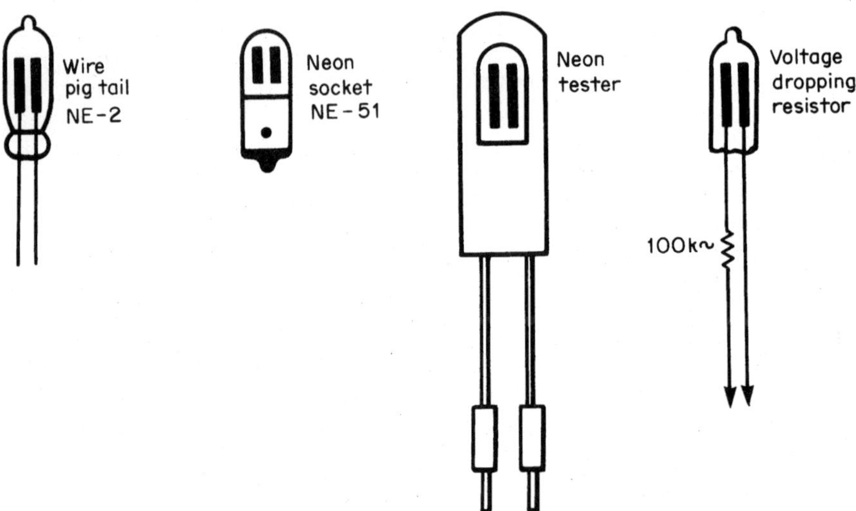

7-15 A voltage-dropping resistor is required to drop the 120-volt line voltage for the NE-2 neon lamp.

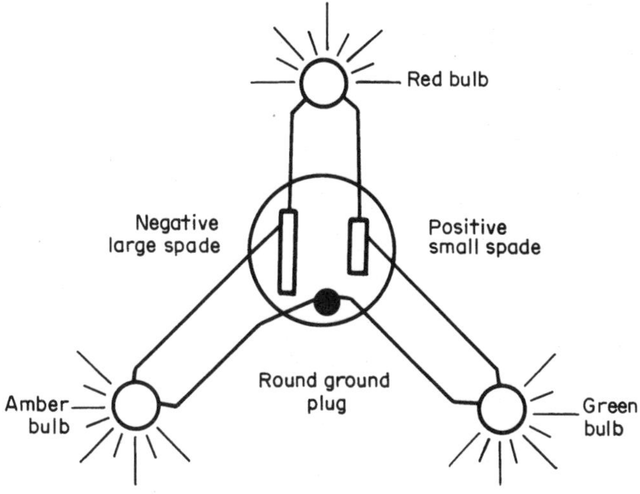

7-16 You can make a polarity and 120-Vac line voltage tester with three neon bulbs and a three-prong plug. The red bulb indicates normal line voltage, amber indicates good ground, and green indicates poor ground.

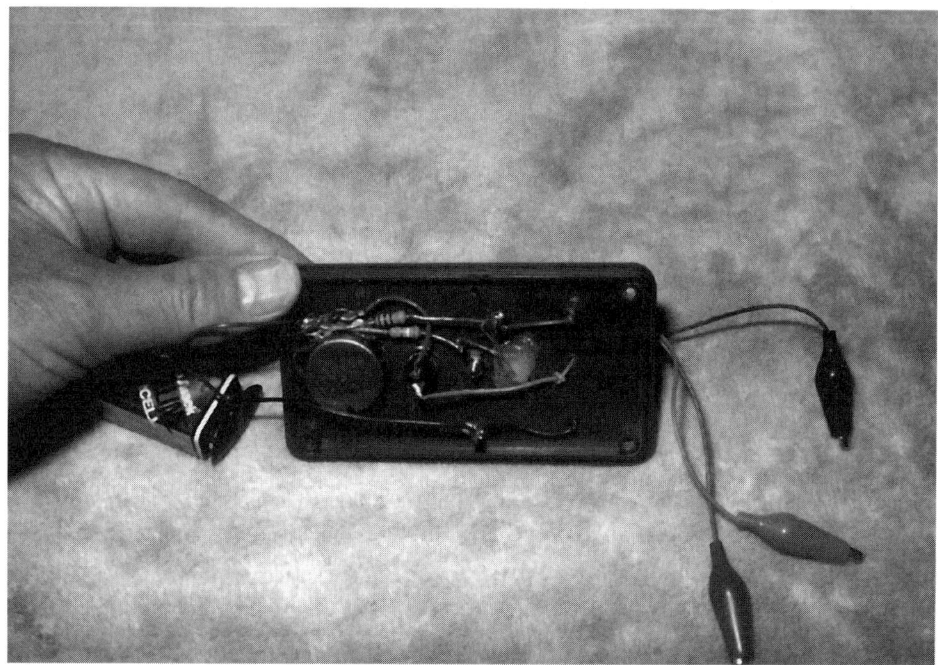

7-17 The LED pilot light can be used in a transistor-diode battery-operated project.

Testing

Check the light bulb in a light-bulb tester or with a low-ohm continuity resistance measurement with the DMM. Simply replace the light bulb when the glass is dark and the bulb still lights. This bulb has very little life left and should be replaced, especially while the electronic product is in for repair and torn apart on the service bench.

VCR sensing lamps

What and where

The lamp bulbs found in the VCR might contain a tape-end sensor, reel sensor, dial lamps, and infrared LED sensing lamps. Within the reel sensor circuit, pulses are sent from the reel sensor attached to the take-up reel base for each rotation of the reel base. These pulses are amplified at Q3002 and the output is applied to pin 11 of microprocessor IC3010 and then to a timer circuit in the SY-2 board (Fig. 7-18).

If there is no pulse at the input to pin 11, a system-down condition is indicated one second later during rewind, fast forward, or forward and reverse search, or three seconds later during the forward mode. No buttons will function except stop and eject, thus protecting the videotape from tape-transport problems.

Testing

Check the reel sensor and phototransistor with the diode test of the DMM. Test the voltage applied to the reel sensor. Check Q3002 and diode D3011 with the diode test of the DMM or with a transistor tester. Measure the voltage applied to the transistor and IC3010 terminals. Scope the pulse at pin 11 of IC3010.

Wilko lamps

What and where

The Wilko lamps are found in microfiche viewers, slide projectors, movie projectors and other equipment where near perfect viewing is needed. Also Wilko lamps are made for miniature fluorescent lamps and tubes. These Wilko projection bulbs come in 120 to 1000 W, 13.8 V at 85 W, 21 V at 150 W, 24 V at 150 to 250 W, and 82 V at 300 to 380 W.

Testing

Check the voltage applied to the lamp. Take a quick continuity measurement of the bulb. Replace the Wilko lamp with the same voltage and wattage.

Laser diodes (CD player)

What and where

The laser diode is a semiconductor diode, often made up of gallium-arsenide material, which emits a nonvisible light when voltage is applied to its terminals. The laser diode within the compact disc player is located under the playing disc (Fig. 7-19).

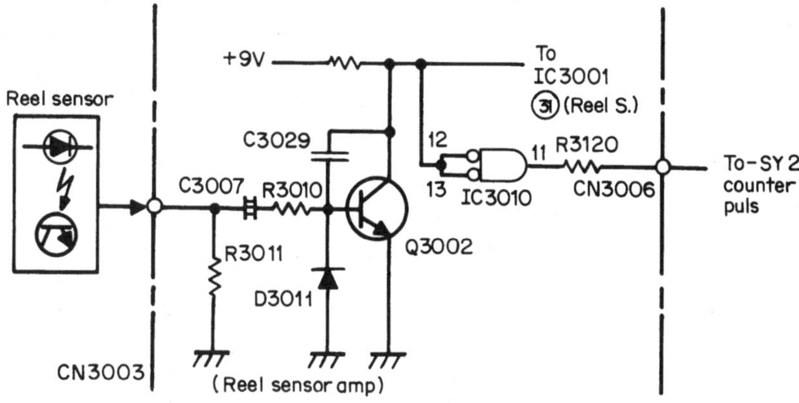

7-18 The VCR reel sensor circuit combines an LED and phototransistor to detect the end of the tape.

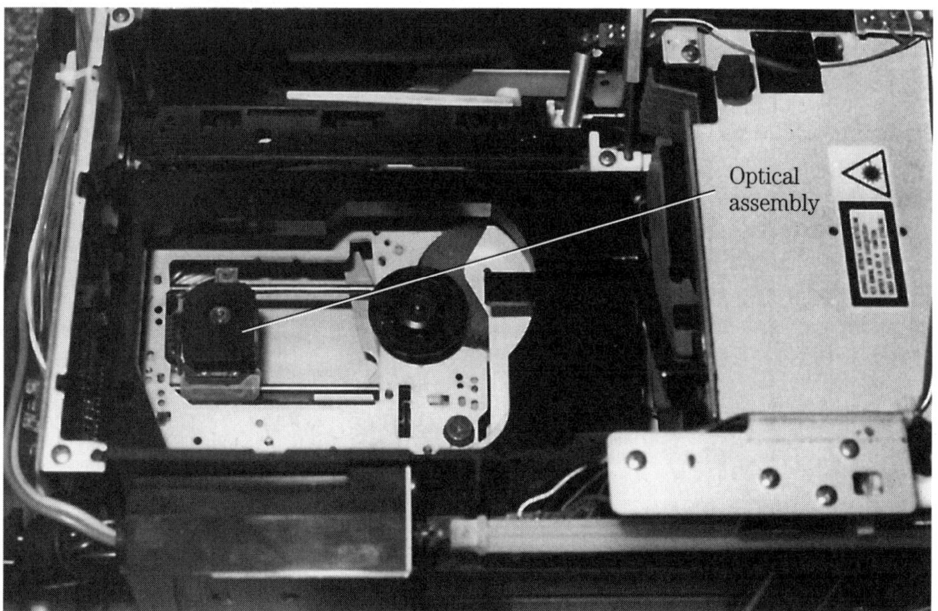

7-19 The laser diode is found in the optical lens assembly of a compact disc player.

The laser diode is located within the CD optical pickup block that travels from the center of the disc towards the outside edge of the disc with a slide, SLED, or carriage motor.

Be careful when working around the CD player to avoid laser beam radiation exposure. You can damage your eyes if you stare at the bare optical line assembly while the player is operating, so keep your eyes 30 cm away from the area. Keep eyes 2 feet away from the laser pickup assembly. The laser beam warning label is usually fastened to the back of the laser optical assembly citing danger. Always keep a disc on the spindle while servicing a CD player. Remember, the laser beam is not visible like that of an LED or pilot light.

The laser beam is formed on the pits and slants of the bottom side of the disc, and the digital signal is picked up by the photodetector diodes. The photodetector diodes sense the EFM signal from the disc and pass it on to the RF amplifiers. Besides supplying the EFM signal, the photodetector diodes (A, B, C, D) provide a tracking-error signal (Fig. 7-20). Because the EFM signal is weak, a preamp is connected to the photodetector diodes. In some pickup assemblies, the photodetector diodes are called HF sensors.

Testing

The laser diode can be tested with a voltage test, an infrared indicator, and a light power meter. The average laser output is from 0.25 to 0.9 mW. The laser power meter is used to measure the laser diode output and infrared remote control unit. The laser meter is particularly suitable for the service of compact disc and laser disc players because of its narrow, tiltable probe. Place the meter probe over the pickup lens assembly. Move the probe around for the greatest measurement on the power meter. The power meter might have three measuring ranges of 0.3 mW, 1 mW, and 3 mW that check all laser light sources within the compact player. The laser interlock switch must be shunted before a laser beam can be measured in most CD players. If the meter indication is less than 0.1 mW, replace the optical pickup assembly.

When a power meter is not available, check the voltage and current of the laser diode drive transistor with the low-voltage ranges of the DMM. Actually, the current of the laser diode is taken by a voltage measurement across a fixed resistor in the laser drive transistor circuit (Fig. 7-21). The normal laser current is given upon the laser pickup assembly. For instance, a Realistic CD-1000 compact disc player has a normal current of 40 to 70 mA. Check the laser driving current by measuring the voltage across R623 (12 ohms). If it is over 100 mA, the laser unit could be defective. The voltage value of 0.48 to 0.84 V equals 48 to 84 mA of current (a safe level).

LCD display

What and where

The liquid-crystal display (LCD) is a readout device for calculators, counter, digital clocks and digital test meters (Fig. 7-22). A digital multimeter (DMM) has a digital display showing the resistance, voltage and current measurements. The LCD display is found in camcorders, VCRs, TVs, CD players, receivers, microwave ovens, and shortwave receivers. Instead of an electronic viewfinder that is built around a small CRT, the liquid display is found as a viewfinder in the latest camcorder. The LCD display might fold out from the camcorder showing the picture that is being taken (Fig. 7-23).

Testing

Scope the signal going to the LCD display. Measure the supply voltage to the LCD display IC. Check the cable going to the black light unit for correct sync, B+,

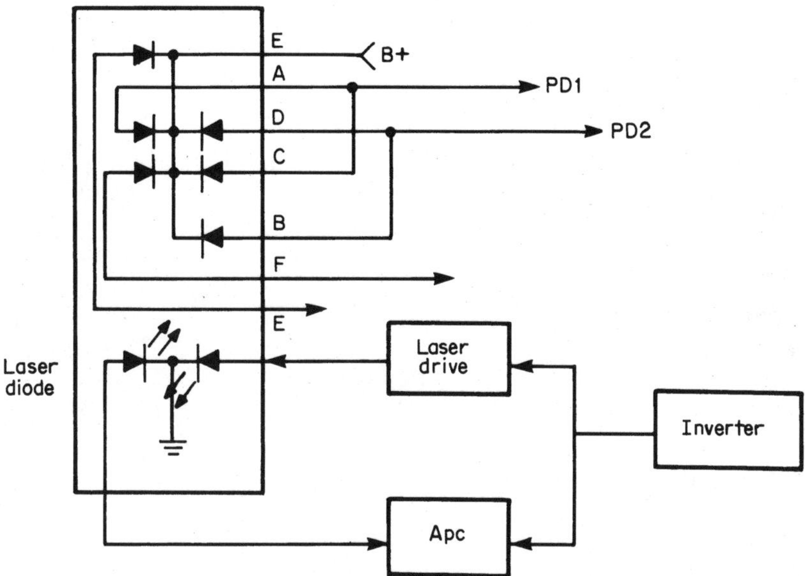

7-20 Both laser diodes and photo diodes operate within the optical lens pickup assembly.

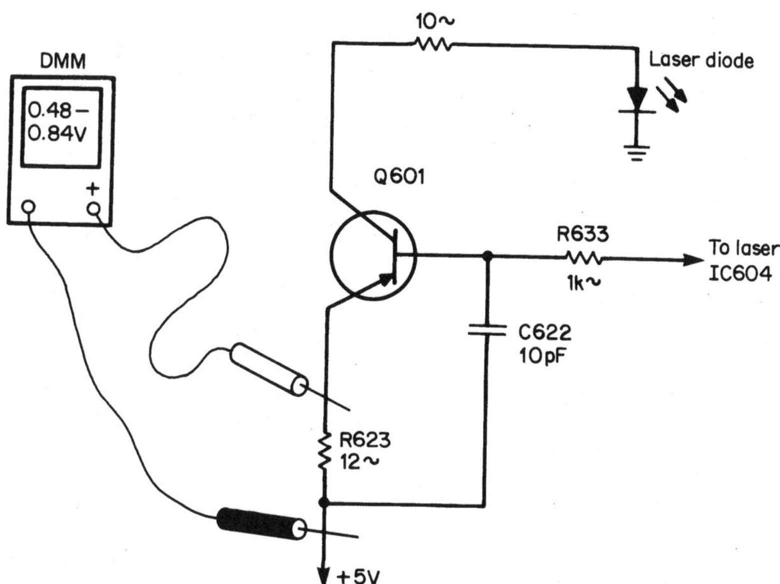

7-21 Check the laser diode operating current with a voltage measurement across the Q601 emitter resistor.

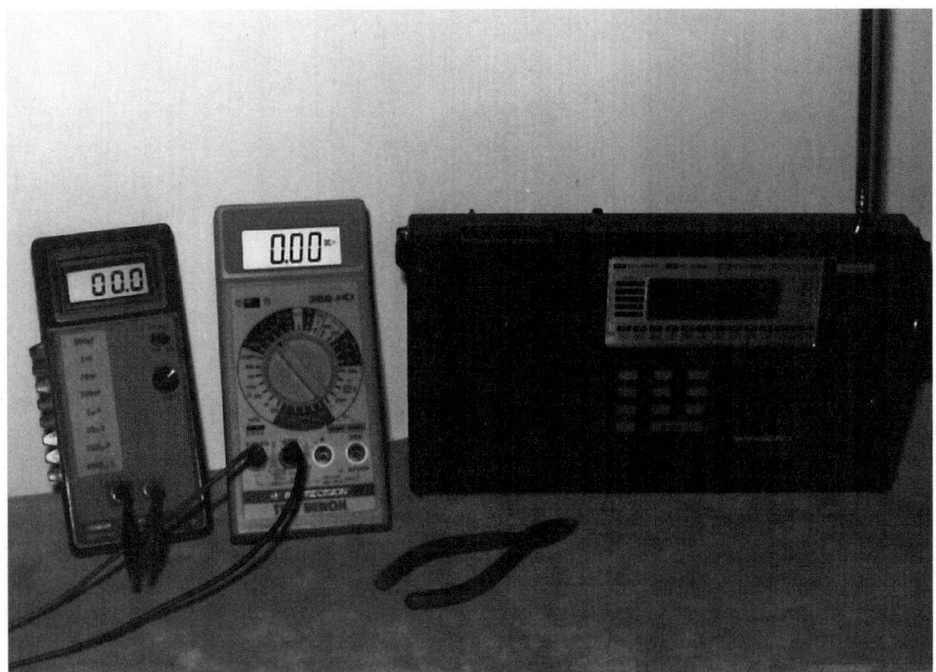

7-22 The DMM and shortwave receiver employ LCD displays to indicate what is measured and played.

and good ground connection. Check all sockets and connectors going to each assembly. Measure the B+ at the black light unit. Scope the sync input circuit to the black light unit. Most LCD assemblies are replaced when found defective (Fig. 7-24).

LED lamps

What and where

The light-emitting diode (LED) is another semiconductor that emits visible light when forward biased. The LED indicator was found in the early electronic products as pilot lights or indicators. These LEDs came in miniature, jumbo, and giant sizes. The LED might be a super-red, high-brightness or super-bright type. The LED might have long leads with a panel mount, have dual-colored LEDs, and appear in a bargraph display (Fig. 7-25). They might be round, flat, or appear in a rectangular display. The blinking LED comes off and on with correctly applied voltage.

Testing

Each LED device can easily be checked with the diode test of the DMM. A leaky LED will show a resistance measurement in both directions. The normal LED will indicate a forward-bias resistance measurement with the positive probe of the VOM on

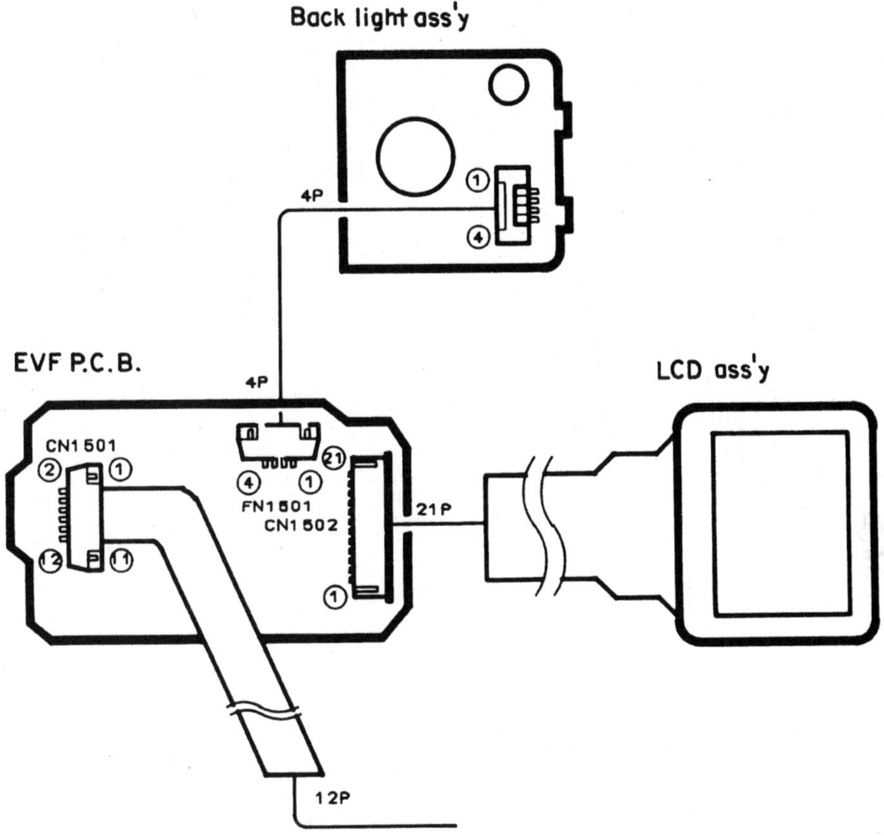

7-23 The LCD display in the camcorder shows what picture is being taken.

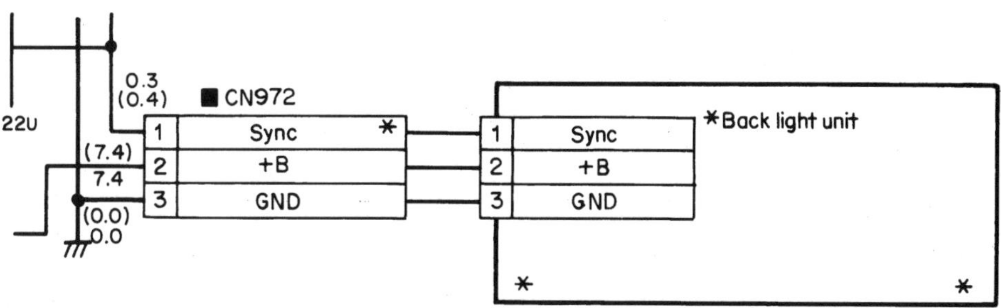

7-24 Check the B+ voltage and scope the sync signal of the black light unit to determine if the LCD is defective.

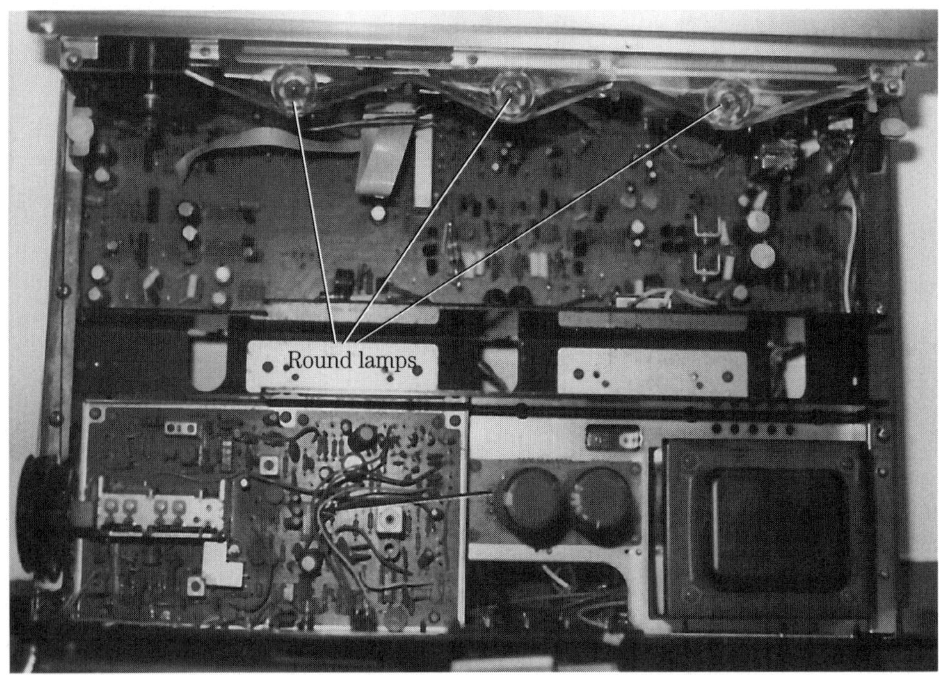

7-25 LED lamps are found in round, rectangular, and blinking indicators.

the anode (+) terminal and negative probe to the cathode (K) terminal. Likewise, the positive probe is placed on the negative (−) anode terminal and the black lead to the cathode terminal of the digital multimeter. The normal LED will have a resistance measurement in only one direction. Notice a black bar line represents the cathode terminal upon an SMD LED, while the shortest wire leg terminal is the cathode of the regular LED (Fig. 7-26).

Dual LEDS

What and where

The LED might operate from 1.6 V at 20 mA, 1.7 V at 20 mA, 2 V at 10 mA, 2.18 V at 20 mA, 2.4 V at 20 mA, 3 V at 20 mA, and 3.4 V at 20 mA. A dual super-bright LED has two different LEDs inside one container with red and green and red and yellow colors. The center terminal is the common cathode terminal and can be tested with the diode range of the DMM (Fig. 7-27). The two-color anode measurement should be comparable in resistance.

Testing

Check the resistance of each color LED in the dual-LED package with the diode test of the DMM. Place the negative probe to the cathode (+) terminal of diode and switch the positive probe to either anode terminal for a resistance measurement. When test leads are reversed, you should have an infinite reading with a normal LED. Notice the common cathode lead is the longest terminal in the dual-flat LED.

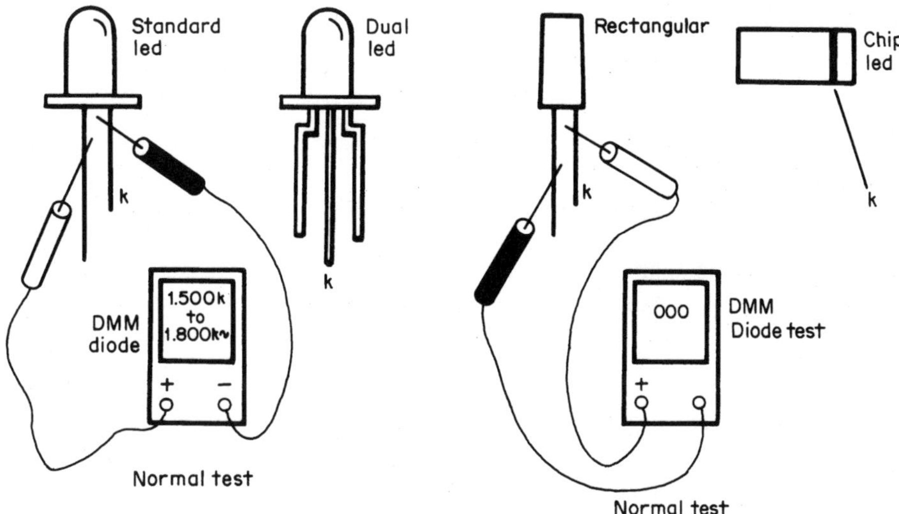

7-26 Check the LED lamps with the diode test of the digital multimeter (DMM).

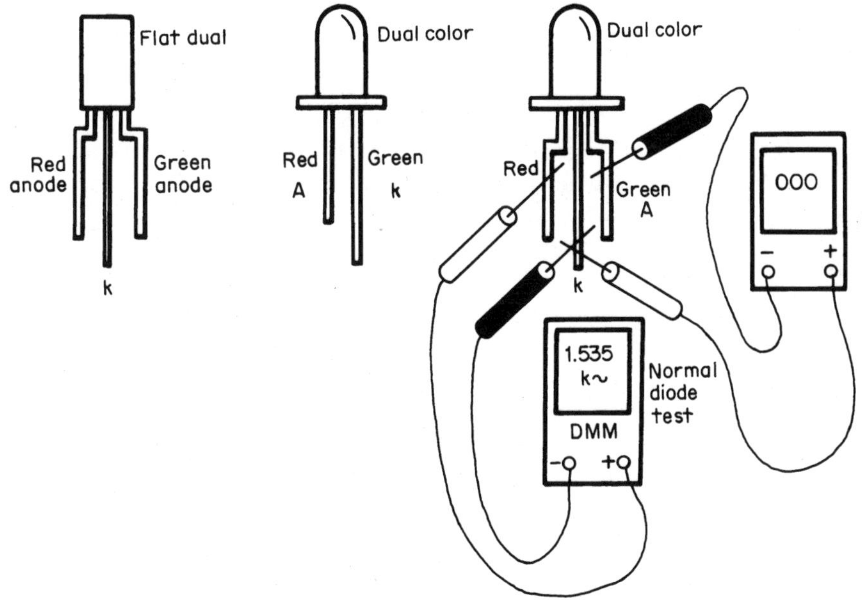

7-27 The dual LED can be checked with the negative terminal at the cathode (K) and the positive terminal at either green or red.

Infrared LED

(See Infrared detectors)

Segment LEDs

What and where

The multiple LEDs might have a seven-segment or ten-element LED in a bar-graph display. The LED bar-graph displays might be found in instrumentation, industrial controls, office equipment, computers, light bars, microwave ovens, and audio displays in cassette players and receivers. The display might come in red, green, and orange colors. The LED bar graph displays might have square or rectangular light bars.

Testing

Check each LED in a bar-graph display with the diode test of the digital multimeter. A defective LED will have a low resistance measurement in both directions. When one LED is found defective, replace the whole display.

Line transformers

What and where

The line transformer is an electromagnetic induction that transfers electrical energy from one circuit to another and isolates one circuit from another. Line transformers are used in PA systems, telephone-matching, and coupling devices. The CD player or amplifier might have a separate line output to connect to a separate amplifier and also a headphone jack for personal listening. The line output transformer used in matching the audio output of an amplifier to several speakers might consist of 4 or 8 ohms to a 70-volt line. Taps of 0.625, 1.25, 2.5, 5, and 10 watts are found on the secondary windings. A telephone coupling transformer might be considered a line output transformer with impedance of 600 to 600, 600 to 600 CT, 600 to 470, 600 to 346, and 600 to 500 ohms.

Testing

Check each tap and winding with the low-ohm scale of the DMM for open connections. The 4- or 8-ohm taps will have a very low ohm measurement compared to the 70-volt line side. You will find a higher ohmmeter reading when measured with the VOM.

Magnetron tubes

(See Electron tubes)

Metal oxide varistor (MOV)

What and where

The metal-oxide varistor (MOV) is a voltage nonlinear dependent resistor whose value will change with different voltage applied across its terminal. The metal-oxide resistor has a film of tin oxide deposited upon a substrate. A varistor is a surge absorber and has excellent clamping voltage characteristics and flat response time. It eliminates the discharge leg that is induction of a gas-type arrestor. ZNR arrestors can be used in ac and dc circuits for protection of positive and negative transients. It is made of zinc oxide. The ZNR has a bilateral, symmetrical, characteristic curve and can be used in circuits in place of back-to-back zener diodes. This provides circuit-clamping protection in either direction (Fig. 7-28). The varistor impedance changes from a very high-standing value to a very low-conducting value, thus changing the transient voltage to a protective level.

The metal-oxide varistor is found in the ac power line to protect microwave oven circuits, consumer appliances, semiconductor components, industrial equipment, telephone lines, railroads, communications, and security and fire-alarm equipment. It is used in automotive environments, ignition pulses, alternator fields, mutual harness couplings, inductive loads, switching fans, heaters, and booster starts. Within voltage-regulator circuits, the MOV protects against destructive high voltages and lightning damage.

Testing

Notice if the varistor has damage or blown marks. Check to see if the unit is blown in two or if one leg has come loose. Simply take a normal continuity measurement with the high range of the VOM or DMM (above 20 megohms). Replace the varistor if the resistance measurement is below 20 megohms.

Meters (panel)

What and where

You might find a panel meter mounted in receivers, VOMs, testers, auto transformers, ac clamp-on meters, isolation testers, satellite meters, signal tracers, EMF meters, CRT restorers, distortion meters, stereo-balance meters, battery testers, and dc power supplies (Fig. 7-29). Some panel meters have a center zero balance and mirror-backed scale. The most popular replacement meter is the 0-1 mA. The replacement meter starts with a 0-50 uA dc, 0-100 uA dc, 0-1 mA dc, 015 A dc, 0-30 A dc, 0-15 Vdc, 0-30 Vdc, and 0-300 Vdc.

Testing

Check the meter continuity with the low-ohm RX1 scale of a DMM. If normal, the suspected meter hand will move up or down with resistance or voltage tests. Check for burned off or broken terminal wires. Suspect a warped meter scale when the meter hand is stuck in one spot. Remove the front cover. Reglue the meter scale. Reverse the leads upon the meter when the hand goes backwards but does not move forward.

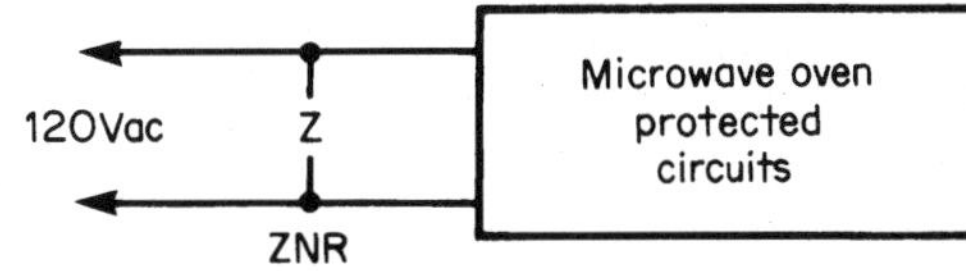

7-28 The metal oxide varistor is placed in the 120-volt ac line to protect panel circuits in the microwave oven.

7-29 The panel meter measures the power supply voltage connected to the motor terminals.

Microphones

Condenser microphone (electret)

What and where

The condenser microphone is similar to an Electret mike. A tightly stretched metal diaphragm is found between two contacts and a dc-bias voltage is applied. When sound waves cause the metal to move or vibrate, this varies the capacity and produces alternating output current. The electret microphone has a dielectric disc or slab that vibrates with sound waves and produces an output voltage. The electret condenser microphone is used in walkie-talkies, tape recorders, cassette players, intercoms, telephone answering machines, hearing aides, toys and sensors (Fig. 7-30).

Testing

The electret and condenser microphones must have a dc voltage applied to them to make them work. The maximum operating voltage is 10 volts, while the average voltage is between 1.5 to 4.5 volts dc. Check the microphone right in the circuit and measure the dc voltage applied across the microphone terminals (Fig. 7-31). If no voltage, check the resistance of voltage-dropping resistor for an increase in resistance. Place a metal screwdriver blade or test probe alongside the coupling capacitor and listen for a loud hum with the volume control wide open. Repair the amplifier circuits with no pickup hum or sound.

7-30 The dual-electret microphone picks up the sound to be recorded in a portable cassette player.

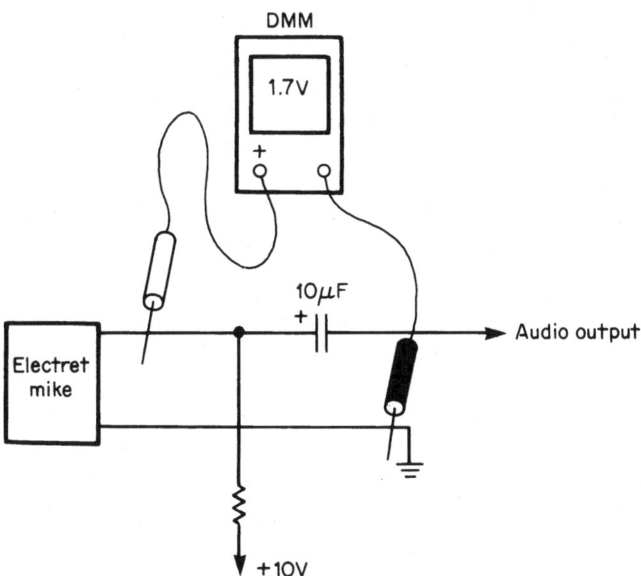

7-31 Check the voltage at the electret mike terminals to determine if the microphone or circuit is defective.

Crystal microphone

What and where

The crystal mike has a transducer made up of piezoelectric crystal. When sound waves strike the crystal transducer, it produces an output voltage. The crystal microphone is a very high impedance microphone used in PA systems, communications, churches, telephones, and wireless equipment. The crystal or ceramic microphone might have a 30-Hz to 15-kHz frequency response.

Testing

Make sure the amplifier circuits are normal by taking audio tests or subbing another microphone. If the microphone is defective, make sure the cable wires are not broken between connector and crystal unit. Remove cable connector and press a finger upon the center terminal and you should hear a loud sound, indicating the microphone is defective. Check for continuity with the RX1 range of the DMM. Replace the crystal unit if it is defective. Most crystal microphones are damaged or broken by dropping the microphone upon a hard surface.

Dynamic microphone

What and where

The dynamic microphone is built somewhat like a speaker. A small coil is mounted on a vibrating diaphragm or cone that moves through a magnetic field, producing an ac output voltage. The dynamic microphone has a very low impedance (600 ohms) and can be used where pickup hum might be a problem. The dynamic microphone might have an 80-Hz to 12-kHz frequency range. These microphones are found in churches, chambers, auditoriums, and for TV announcers.

Testing

Check the continuity of the dynamic microphone with the RX1 range of the DMM (Fig. 7-32). A distorted dynamic mike might have a dropped or frozen cone against the magnetic pole piece. Distortion can also be caused with holes or damage to the cone or diaphragm of the microphone. Try the dynamic mike on another amplifier. Expensive dynamic microphones can be sent in for repair at audio electronic stores or to manufacturer depots.

Electret microphone

(See condenser microphone)

Piezo buzzer or speaker

What and where

The piezo buzzer or transducer has the same type of piezo crystal material such as quartz, Rochelle salt, tourmaline, or various synthetics. This buzzer is used as an indicator. The piezo buzzer might have a high-pitched sound with low-output voltage, low power consumption, long life, and a wide range of frequency (2 kHz–4.4 kHz). The buzzer operates between 3 to 30 volts dc. These inductors are found in microwave ovens, various buzzers, telephone ringers, smoke detectors, alarms, radio beepers, electronic games, cameras, and home appliances (Fig. 7-33).

Testing

Connect a 3- to 10-volt source to the piezo buzzer and listen for a high-pitch squeal. Measure the voltage applied across the buzzer terminals when the buzzer should sound off, and replace the defective piezo buzzer instead of trying to repair the speaker unit.

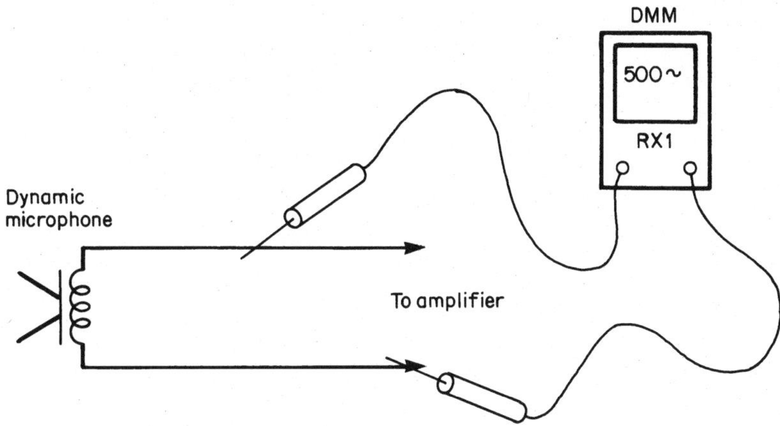

7-32 Check the dynamic microphone continuity with the RX1 scale of the DMM.

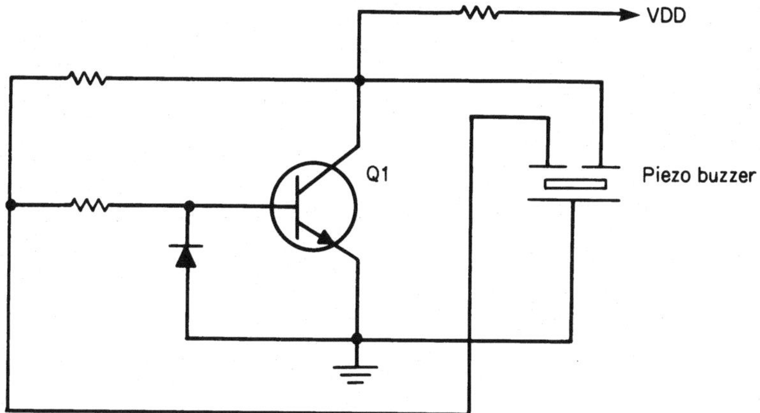

7-33 A piezo buzzer indicates when the cooking cycle is finished in the microwave oven.

Microprocessors

(See Semiconductors)

Motors

Camcorder motors

What and where

The camcorder might have a loading, capstan, drum or cylinder, auto focus, zoom, and iris motors. Some camcorders use mechanical loading of cassettes, while others have a loading motor. The auto focus, iris, and zoom motors are located on the lens assembly (Fig. 7-34). The loading motor can eject, load, and unload the videocassette. A capstan motor provides tape movement in play, record, rewind, fast forward, and search modes. The drum or cylinder movement rotates the video heads. An auto-focus motor controls focusing of the lens assembly, and the zoom motor brings the view up close or far away from the lens assembly.

Testing

Check the voltage across the loading motor. Take a continuity resistance measurement of the motor winding, and scope the loading signal from the IC MI-COM to the loading motor driver and up to the motor windings (Fig. 7-35). Check the supply voltage on the loading motor driver IC and the MI-COM processor when no waveform is noted. Applying a dc voltage to the motor windings can test the small loading motor.

Drum or cylinder problems are indicated when the tape is inserted and, after a few seconds, it unloads. There is no take-up action and the cassette is ejected. These symptoms can indicate that there is no drum or capstan motor rotation. Inspect the motor belt for cracks, breaks, or signs of slippage. If the motor is operating and there is no action, suspect a defective capstan motor belt.

Check the voltage applied to the motor terminals. No voltage can indicate a defective driver or drive signal from the microprocessor. Measure the regulated 12 volts at pin 7 of the capstan motor driver IC603 (Fig. 7-36). Scope the drive signal from the microprocessor to the motor driver IC to the motor windings to determine what stage is defective. All motor circuits within the camcorder can be tested and repaired in the same manner, and the defective motor should be replaced with the exact part number.

CD motors

What and where

Most compact disc players have a loading, SLED, slide, disc, or turntable motor. The turntable CD player might have a carousel, turntable, roulette, up-and-down, magazine, or chucking motor, besides a tray or loading, disc, and SLED motor. A loading motor loads and unloads the cassette, while the SLED or slide motor moves the laser assembly out towards the outside of disc, and the disc or turntable motor rotates the CD at a different speed (Fig. 7-37).

7-34 The auto-focus and zoom motors are found mounted on the lens assembly of the camcorder.

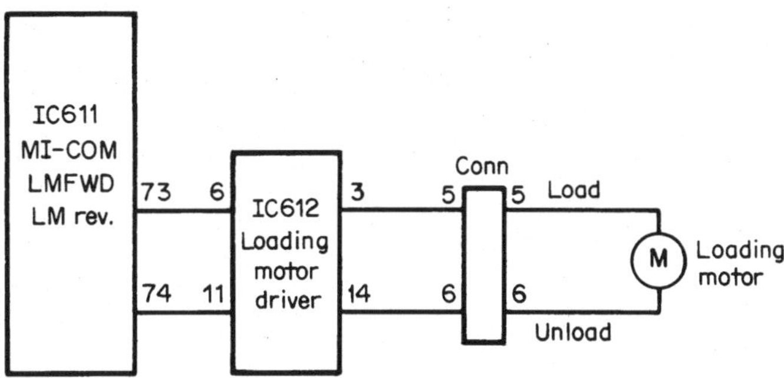

7-35 The loading motor is driven by a voltage and signal from the loading motor driver IC and MI-COM microprocessor.

The roulette or carousel motor rotates the five- or six-turntable tray and stops at the correct disc selection. The disc select motor selects the correct disc in the tray. Most of these motors have a 10- to 20-ohm winding resistance. A Sony portable SLED motor has a resistance of 12 ohms, and a Magnavox CD changer (CDC-745) has a turntable motor resistance of 18 ohms.

Testing

Check the continuity or motor resistance if the voltage is applied to the motor terminals and there is no rotation of the motor (Fig. 7-38). Scope the motor signal

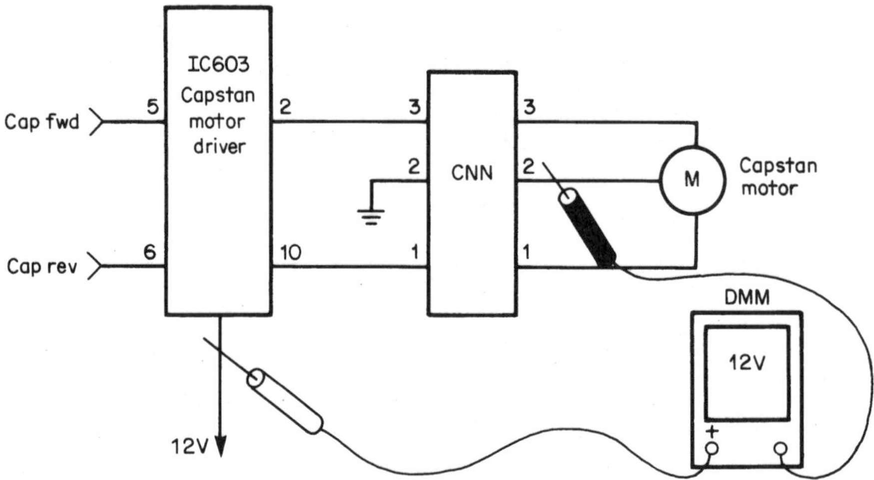

7-36 Measure the supply voltage at the capstan motor driver IC603 when the capstan motor will not rotate.

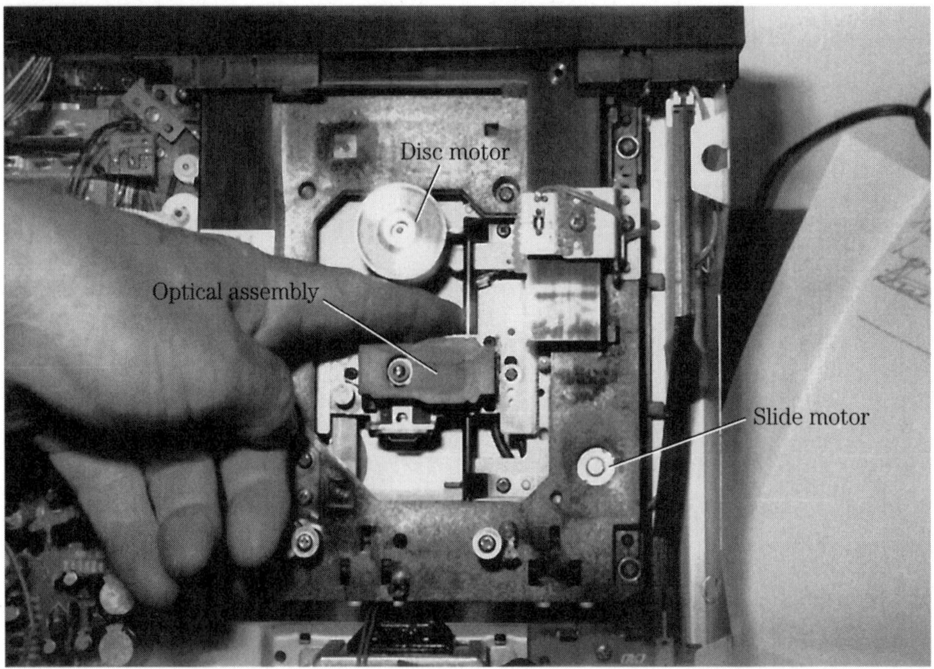

7-37 The slide or SLED motor moves the optical pickup assembly towards the outside edge of the rotating disc.

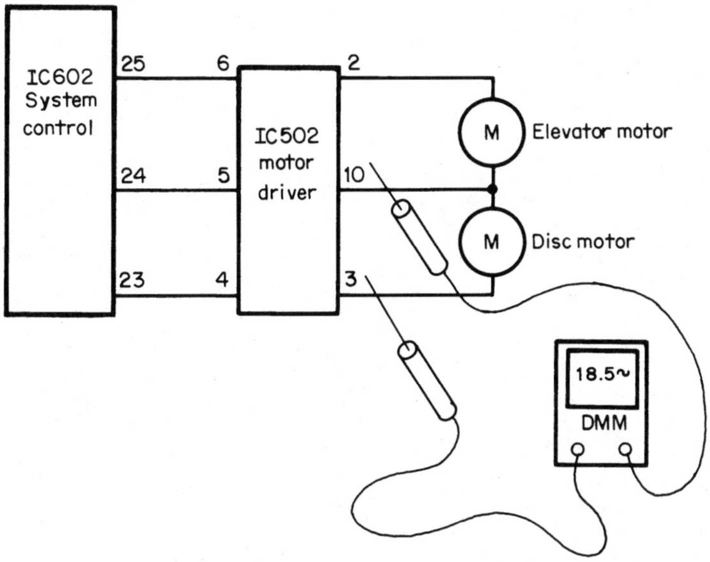

7-38 Check the motor resistance continuity when voltage is found at the disc motor terminals.

from system control to the motor driver IC and to each motor terminal to see what stage is defective. Check the supply voltage found on the supply pin of the motor driver IC and the system control IC. Replace with a part that has exactly the same number.

Cassette motors

What and where

The cassette motor within the cassette player provides tape movement in play, record, fast forward, and reverse modes (Fig. 7-39). The portable cassette player might operate from a 12-volt battery source while in ac power supply; the dc voltage might be from 10.5 to 18 volts. Some deluxe cassette players have a regulated motor speed-control circuit.

Testing

A defective motor might be dead, operate intermittently, or change to different speeds. Monitor the dc voltage across the motor terminals (Fig. 7-40). Check for broken motor wires or defective power circuits if there is no dc voltage. In some models, one motor wire might go to the pause control. When the pause control is pushed down, it opens the motor connections. Inspect the pause control switch for broken wires. A dirty leaf or play function switch might prevent voltage from reaching the motor terminals.

The dead motor might start up after the motor pulley is rotated by hand. Sometimes tapping the shell of the motor with a screwdriver handle will make the motor

7-39 The cassette player has a separate motor for each cassette compartment.

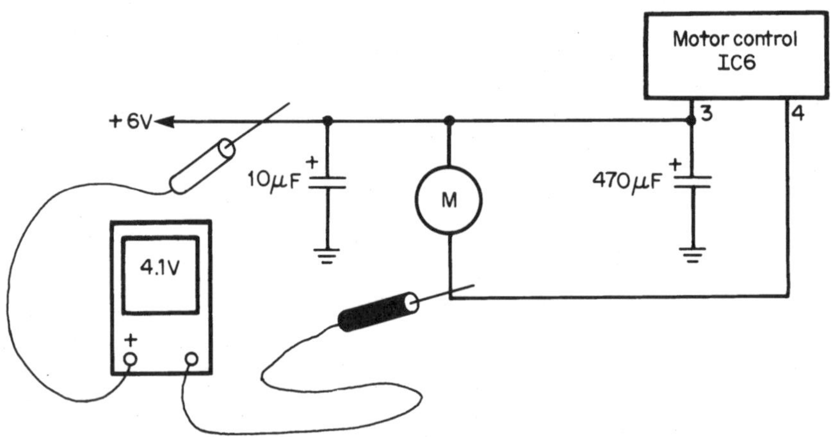

7-40 Check the voltage applied to the cassette motor to determine if motor or motor control circuits are defective.

change speeds. Moving the motor lead connections might make the motor change speeds. If the cassette is running too fast, inspect the capstan for tape wrapped around it. Check if the motor belt is operating high on the motor pulley to increase the speed. If it is high on the pulley, replace the motor belt. Always replace the defective cassette motor with a part that has the original part number.

Ac-dc motors

What and where

The ac-dc motor is found in many consumer electronic products that people operate every day. These small motors are found in drills, routers, hand tools, shapers, leaf blowers, and sanders. The ac-dc motor has two field coils with armature and brushes. The universal motors can operate from the ac power line or a dc source.

Testing

The defective universal motor might overheat and run slow, become intermittent, or fail to rotate. Check the brushes for worn areas or improper seatings. Sometimes the motor winding will become leaky between the motor chassis and winding, causing it to run slowly and overheat. Check for 120V ac across the motor terminals (Fig. 7-41). Take a quick continuity resistance measurement of the field coils and armature. Take a high-resistance measurement from motor winding to metal end bell for a leakage test. Inspect the motor for dry or sluggish bearings. When removing the end bells to check the internal armature and windings, make a metal punch marking across from each metal piece so it can go back together at the precise mounting.

Fan motors

(See Fan blowers)

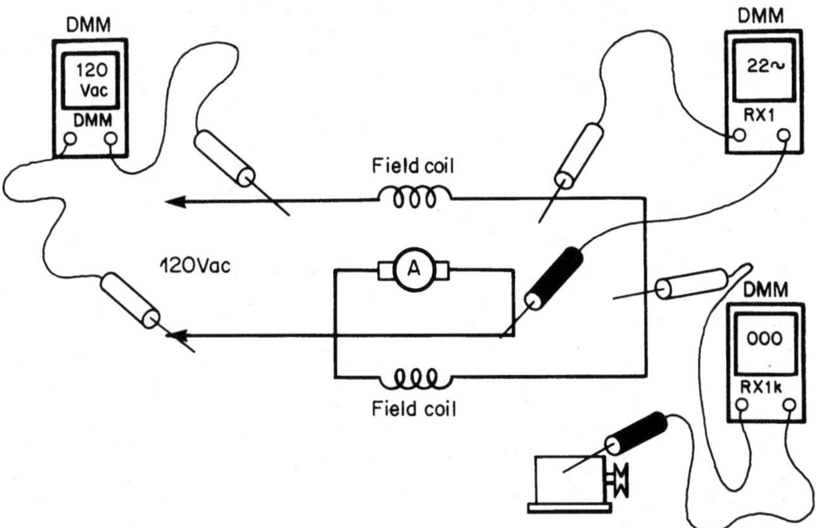

7-41 Check the 120 Vac applied to the universal motor, the continuity of the motor windings, and leakage from the winding to the metal bell of the motor.

Microwave oven motors

What and where

The microwave oven might have only one fan blower in the early ovens. Today, the convection oven might have a fan blower, stirrer or turntable, convection blower, and damper door motor (Fig. 7-42). The fan blower moves the air across the magnetron and out the back vent. A stirrer motor rotates or fans out the RF waves to provide even cooking. Also, the turntable motor rotates the food to eliminate dead spots and promote even cooking. A convection blower moves the hot air in the oven cavity, while the damper door motor opens the damper door to let out hot air.

Testing

Check for 120 Vac across the motor terminals. Measure the motor continuity for open winding and poor connections, and suspect dry or gummed-up bearings when the motor slowly rotates. Wash out bearings and lubricate.

VCR motors

What and where

The VCR machine might have a loading, drum, or cylinder and capstan motor. A loading motor loads the cartridge, while the drum or cylinder motor rotates the video heads (Fig. 7-43). The capstan motor controls the speed of tape moving past the

7-42 The stirrer motor within the microwave oven is located at the top of the oven.

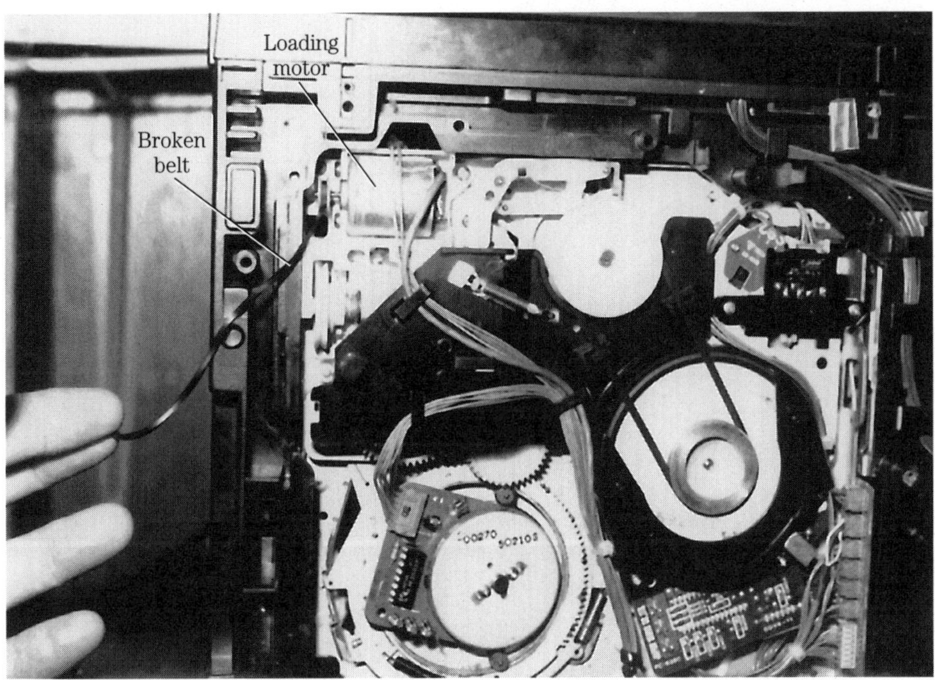

7-43 Notice that the loading motor has a broken belt that prevented cartridge loading in the VCR.

heads and moves the tape. The capstan servo assembly affects both audio and picture. When both picture and sound are affected, you have capstan problems. The capstan motor might be controlled by speed- and phase-control circuits.

Testing

Check the voltage across the loading motor. Take a low-resistance measurement of the loading-motor winding. Scope the input and output of the loading-motor driver IC and on the motor terminals. Check the supply voltage on the motor driver IC (Fig. 7-44). Check for a jammed loading assembly or drive gears, and inspect the loading motor drive belt for cracks and enlarged belt.

Notice if the capstan and drum or cylinder motor are driven by a digital servo or microprocessor. Measure the voltage applied to the capstan motor. Inspect the capstan belt for breakage or a loose belt, and take a low-resistance continuity measurement of the capstan motor. Likewise, check the drum or cylinder motor in the same manner and scope FG and PG motor pulses (Fig. 7-45).

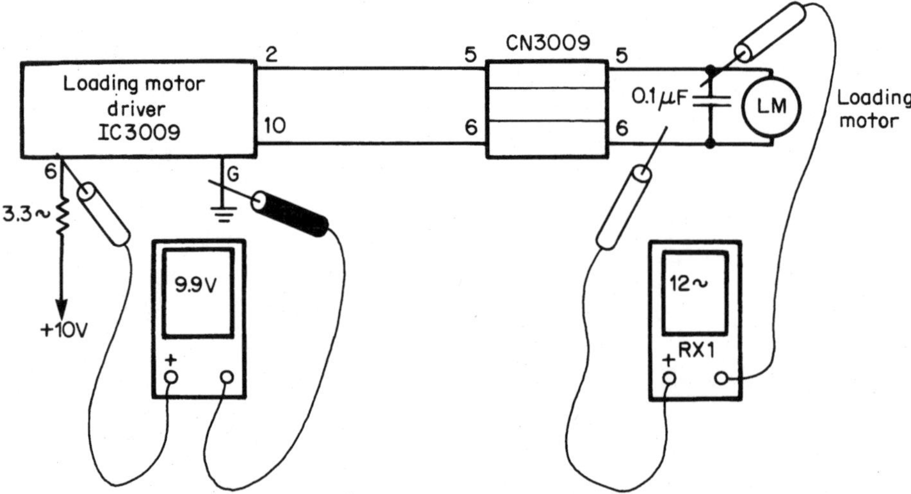

7-44 Check the motor resistance and supply voltage of the loading motor driver IC to determine if the motor or driver circuits are defective.

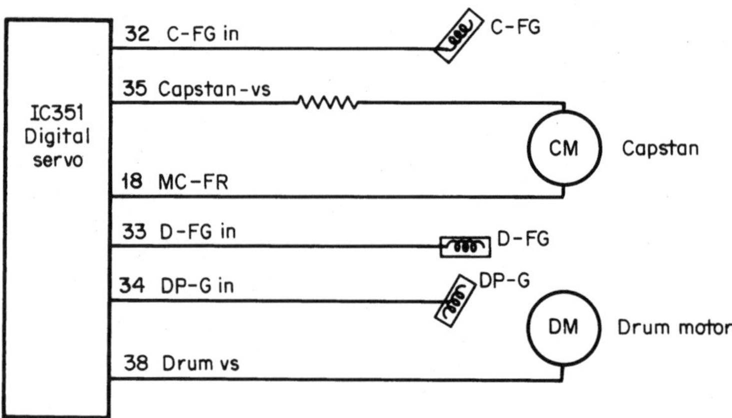

7-45 Scope the different signals from the drum and capstan motors to determine if the digital servo IC351 is defective.

8
CHAPTER

Testing electronic components O to R

(Optoelectronics to radios)

The opto-isolator is found mainly in the VCR and the latest TV chassis. A photo detector transistor is used in audio and video circuits, sensors, cameras, and many other consumer electronic products. The various oscillator circuits are found in AM-FM-MPX receivers, TVs, and transmitters. The many different potentiometers might contain a slide or variable control to provide easy audio command with a radio, CD, or cassette player. Just about every electronic product operating from the power line contains some type of power supply, while the preamp circuits are found within the phonograph, cassette, CD, and audio circuits. How to test opto-isolators and radio receivers is covered in Chapter 8.

Opto-isolators

What and where

The opto-isolator components are found in the VCR sensor circuits and the latest TV power supplies. The opto-isolator is a coupling component that connects circuits with a light beam. In the Sylvania C9 chassis, the opto-isolator (IC404) separates the hot circuits from the cold ground chassis of the switched-mode power supply (SMPS). IC404, besides isolation, helps regulate the input and output voltage applied to the switch mode transformer (Fig. 8-1). The opto-isolator component adds to or takes away voltage when the output voltage changes.

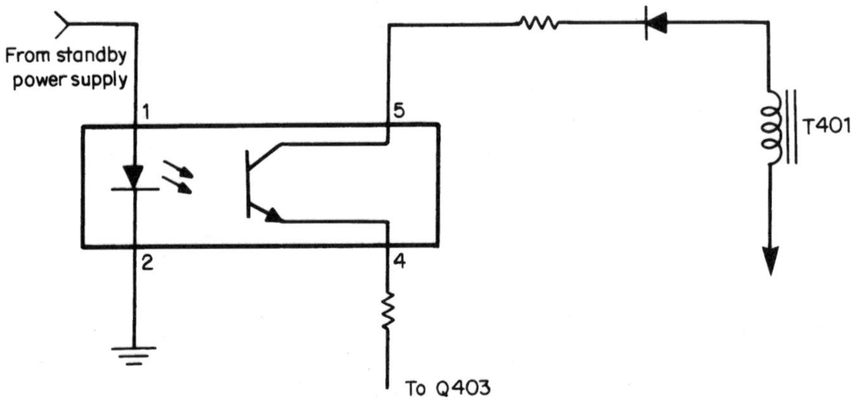

8-1 The opto-isolator is found in Sylvania's C9 TV switch-mode power supply (SMPS).

Testing

The opto-isolator can be checked with the diode test of a digital multimeter. Rotate the DMM to the diode junction test. Place the positive probe to pin 1 and the negative (black) probe to pin 2. Check the LED diode with the diode test. You should have continuity with this measurement. Reverse the test leads, and a normal measurement should have infinite resistance. Replace the opto-isolator with a resistance measurement in both directions. Test the resistance measurement between pins 4 and 5 with the diode test. A normal phototransistor should have infinite measurement in both directions (Fig. 8-2).

Photo detector transistor output

What and where

The photo detector transistor is found in audio and video circuits, VCRs, remote controls, sensors, light sensors, cameras, door openers, alarms, industrial controls, instrumentation, solid-state switches, power IC relays, and telephone lines with pulse functions and ring detection. A photo detector is an optoelectronic coupler. The phototransistor might appear in transistors and Darlington-type circuits (Fig. 8-3). The isolation voltage might be from 1.0 V to 5.3 KV. The photo detector diode is found in a typical TV remote control (Fig. 8-4).

Testing

Connect the diode test of the DMM to pin 1 and 2 of the phototransistor. You should have a low-resistance measurement with positive lead at pin 1 and negative probe at pin 2. No measurement should be read with reverse test leads. The transistor terminal can be tested with the diode-junction test of the DMM or a transistor tester on pins 4, 5, and 6. Check the phototransistor with a leakage test between emitter (4) and collector (3). Scope the signal in at pin 35 of IC901 (Fig. 8- 5). Check the B+ voltage applied to the remote phototransistor preamp.

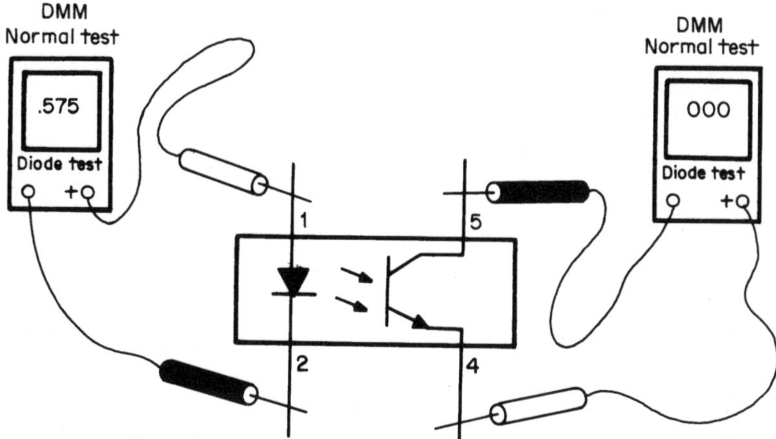

8-2 Check the opto-isolator with a resistance, diode, and transistor test.

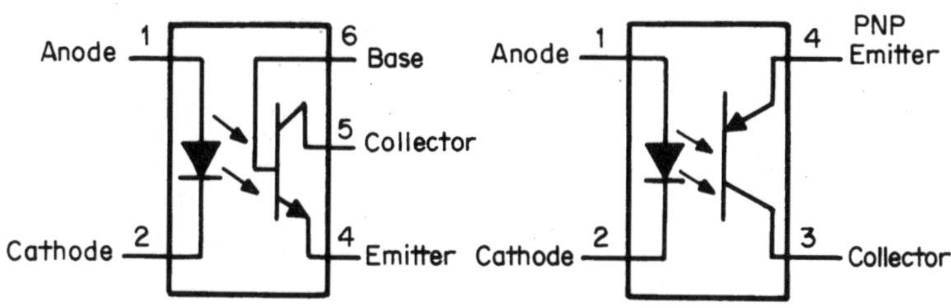

8-3 The phototransistor might appear in an LED-Arlington transistor-type circuit.

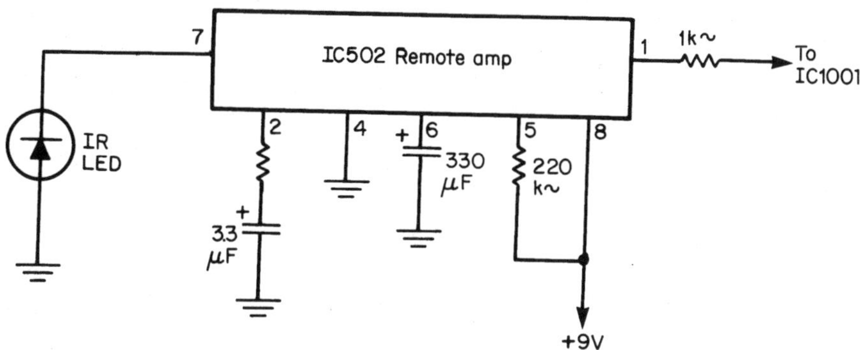

8-4 A photo-detector diode can be found in a typical TV remote control.

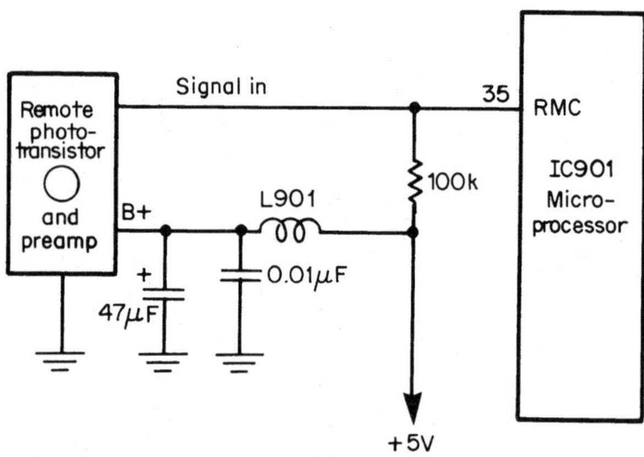

8-5 Scope the signal in at pin 35 of IC901 to determine if the phototransistor is operating.

Photo Darlington output

What and where

A photo Darlington transistor output device is a phototransistor with a Darlington transistor amplifier for higher output current. The opto-isolator's output might appear in the phototransistor output, opto-MOS solid-state switches, optical sensors, micro-electronic power IC relays, optical sensors with built-in amps, opto-SCR and trice output, logic output, photonic detectors, high-voltage transistor output, high-speed phototransistor output, dual high-speed photo Darlington output, high-voltage photo Darlington output, high-gain split-photo Darlington output, and optocouplers.

The photo opto-isolators might appear in transistor or IC configurations. IC types might be found in instrumentation and control equipment, ac line/digital logic, digital logic, twisted-pair line receivers, and telephone/telegraph line receivers. The photo Darlington output comes in dual, triple, and quad Darlington outputs (Fig. 8-6). The isolation voltage might be from 3.75 KV to 5.3 KV.

Testing

Test the photo Darlington output device with the diode test of the DMM. Check the anode (2) and cathode (4) terminals like any diode. Tests for high leakage between emitter (1) and collector (3) terminals with the diode-junction test. Replace the unit with leakage between terminals 1 and 3.

High-gain Darlington output

What and where

The high-gain Darlington output device might appear in an IC package with three or four photo Darlington output transistors. The quad-photo Darlington output device has four separate Darlington transistors with separate terminals (Fig. 8-7).

The quad-photo Darlington output might have two detector diodes in parallel in the 16-terminal chip.

Testing

Check each photo LED diode with the diode test of the DMM. The quad-Darlington output from pins 9 through 16 can be tested with a leakage test between emitter and collector terminals with the diode-junction test of the DMM. An infinite reading should be obtained with reverse test leads. Replace the entire unit when leakage is observed.

SCR and triac driver output

What and where

The silicon-controlled rectifier and triac assembly are found in the output of an opto-isolator device. These SCR and triac devices control isolated circuits through a light beam from an enclosed LED source. The isolated voltage might be around 660 V, while the trigger current is from 200 to 400 V with an output current of 300 mAs.

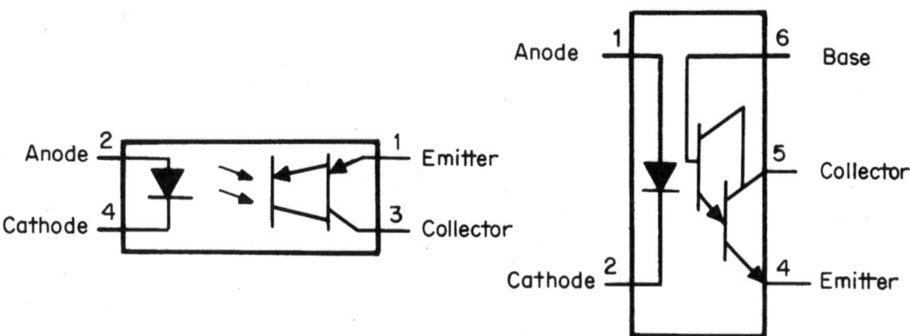

8-6 You might find the photo-Darlington transistor in a dual, triple, or quad output.

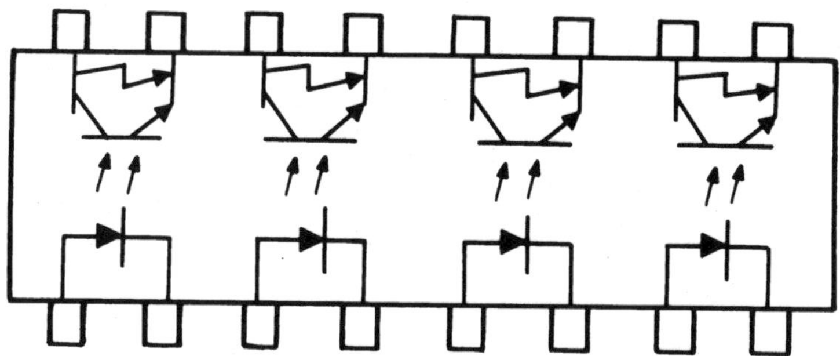

8-7 A quad-photo Darlington has four separate Darlington transistors and diodes in one component.

Testing

Test the LED diode (terminals 2 and 3) as any diode with the diode test of the DMM (Fig. 8-8). Since the SCR terminal connections are found in 4, 5, and 6, check the silicon-controlled rectifier with an SCR or triac tester. The TRI-CHECK II is a new test instrument designed by Electronic Systems, Inc.

Disconnect all wire terminals to the SCR. Turn on the TRI-CHECK II, and the middle light will blink. Connect the black lead (T1) to the cathode terminal, clear lead to (T2) the anode, and red (G) to the gate terminal. Press the test button. The SCR light should light when the button is pressed and turn off when the button is released. Both lights will light with a shorted SCR. No lights are on with an open SCR.

Check the triac output portion of the opto-isolator with the TRI-CHECK II test instrument. Disconnect all wires to the triac. Turn on the TRI-CHECK II and the middle light should blank . Now connect the TRI-CHECK II to the triac. Place the black lead (T1) to MT1, clear wire (T2) lead to MT2, and red (G) lead to the gate terminal. Press the test button.

If the triac is normal, both lights should light when the test button is pressed, and they should turn off when the button is released. Both lights will always light with a shorted triac. No lights are on with an open triac. Always turn the tester off when not in use. Notice that the triac output opto-isolator has no gate terminal, as the triac is triggered by the light from the LED.

The triac section of the opto-isolator can be tested with resistance measurements with the DMM. Check the continuity between any triac terminals with the VOM, VTVM, digital VOM, FET-VOM, or diode test of the DMM. Switch the DMM meter to the diode test and read the resistance between MT1 and MT2. Infinite measurement should indicate a normal reading even with reversed test leads. Replace the entire assembly when a 5-megohm or lower measurement is noted between MT1 and MT2 (Fig. 8-9). A normal measurement should be above 10 megohms.

Photo cells

What and where

A photocell converts light into electrical energy. The photovoltaic cell is used to charge batteries. The photovoltaic cell, silicon cell, selenium cell, solar cell, or sun battery are known as photocells. A light-sensitive resistor might be called a light-sensitive diode, photoconductive cell, and selenium cell (Fig. 8-10).

The photoconductive cells (DCS) might be made up of cadmium-sulfide, whose resistance changes with visible light, infrared, or ultraviolet light. These photoconductive cells are found in exposure meters, electronic shutters, optical switches, photoelectric lamps, and color temperature measurements.

Testing

Check the resistance of a photoconductive cell without any light (500 K to 20 M) or with light in front of the cell (3 K to 100 K) ohms (Fig. 8-11). Measure the output voltages from a photovoltaic cell with the low voltmeter range of the DMM, VOM, or FET-VOM. Each photovoltaic cell has an output voltage of 0.45 to 0.5 volts.

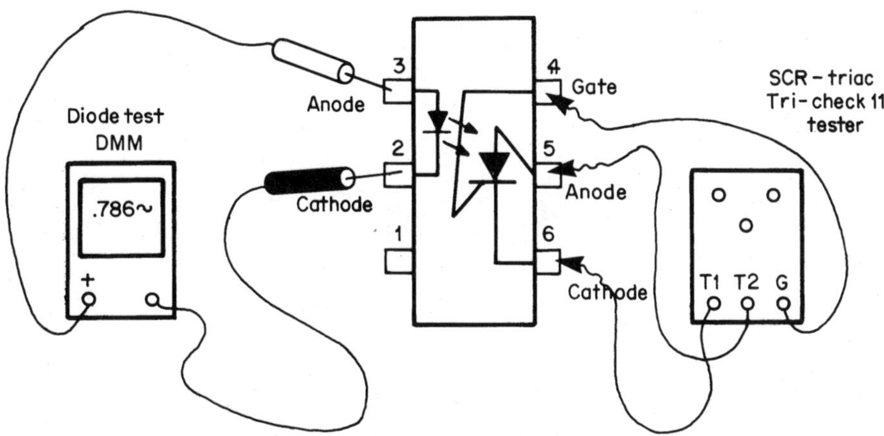

8-8 Check the LED diode with the diode test of the DMM, and test the SCR-triac section with the SCR-triac tester.

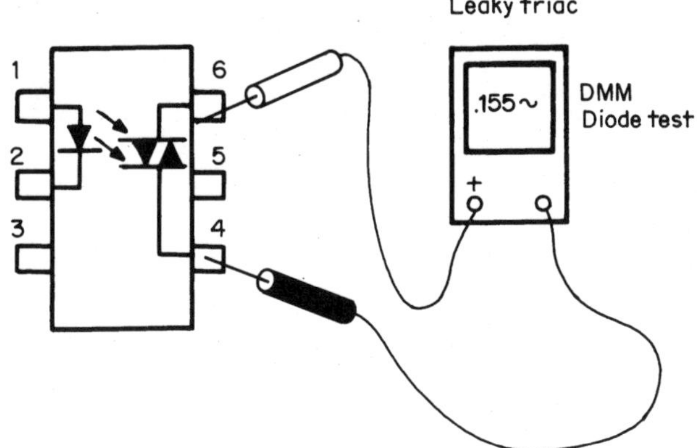

8-9 A leaky triac assembly will have a low-resistance measurement between the MT1 and MT2 terminals.

LEDs

(See LEDS)

OP-AMP

(See Semiconductors)

Oscillators

AM radio oscillators

What and where

The oscillator circuit found in the AM radio might be called a converter stage, with Q3 as oscillator and RF amplifier (Fig. 8-12). The output of the oscillator primary

8-10 The photocells wired in series provide a voltage bank to charge batteries.

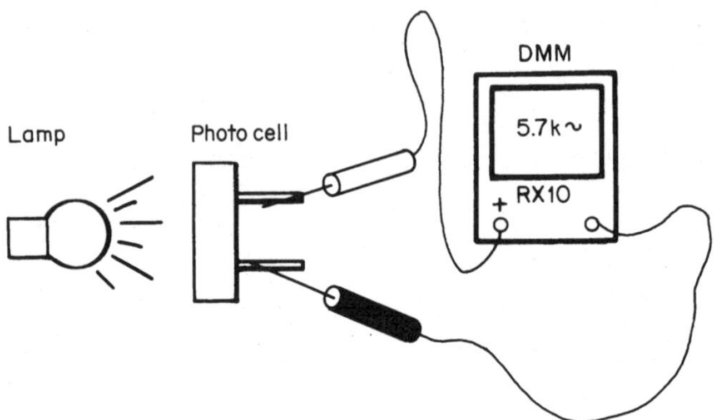

8-11 Check the photocell resistance with and without a light in front of the cell.

winding connects to the 455-kHz IF stages. The RF amp coil L2 and oscillator coil (L3) has permeability-type tuning, in which an iron core slides in and out of a tightly wound coil.

Testing

You can quickly test the AM radio oscillator circuit by placing another receiver nearby (Fig. 8-13). Turn both radios on and rotate the suspected radio tuning until a whistle is heard in the other radio. If no tunable noise is heard, check the AM oscillator circuit. Measure the voltage on all three transistor terminals. Notice in this circuit that a PNP oscillator transistor (Q3) is used with very low collector and very high base and emitter voltage. Test Q3 with an in-circuit transistor tester or diode-junction test of the DMM (Fig. 8-14). Check each coil for continuity. When

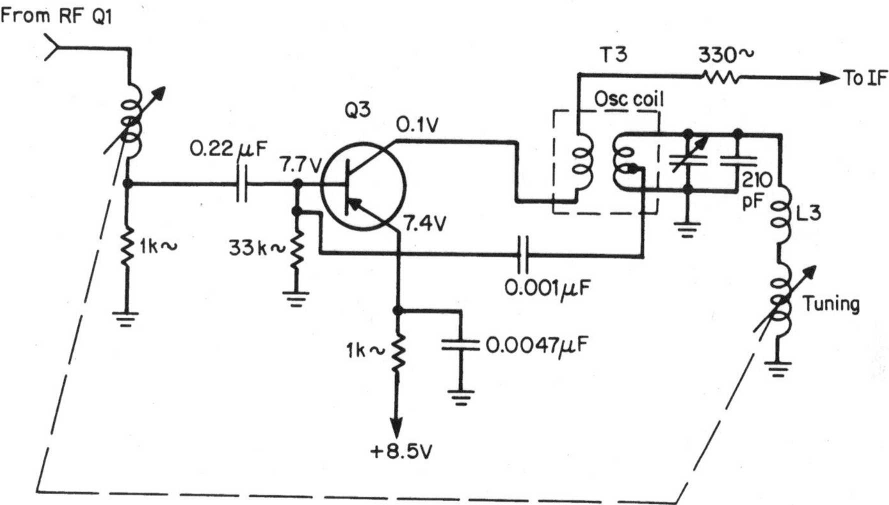

8-12 The converter stage in the early radio consists of permeability tuning.

8-13 Check the oscillator circuits by tuning in another radio nearby and listen for a squealing noise when tuned to the exact frequency.

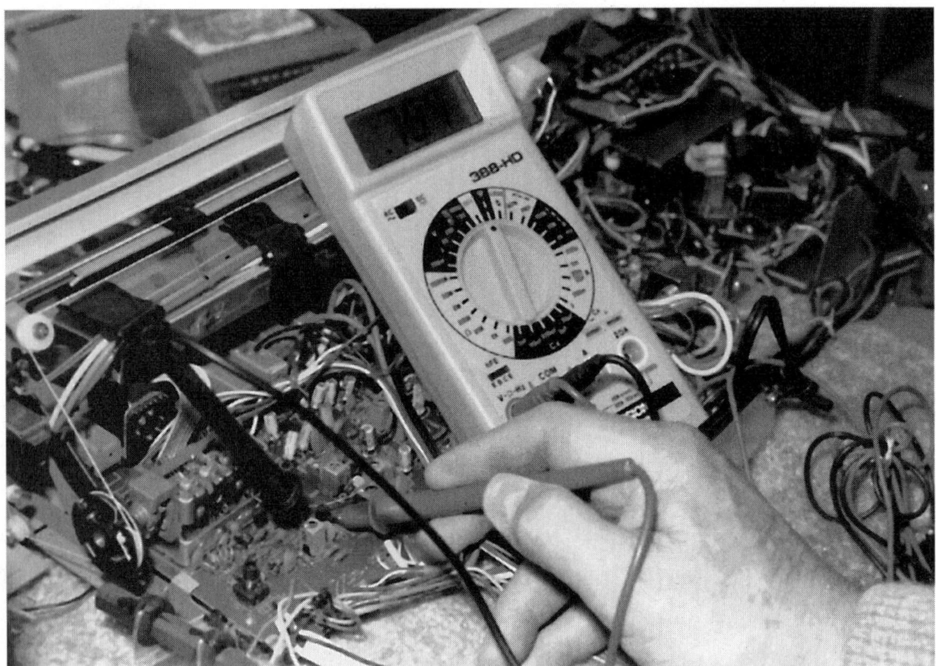

8-14 Check the suspected oscillator transistor with the diode-junction test of the DMM.

stations will change without touching the tuning dial, suspect a broken oscillator metal core in L3.

Crystal oscillator

What and where

The frequency of the crystal oscillator is determined by the frequency of a piezo-quartz crystal. A crystal oscillator might be found in radio and shortwave receivers, clock circuits, timing circuit testers, TV sets, VCRs, and CD players (Fig. 8-15). The crystal oscillator is a very stable and reliable circuit.

Testing

The crystal oscillator circuit can be tested at the output with a scope waveform and frequency meter. Check the suspected crystal within a crystal tester. Measure the voltage on the crystal oscillator transistor (Q1). Suspect a defective PF capacitor when the oscillator circuit is not stable. Check each bypass capacitor on the small capacitor tester for correct capacity (Fig. 8-16).

FM oscillator

What and where

Often, the FM oscillator output is fed to a separate mixer stage. The FM oscillator circuit is found between FM-RF transistor and mixer circuits. The frequency modulation (FM) band is between 88 to 108 MHz. Special transistors and IC components are found in the critical FM circuits. The FM oscillator is found in the FM-AM-MPX, FM-AM portable radios, and auto receivers (Fig. 8-17).

Testing

The FM oscillator can be tested in the very same manner as the AM circuits. Suspect the FM oscillator when a loud hissing noise is heard without any stations tuned

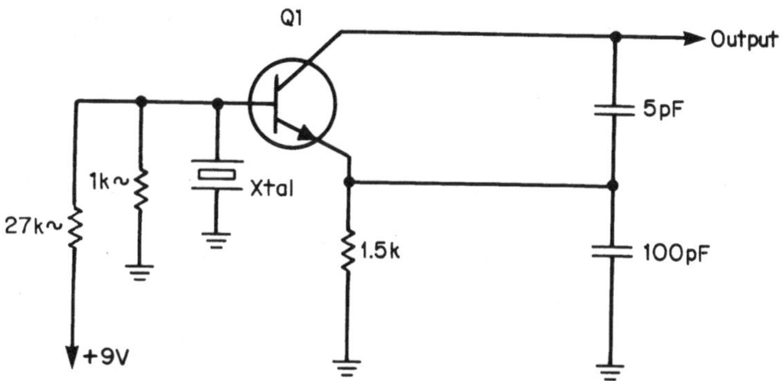

8-15 The crystal oscillator circuit is found in electronic clocks, radios, shortwave receivers, and timing circuits, to name a few.

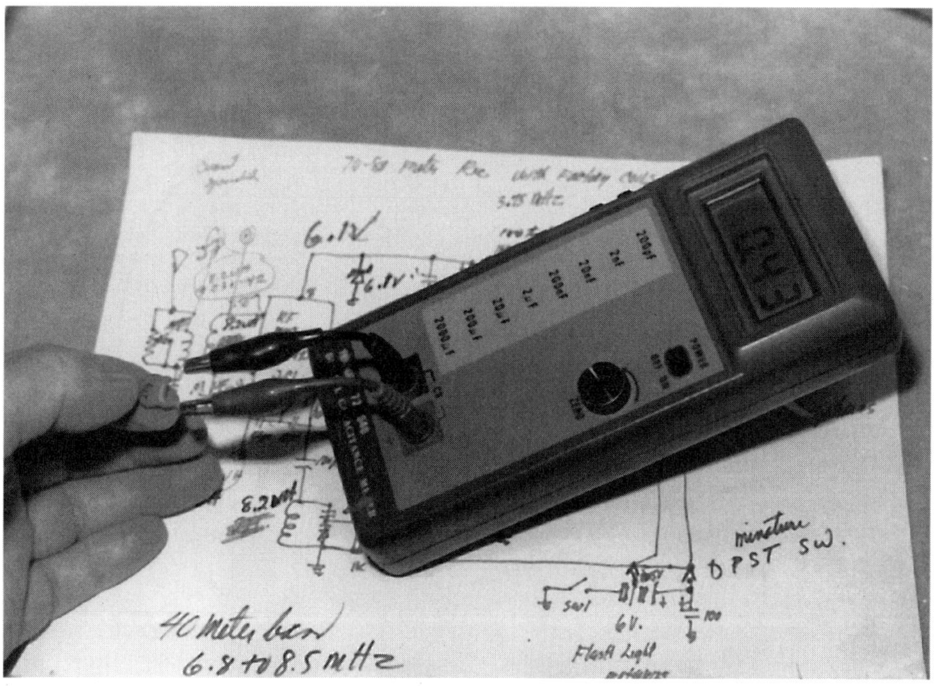

8-16 Check the small bypass and coupling capacitors with a capacitor tester.

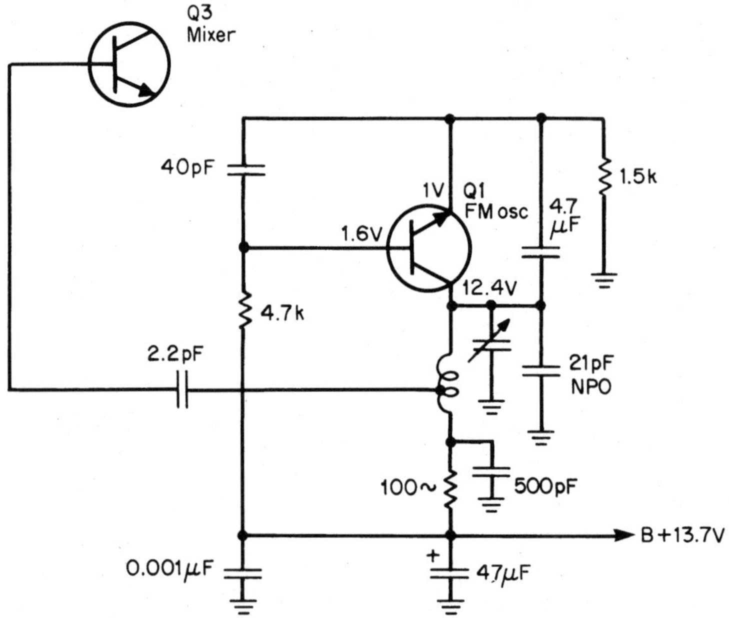

8-17 The FM oscillator circuit found in the AM-FM-MPX radio.

in. Check the FM oscillator mixer stage if there is no tunable section, but there is a normal AM section. Measure the voltage on the FM front-end IC if there is no FM reception. Check the RF-FM transistor or stage when only a local FM station can be heard. Take critical voltage and transistor tests of the oscillator transistor (Q1). Suspect a voltage regulator transistor or circuit when voltage is low at the FM oscillator transistor.

Cut the leads on the replacement transistor the same length as the original, and mount in the same position. Check each FM coil with the RX1 scale of the DMM or FET DMM (Fig. 8-18). Test each coil connection at the foil board connections. Push and prod around on FM components when FM reception appears intermittent.

PA systems

What and where

The public address (PA) system can be found in churches, auditoriums, outside meetings, and at county fairs. The public address system might consist of a high-powered amplifier, several speakers, microphone, and tapes (Fig. 8-19). Usually, the PA system is used to announce events, etc.

Testing

Sub another microphone or connect the suspected microphone to another amplifier to determine if microphone is okay. Likewise, sub another speaker if one is found dead or noisy. Check all connecting microphone cables and speaker wires.

8-18 Take continuity tests of the coils and PC wiring in the FM circuits.

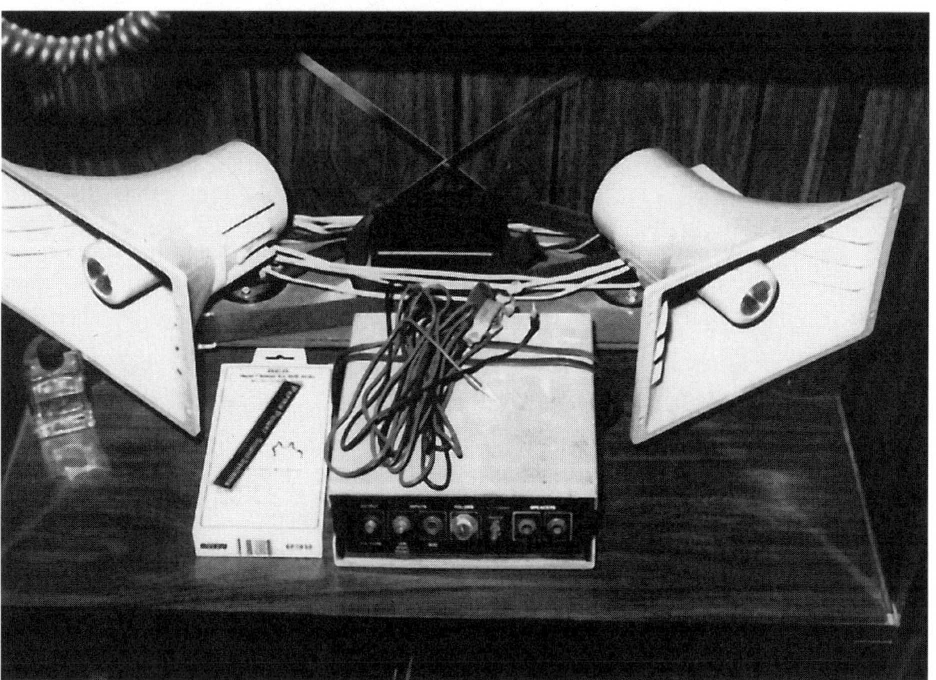

8-19 The public address (PA) system might have rain-proof outdoor speakers.

Repair the amp circuits like any audio stage with sine, square waves, and scope as indicator. Replace solid-state components with original or high-wattage parts.

Photo cells

(See Optoelectronics)

Positors

What and where

A posistor is a thermal-resistor device that is found in amplifiers, transformers, and TV degaussing circuits. A thermal posistor is used in computer monitors and many popular TV degaussing coil circuits. The posistor found in amplifiers, protects high-wattage power output transistors from overheating. A posistor might be found on the power transformer to determine if the transformer is overheating. A posistor operates within a shut-down circuit that disables the unit when the output transistor and transformer run too hot in the large high-powered amplifiers.

The posistor in the amplifier power supply prevents overheating of the output transistors, ICs, or power transformer. Here the receiver has two built-in overload thermal protection circuits in case of abnormal operation. When the temperature of the posistor on the transformer (TH603) and the power output transistor (TH601) heat sink changes abnormally, the resistance of the posistor becomes larger and pin 1 of IC702 will increase and cause the protecting transistor TR605 to turn off. As a result, pin 2 of IC701 is cut off from the power supply (Vcc) line, and the power transistors are protected (Fig. 8-20). Often, the posistor is cemented to the heat sink, transistor, and power transformer shell.

Testing

Check the resistance of the posistor when cold. Then apply a soldering iron or heat near the posistor and notice if the posistor resistance changes. A thermal posistor resistance is low at the beginning and increases in resistance as it heats up. Check for a broken or burned-off terminal at the posistor body. Replace with the exact part number when found in the amplifier and transformer circuits.

Potentiometers

Carbon pots

What and where

A carbon resistor or potentiometer is made from carbon, graphite, or carbon-related material. A potentiometer varies the resistance in a circuit. The average volume or balance control is made up of a carbon substance. These carbon controls are found in radios, receivers, cassette and CD players, VCRs, TVs, and just about any commercial entertainment product (Fig. 8-21). The carbon volume control might be a rotating or sliding-type control. Carbon controls are found in cermet, multiturn,

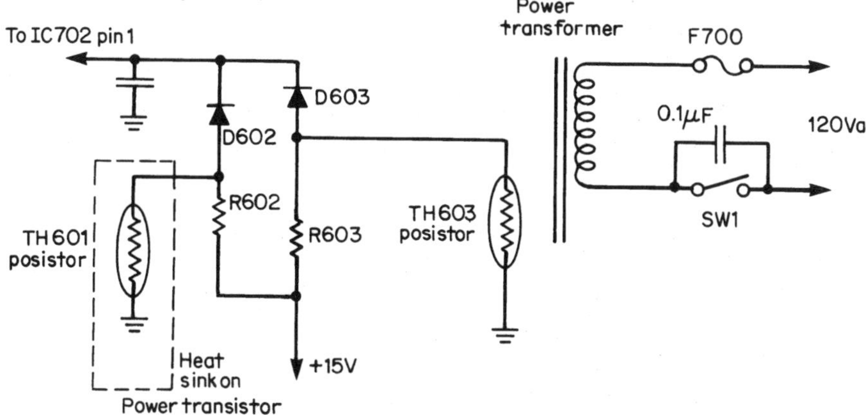

8-20 Small thermal posistors protect overheated transistors and transformers found in expensive amplifiers and receivers.

8-21 The large cassette portable might have carbon tone, volume, and balance controls.

single turn, PC mount, slide, SMD, thumbwheel, and trimmer types. The replacement carbon potentiometer might appear in ⅛-, ¼-, and ½-watt types.

When

The defective carbon control might cause a hum in the sound, intermittent audio, noisy sound, or no sound at all. The carbon control can short internally, and cancel the audio to the metal control. A worn or damaged carbon control can cause excessive noise or intermittent audio (Fig. 8-22). Poor terminal connections can produce intermittent or no sound. The control might come in linear and audio taper. Replace the audio volume control with an audio taper control and the linear taper with a linear control.

Testing

Check the resistance of the carbon control with the ohmmeter range. A DMM ohmmeter range is more accurate than a VOM. The replacement control may have a 1-K, 2.5-K, 5-K, 10-K, 25-K, 50-K, 100-K, 250-K, 500-K, and 1-megohm resistance. Measure the resistance between each terminal and to the metal outside of control for possible leakage. Most noisy variable pots can be cleaned up with cleaning spray. Replace the control when it is excessively worn or has a broken internal connection. The universal carbon pot might have a pick-a-shaft where the different shafts can be inserted.

Cermet control

What and where

The cermet potentiometer might appear in 2-watt, square, single-turn, or miniature trimmer controls. Cermet is an alloy of ceramic, such as nickel, metal, surface mount, or titanium carbide. A thin film of cermet is used as a variable resistance. The miniature cermet control may have a resistance tolerance of 10 to 20 percent.

Testing

Check the variable resistance with a FET-VOM or DMM for correct resistance. The defective control can be replaced with universal replacements.

Motor-driver control

What and where

The motor-driven volume control can be rotated with a small motor within a motor control circuit (Fig. 8-23). The motor-driver control circuits are found in the deluxe receiver, phonographs, CD players, camcorders, and VCRs. The motor-driver and system control circuits might be transistor or IC driven. System control (IC550) provides volume up and down signals to a volume control driver (Q551). The volume control motor rotates the control up and down with a forward and reverse voltage.

Testing

Suspect a defective motor when dc voltage is found at the motor terminals and no rotation (3.5 to 6 volts dc). Reverse the motor control and notice the change in polarity of voltage at the motor terminals. Check the motor continuity for open winding of the dc motor. Replace the motor with the exact part number. Test all transistors within the circuit. Double-check the supply voltage at the driver transistor and system control IC.

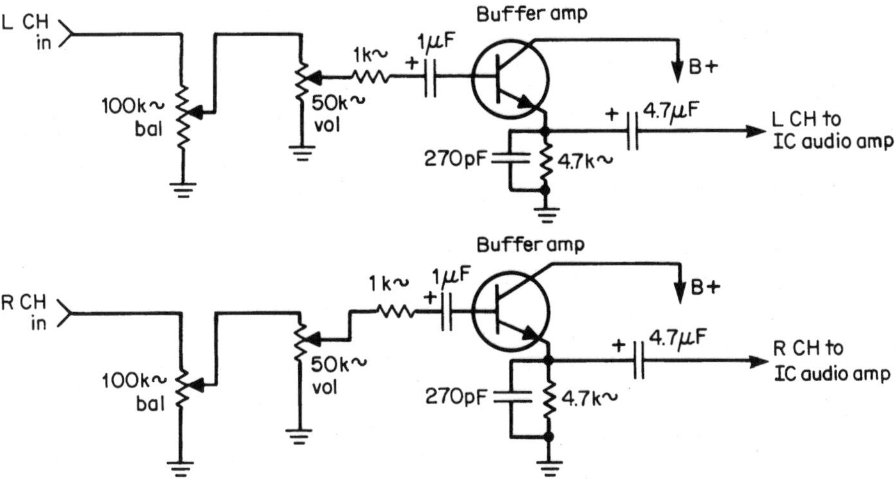

8-22 For a dirty or worn control, spray each volume control in the stereo amplifier section.

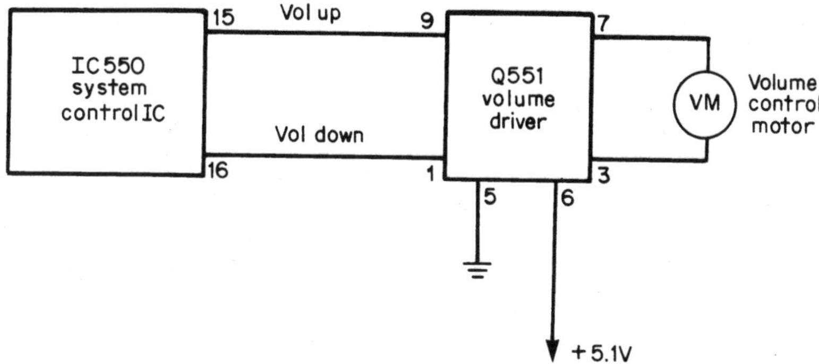

8-23 You might find a volume control driver motor circuit within the remote control receivers.

Multiturn and single-turn controls

What and where

The single-turn control completes its resistance in one turn, while a multiturn control might have a 10-, 12-, 20-, or 25-turn rotation. A multiturn control shaft must be rotated through several revolutions to complete the resistance. The multiturn control appears in wire-wound, cermet, trimmer, and SMD types.

Testing

Clip the ohmmeter leads across one leg and the center terminal. Rotate the control through a complete rotation and look for a flicker or break in the reading. Replace the control if open or erratic.

PC mount controls

What and where

The PC-mounted control might be made up of carbon, wire-wound, cermet, and miniature carbon composition. They come in trimmers, miniature snap-in, square and rectangular, nonconductive body and shaft, modular style, thumb wheel, and SMD trimmer potentiometers. The PC controls might mount vertically or horizontally on the PC board.

Testing

The small trimmer controls have a tendency to become noisy in audio bias circuits. Sometimes the rotating contact point becomes oxidized and makes a poor contact. Clean up and rotate the control to prevent noisy audio. Check the bias PC board control with a resistance measurement right in the circuit (Fig. 8-24). Replace the control if the resistance changes or has a worn spot. Reset for correct bias voltage.

Slide controls

What and where

Slide controls might be found as volume, balance graphic equalizers, and tone controls in cassette and phono players, radios, and CD players (Fig. 8-25). The balance control is set at the center and can slide either way to balance the sound. Audio controls can be adjusted to the right for more volume and lowered to the left side of the sliding control knob. The sliding control also comes in the PC mount with a linear taper.

Testing

Check for correct resistance of the sliding control on the ohmmeter range of the DMM or FET-VOM. Make sure you have the correct terminals. Spray cleaning fluid inside, along the sliding lever, to clean up a noisy or dirty volume and tone control. Replace the control with worn or erratic operation.

SMD control

What and where

SMT (surface-mounted potentiometers) might appear in multiturn, trimmer, and square single-turn cermet controls. The standard replacement range of an SMT cermet potentiometer is 10 to 2 megohms. The SMT trimmer control might range from 220 ohms to 1 megohm (Fig. 8-26). The SMT pots are linear-type screw-slotted controls. The SMT trimmer control might be used as bias controls in audio circuits.

Testing

Check the control on the PC wiring with the ohmmeter scale of a DMM or FET-VOM. Rotate the trimmer and notice if an erratic reading is obtained, causing intermittent bias and sound. Remove the SMT trimmer from the PC circuitry with a soldering iron and solder the wick material if open or erratic.

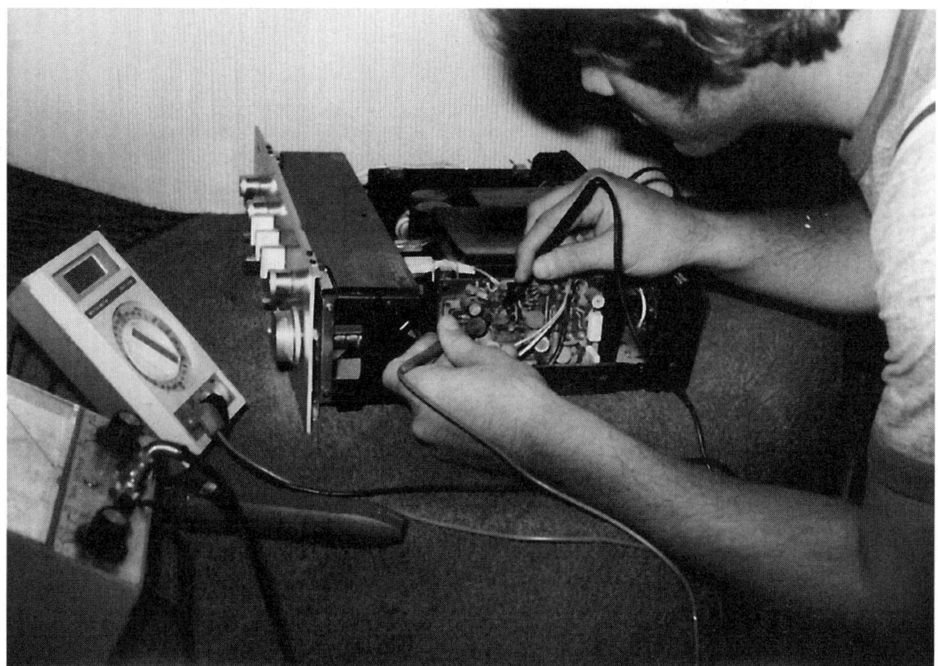

8-24 Check for a worn or noisy bias control in the high-powered amplifier.

8-25 The balance and equalizer controls are sliding components in the G.E. portable dual-cassette player.

8-26 The SMT trimmer control can be replaced by unsoldering tabs on the PC wiring.

Thumbwheel controls

What and where

The subminiature potentiometer with thumbwheel knobs was first used in pocket-sized radios and cassette players. The serrated-edge knob can easily be rotated by the thumb or finger. You might find a small off-and-on switch attached at the off-end rotation. These replacement controls come in linear and audio taper (Fig. 8-27). The linear taper resistance starts at 500 ohms to 1 megohms, while the audio taper begins at 1-K ohms to 1 megohm.

Testing

Carefully inspect the control for broken or loose contacts on the PC board. Measure the resistance of the control. Replace the thumbwheel control when worn or cracked. Spray internally for a minor scratching or noisy control.

Trimmer controls

What and where

The trimmer or screwdriver-slot control is found in preset circuits, amps, bias controls, power supplies, industrial circuits, auto-motor adjustment equipment, TVs, VCRs, and CD players. They might appear as cermet trimmers, multiturn, single-turn, or carbon, and can be mounted horizontally or vertically on the chassis. The surface-mounted (SMD) is mounted directly on the PC board with a screwdriver adjustment (Fig. 8-28).

Testing

Take a resistance measurement on the outside terminals of the control. Remove any trimmer control with broken contacts and element.

Wire-wound controls

What and where

The wire-wound control is usually found in higher current wattage circuits. They might appear in 2-, 4-, and 5-watt types. Wire-wound controls are used in audio amps, speaker circuits, TVs, and radio applications (Fig. 8-29). These wire-wound controls might be single, three-turn, ten-turn, or wire-wound trimmers with the contact arm at

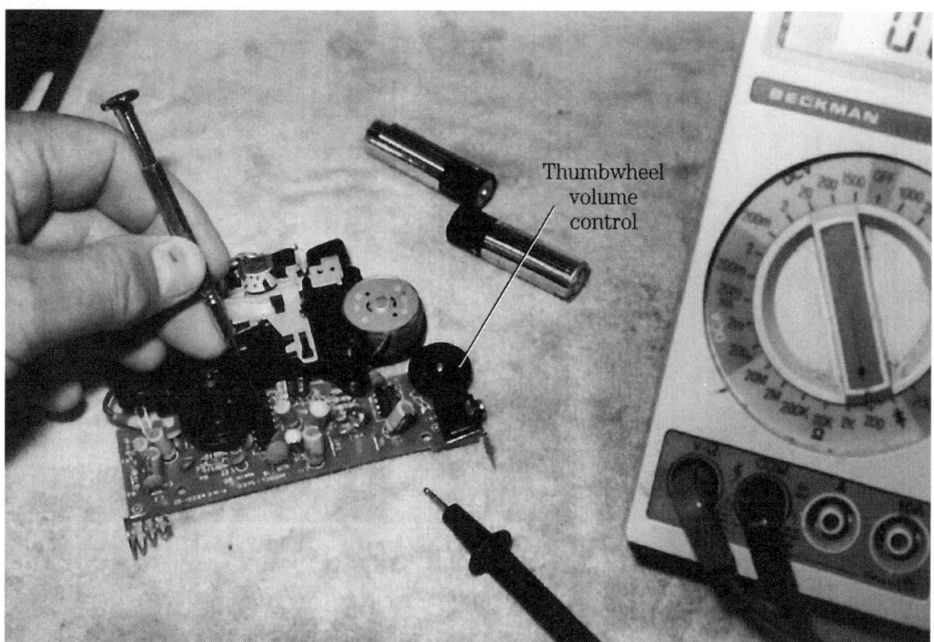

8-27 The thumbwheel volume control is found in portable and hand-held cassette players.

8-28 The small bias-trimmer controls are mounted directly on the PC board in the high-powered amplifier.

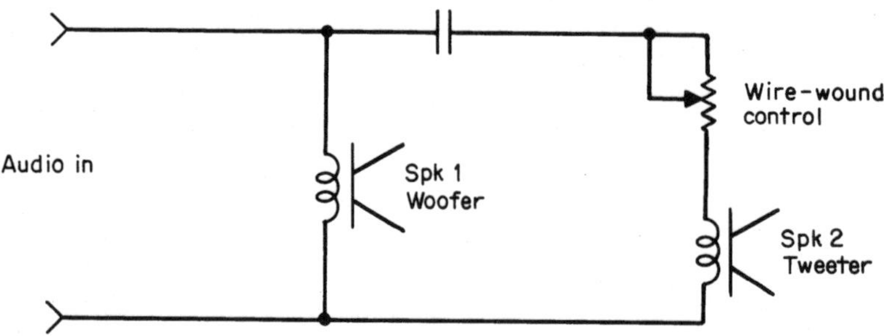

8-29 The wire-wound control might be found in the speaker, audio amp, and TV circuits.

ground potential. The wire-wound control might have a resistance of 100 to 100K ohms. Usually, the wire-wound control is wound with nichrome wire.

Testing

Wire-wound controls have a tendency to oxidize contacts and become noisy. Sometimes, spraying the control with cleaning fluid and readjustment of the control is all that is needed. Check for correct resistance with a DMM or FET-VOM ohmmeter.

Power adapters

(See Adapters)

Power cord

What and where

The ac line cord is found in all consumer electronic products that operate from the power line. The cheater cord is found upon the early TV receivers (Fig. 8-30). A cheater cord is used to power the TV chassis when the back cover is removed. Most power cords are two- and three-conduction types. The three-conductor cord has a ground terminal and is connected to the common ground terminal of power line products. The microwave oven cord has a three-conductor power line cord with a large prong as the ground terminal. The green-ground wire is bolted to the metal case of the microwave oven.

Testing

Inspect the power cord for cracks or breaks and frayed areas. Replace these damaged cords to prevent fires. Do not use an extension cord to operate a microwave oven. Replace the entire cord when electronic products are hit by lightning. Run a continuity test of each conductor with the RX1 range of the ohmmeter from plug to internal connection. Flex the ac power cord for possible breakage and intermittent connections. Replace the damaged or broken plug with a new polarized plug (Fig. 8-31).

Power inverters

(See Inverters)

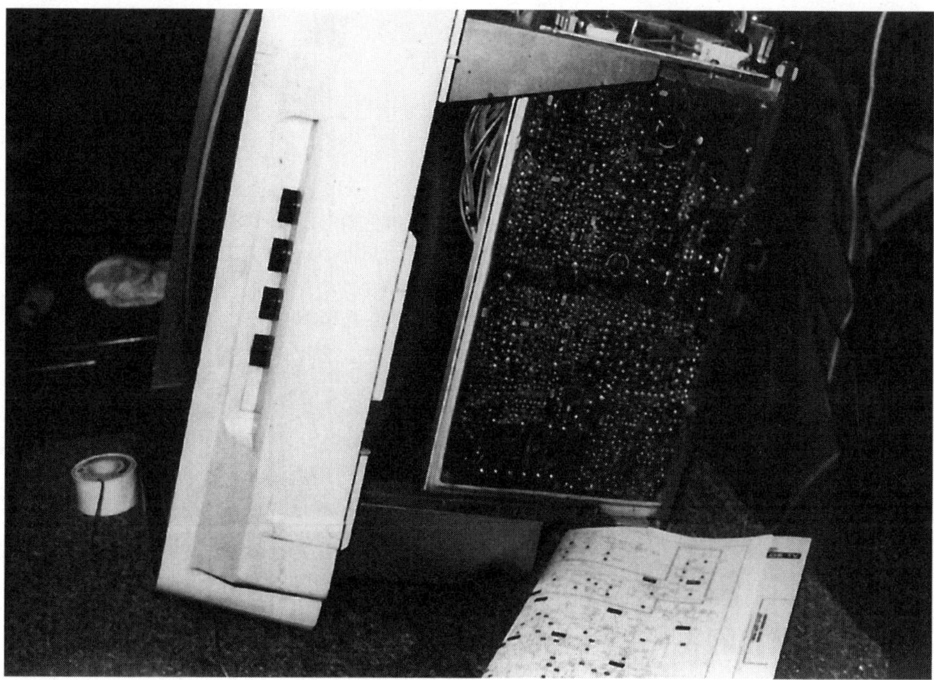

8-30 The cheater cord is used to power the TV chassis with the regular power cord removed.

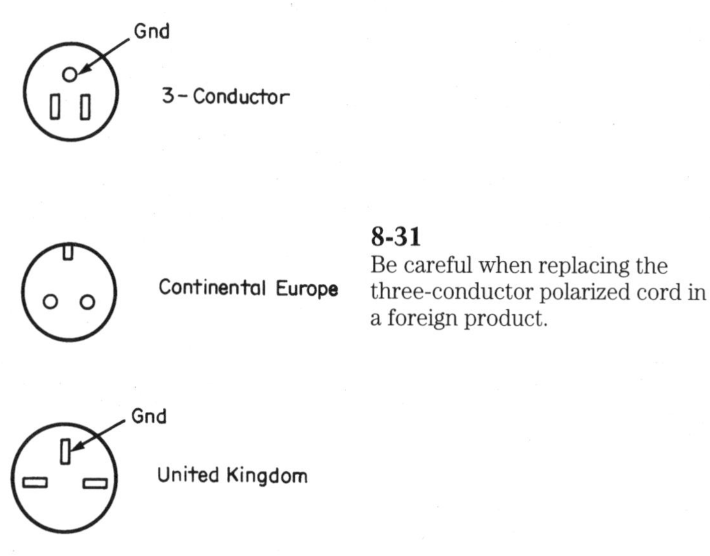

8-31
Be careful when replacing the three-conductor polarized cord in a foreign product.

Power supplies

Commercial power supplies

What and where

The power supply provides a voltage source to the electronic product. A power supply might consist of a halfwave, fullwave, or bridge rectifier circuit receiving voltage from a power transformer or power line. The power rectifier circuits might contain filter capacitors, resistor or choke filtering, and IC or transistor regulation circuits. The commercial switching power supply might provide a voltage source while servicing receivers, camcorders, computers, games, and TV/VCR products (Fig. 8-32).

A general-purpose power supply may furnish 5–15 volts with 1-amp adjustable output. The commercial power supply might provide a 0–30 Vdc 3-amp triple output power supply of two 0–30 Vdc 2-amp sections, with one at 4–6 Vdc at 5 amps, and a high-current power supply of 3–14 Vdc at 12 to 25 amps. The computer switching power supply might have a 200-watt power supply with output voltages of +5 V, 20 A; -5 V, 0.5 A; +12 V, 8 A; and -12 V, 0.5 A.

The built-in power supplies found in radios, TVs, VCRs, CDs, and cassette players might consist of halfwave, fullwave, and bridge rectifiers. The very simple power supply found in the early tape deck consisted of a halfwave rectifier (D101), 68-ohm resistor, and 2200-uF electrolytic filter capacitors (Fig. 8-33). Today, the power supply circuits have either a fullwave or bridge rectifier circuit with large filter capacitors. The large AM-FM-MPX high-powered receiver might have three or four different voltage output circuits. The step-down power transformer might have two or more windings, supplying several different voltage circuits. Transistor and IC voltage regulators are found in critical voltage sources (Fig. 8-34).

Testing

Check the output voltage of the suspected power supply circuits. Quickly measure the voltage across the large filter capacitors. Of course, the highest voltage can be found on the largest filter capacitor terminals. A quick voltage measurement across each filter capacitor can indicate the defective voltage source (Fig. 8-35). If low voltage is found across one large-filter capacitor, check for a leaky diode, zener diode, transistor, or IC regulator in that voltage circuit. Leaky silicon diodes can keep blowing the line fuse and damage the power transformer. Do not overlook an open or dried-up filter capacitor with a very low voltage source.

Heavy-duty power supply

What and where

In the early days of auto radio repair, the 12-volt, 10-amp power supply did not have any form of voltage regulation. Today, the 0- to 15-volt, 30- to 40-amp regulated power supply is used for powering automotive electronics and ham products. The power supply might have a built-in dc voltmeter and current meter. Overcurrent protection automatically shuts off current flow when maximum output is exceeded (31 A).

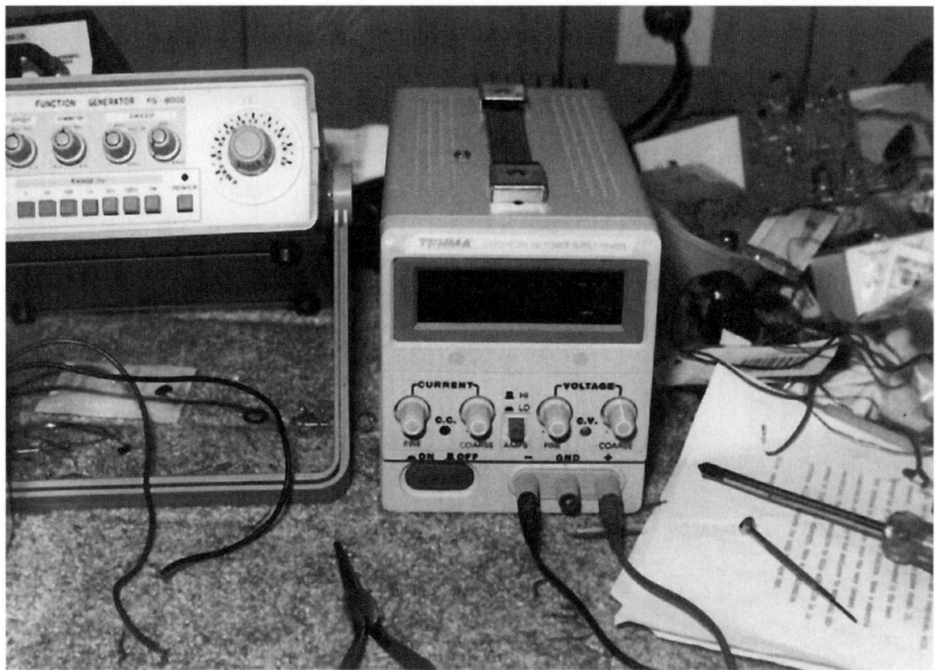

8-32 The commercial power supply might have two or more variable power sources.

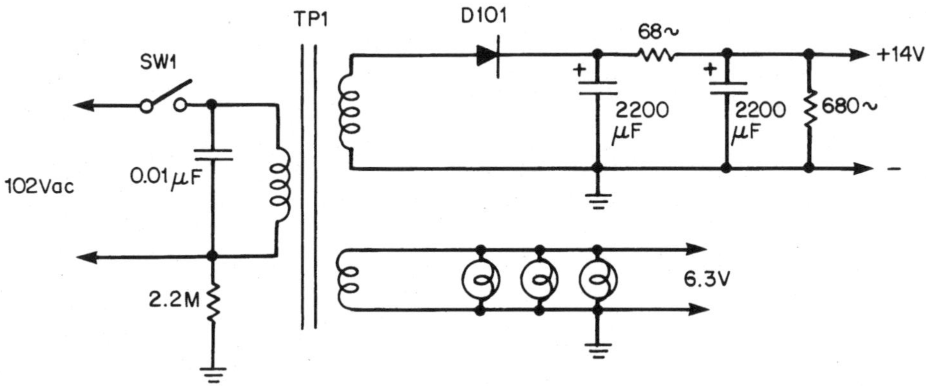

8-33 A halfwave power supply circuit found in the early table and console radios.

Testing

Check the line fuse with no voltage or measured output voltage. Check the overload circuits for defects. Measure the voltage across each large filter capacitor. Trace the no-voltage source to a transistor or IC regulator. Often, a leaky regulator transistor will have a burned zener diode in the base circuit. No voltage out of the transistor regulator may point to an open regulator transistor. Suspect a leaky collector-to-emitter short for higher-than-normal voltage. Take in-circuit

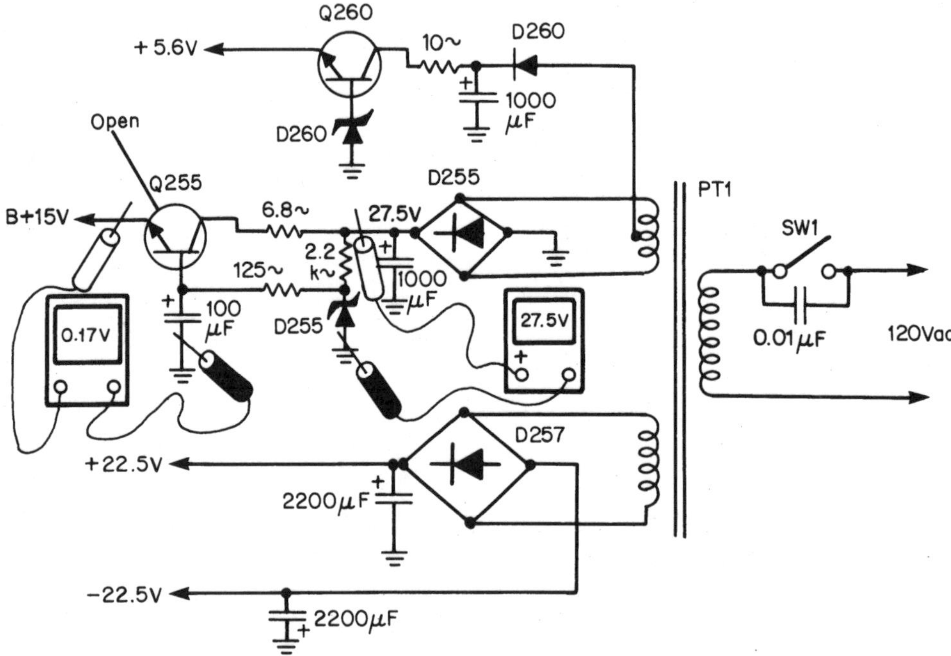

8-34 Check the suspected transistor and zener diode regulator circuits with voltage and diode tests.

8-35 Check the voltage across each large filter capacitor for a low- or no-voltage source.

transistor tests with a transistor tester. Do not overlook defective filter capacitors within the bench heavy-duty power supply when excessive hum can be heard in the connected auto receiver.

Regulated power supply

What and where

You might find transistor, IC, and zener-diode voltage regulation within the regulated power supply circuits (Fig. 8-36). Suspect a defective low-voltage regulator circuit when the audio output circuit and the dc motor is operating, but the receiver or preamp is dead in a cassette player. The voltage-regulated circuits might consist of a single transistor and zener diode or an IC-regulated circuit. The dc voltage from the bridge rectifier feeds directly to the regulating component.

The higher voltage source provides voltage to the audio-output transistors or ICs, and the regulated 12.2-V supplies voltage to the motor. Lower, regulated IC voltage is fed to the receiver, preamp, and AF-amp circuits. Determine what circuits are operating before taking voltage measurements, then measure the voltage source at the defective circuits. Improperly regulated dc voltages are caused by leaky transistors, ICs, or zener diodes.

Testing

Measure the dc voltage across each 1000uF and 6800uF filter capacitor (Fig. 8-37). When the voltage is low or not present, troubleshoot that circuit. Leaky or shorted

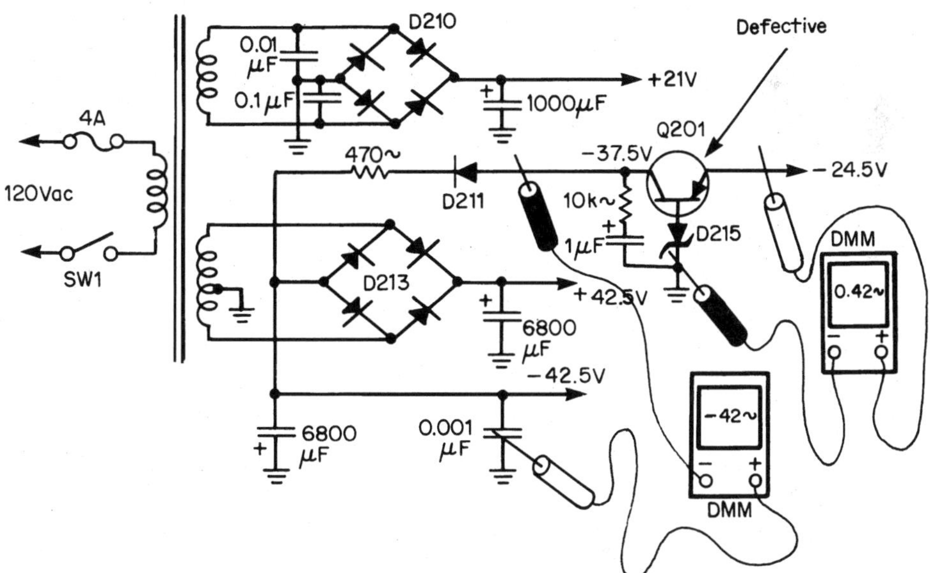

8-36 The regulated-bridge rectifier might contain transistor and zener-diode regulation circuits and can be tested with the DMM.

8-37 Check for correct voltage across each large filter capacitor with the DMM or FET-VOM.

silicon diodes can cause the fuse to blow or damage the step-down transformer. Open or dried-up filter capacitors can cause hum in the audio with lower voltages. A leaky filter capacitor might blow the main fuse and destroy the silicon diodes. The open regulator transistor or IC will have low or no output voltage. A leaky regulator transistor or IC might produce lower or no voltage. Test the transistor within the circuit for leakage or open conditions with the diode-junction test of the DMM or transistor tester.

Power transformers

(See Transformers)

Preamplifiers

Auto preamp

What and where

The audio preamp stage is found in audio amplifiers, cassette and CD players, phono-magnetic, crystal cartridge, and tape circuits. The preamp stages amplify the weak audio and are coupled to the AF and audio output circuit. Usually, the preamp is a voltage amplifier to boost the input signal. These preamp stages might consist of a transistor or IC circuit (Fig. 8-38).

Testing

The audio preamplifier can be tested with voltage, transistor, signal-injection, and signal-tracing methods. Take critical voltage measurements to determine if the correct voltage source is applied to the preamp circuit. Test each transistor with in-circuit tests with the diode-junction test of the DMM or a transistor tester. Clip a 1-kHz sine or square-wave generator signal to the input capacitor and use a scope or speaker as indicator to locate the defective preamp stage. Likewise, switch in the radio, tape, or cassette signal at the input, and signal trace each stage with a scope or external amplifier as indicator. After locating the weak or dead stage, take critical voltage and resistance measurements to locate the defective component.

Cassette preamp circuits

What and where

The preamp stages within the cassette player might consist of one or two stages of transistors or an IC circuit. You might find one IC operating both stereo stages (Fig. 8-39). Often, the R/P tape heads are directly or capacitor- coupled to the input of each stereo channel. Suspect a defective IC or power source when both input channels are weak and dead.

Testing

Clean up the tape head with alcohol and a cleaning stick. A dirty tape head can cause a weak and distorted channel. Take critical voltages on each IC terminal and compare them to the schematic. Take a critical voltage test of the supply source. Insert a 1-kHz test cassette and signal-trace the signal from tape head to the input

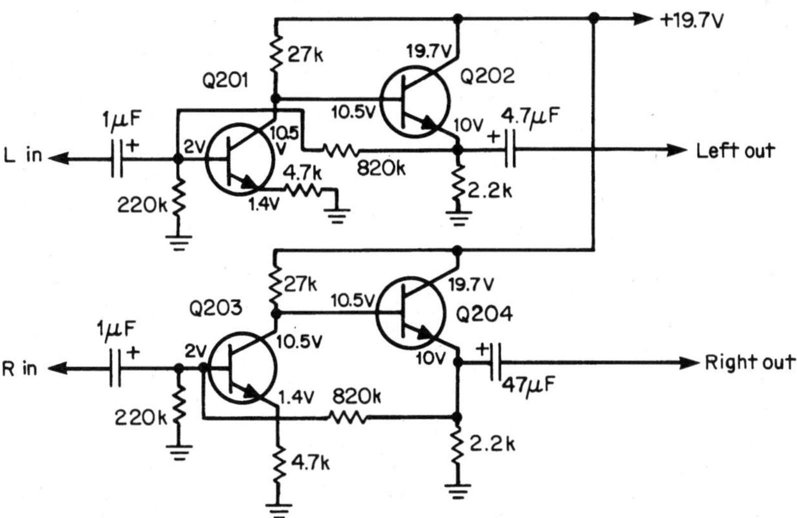

8-38 The preamp audio stages might consist of directly coupled transistors.

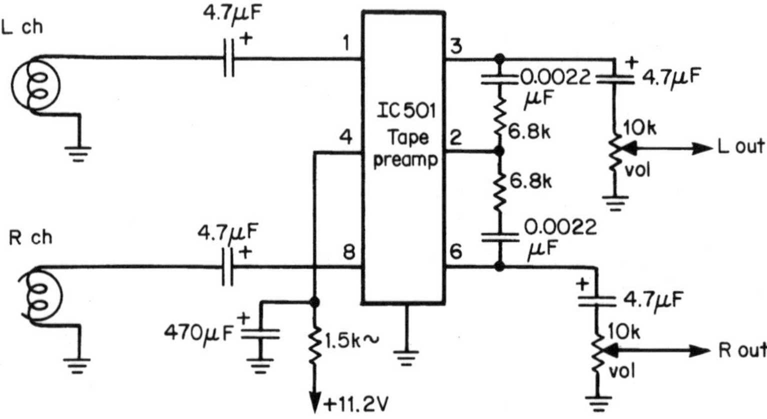

8-39 IC501 provides preamp circuits for the stereo channels of a cassette player.

terminal with the external amplifier. Scope the defective input and output terminal of the dead channel. Compare the input and output waveforms with the normal channel. Suspect a defective IC when a normal waveform is found on the input terminal and no or weak waveform is found at the output terminal. You might find a leaky IC with weak and distorted sound output.

Phono-preamp stages

What and where

In the early phono preamp stages, a tube or transistor amplified the weak signal from a crystal cartridge. The magnetic pickup is a variable-resistance cartridge in which the stylus or needle cause vibrations in a magnetic field. The magnetic cartridge has a very weak output signal and voltage. Thus, a two-stage transistor or IC circuit must amplify the weak signal to be connected to the audio circuits (Fig. 8-40). The weak input signal for each stereo channel is capacity-coupled to the input terminals of IC115, with the amplified audio output at pins 1 and 7.

Testing

Clean off the dirt and dust from the stylus with a small brush. Check the continuity of each magnetic winding of the cartridge with the RX1 scale of the ohmmeter. Both windings should have a similar ohmmeter reading. Measure the voltage on each IC terminal. Mark it down. Compare the voltage measurements with those of the schematic. Signal-trace each stereo channel with a record playing. Compare the input and output signal at pins 3 and 5, and 1 and 7 of IC115, with the scope. Suspect a defective IC or voltage supply source with both channels weak and distorted. Monitor the output with a scope or external audio amplifier when one channel is weak or intermittent. Do not overlook leaky or open electrolytic bypass capacitors tied to pins 2 and 6 of IC115.

Radios

AM-FM radio circuits

What and where

The simple AM-FM radio might consist of separate AM and FM circuits or a combination IC. The front-end circuit might consist of a separate FM-RF, FM mixer, and FM oscillator stage with varactor diode tuning (Fig. 8-41). A large AM-FM MPX receiver might have separate front-end stages with a combination FM IF-MPX and AM radio circuit. The AM and FM tuning is accomplished with a synthesizer and controller processor circuit. Both the AM and FM sections are controlled from the synthesizer control to the varactor-tuning circuit. A signal voltage from the controller selects the station.

When

A weak FM local station symptom might result from a leaky or open RF-FM transistor. Only a rush in the sound can be caused by a defective oscillator circuit.

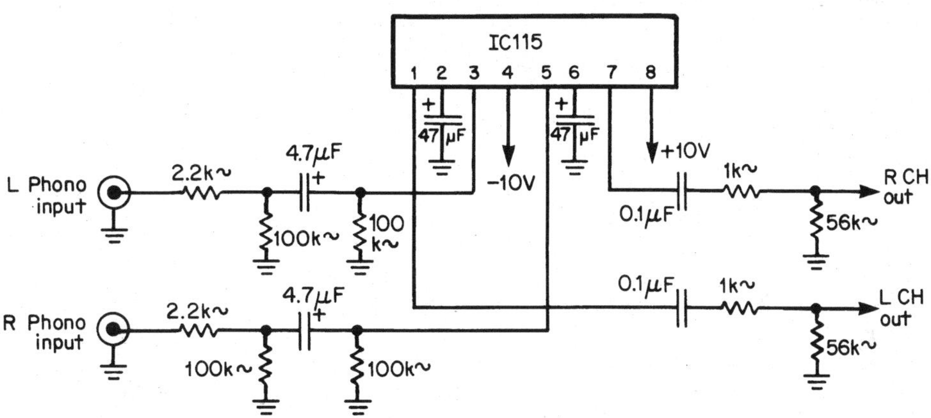

8-40 IC115 provides a stereo phono preamp circuit to feed the audio output stages.

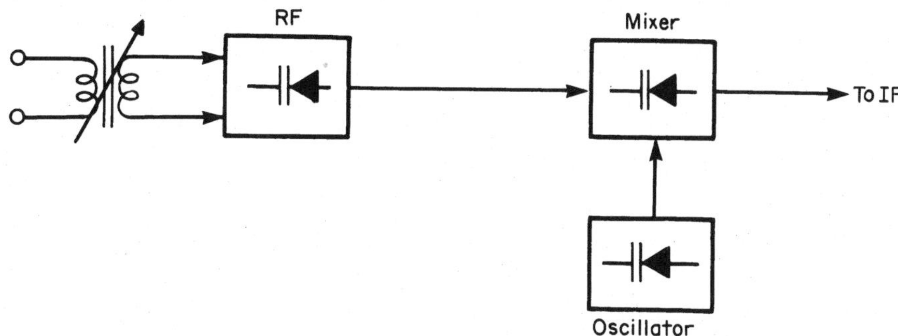

8-41 Varactor diode tuning is found in the FM-RF, mixer, and oscillator circuits.

No tuning of stations might be caused by a defective oscillator transistor or circuit. Try tuning an FM station in with another station nearby. If no squeal or noise is noted on the normal radio as the station is being tuned in, suspect a defective oscillator stage. When both the AM and FM stations cannot be heard, suspect a defective IF section or power source.

Testing

Test each FM transistor within each stage. Take critical voltage measurements. Inject an FM-RF signal at the antenna terminals with the FM-RF signal generator to check the RF, oscillator, and mixer stages. With no response, inject the FM IF signal (10.5 MHz) at the base of the mixer transistor to check the IF circuits. Measure the tuning voltage from the synthesizer control to the varactor diodes of each front-end transistor. Suspect a defective controller circuit or power source with no or low tuning voltage at each varactor diode. Measure the supply voltage pin of the controller processor.

Check the power supply circuit of a conventional dead AM or FM tuning circuit. Notice if the audio section is normal in the radio receiver. If the phono or cassette player is okay, suspect problems within the front-end circuit of the AM/FM tuner in a compact unit. Check the audio at the line output terminals of the tuner. Inject a 10.5-MHz IF signal at the mixer input terminal and notice if you can hear the signal in the speakers. If there is no audio, troubleshoot the IF and MPX circuits. Do not forget to check for a dirty AM/FM switch. Check for a broken AM/FM function switch.

Auto radio circuits

What and where

The deluxe auto receiver and cassette player might have separate FM-RF-IF and FM-MPX circuits, while the AM circuits consist of AM RF-converter-IF that feeds into the MPX circuits (Fig. 8-42). The stereo tape heads connect to a common preamp IC and switch into the first AF audio stages. The power audio output circuits might consist of a dual-power IC or separate power output transistors.

When

Service the front-end and IF circuits of both AM-FM-MPX circuits with the very same method as the table or console receiver. Today most audio functions in the cassette player have been taken over by ICs. Within small low-wattage audio-output circuits, a simple IC might take care of both audio-output channels. In higher-power outputs, separate ICs might be found in each stereo channel (Fig. 8-43). Transistors or ICs might be found in the preamp tape deck audio circuits, or the AM/FM audio circuits are switched in before or after the preamp circuits.

Testing

The dead right or left channel might be caused by a defective output IC, coupling capacitor, or power source. A weak left channel can be caused by an open coupling capacitor, transistor, or IC. The distorted channel can result from a leaky

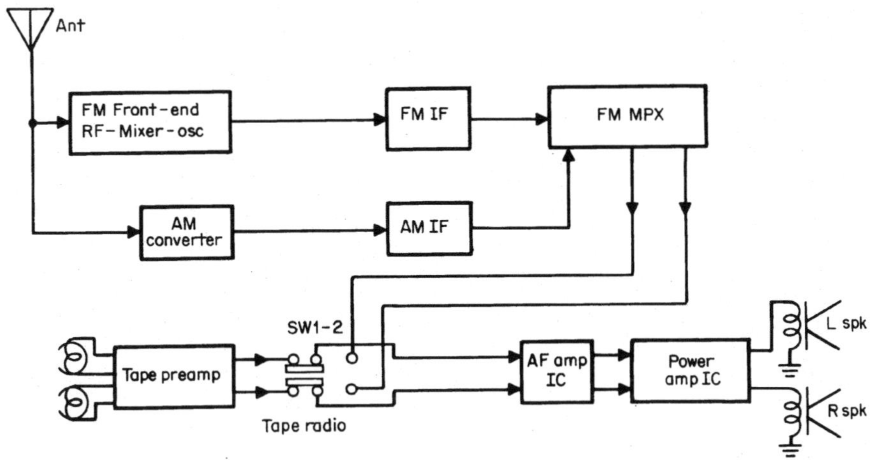

8-42 Separate AM and FM front-end circuits are fed to a common FM-MPX circuit in the AM-FM-MPX auto radio.

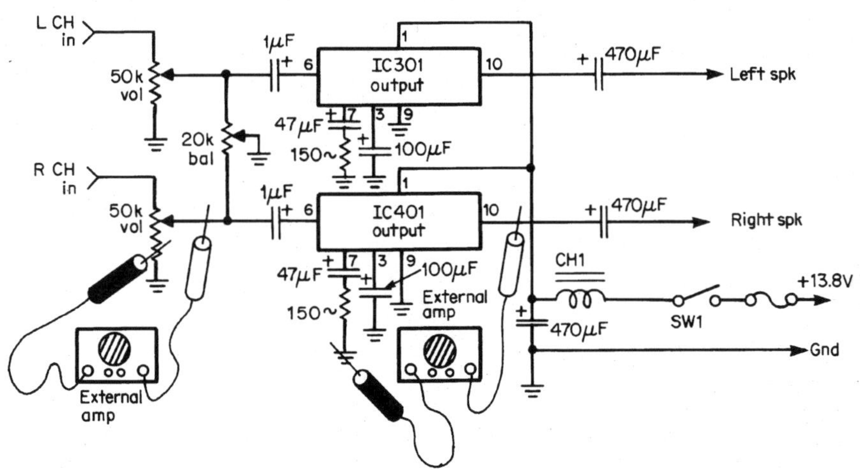

8-43 Troubleshoot the separate audio output ICs with an external audio amplifier or signal injection tests.

driver or coupling capacitor and IC. Suspect a dual-power output IC when both channels are distorted. Intermittent sound problems are difficult to locate. Isolate the intermittent audio to one or both audio channels. A noisy channel can be caused by defective transistors, ICs, and mica or ceramic bypass capacitors.

When audio is found on either IC and there is no output signal, check the voltage supply pin (Vcc) of the IC. If voltage supply pin 1 of IC301 is low, suspect a leaky

IC or improper low-voltage source. Check all components tied to the IC terminals for leaky or open components. Measure the voltage on each pin of the IC and compare to the normal IC terminals. Suspect a defective IC when a signal is found at pin 6 of IC401 and there is no audio output at pin 10.

Troubleshoot the audio section by injecting a 1-kHz audio signal at the input and output to determine what IC is defective. The audio signal can be injected right up to the speakers for a weak or dead audio symptom. Signal trace the audio channels with the external audio amp while a cassette is playing. Start at the preamp and signal trace to the output terminal of audio amp IC for a weak, distorted, or dead symptom with the external audio amplifier. The audio signal should become greater as you proceed through the audio circuits.

Shortwave radio

What and where

Shortwave receivers are constructed like any AM-FM-MPX receiver, except covering the 3- to 30-MHz bands. Like any low-priced radio receiver, the shortwave radio might be constructed from a couple of transistors in a regenerative or direct conversion circuit. There are many different types of shortwave receivers constructed in a simple direct-conversion circuit (Fig. 8-44).

Here a TDA7000 IC serves as an RF amplifier, mixer, and VFO oscillator circuit. The shortwave signal is fed in at pins 13 and 14, with the VFO (variable frequency oscillator) connected to pin 6. The different oscillator coils are tuned in with SW2 and tuned by a tuning varactor diode (TD1). R12 and R13 vary the voltage applied to the varactor diode (TD1) to provide a tuning voltage. R12 provides band-spread tuning. The audio output is controlled by R11 to a small IC amp with headphone reception.

Testing

Service the shortwave receiver as any AM-FM radio. When one band is difficult to tune in, clean up the band-function switch with cleaning fluid. Place the plastic tube right down on each switch contact, and rotate the switch back and forth.

Isolate the circuits by starting with the audio circuits at the volume control. Inject a 1-kHz signal at the volume control to determine if the audio circuits are normal. Troubleshoot the IF circuits by injecting an IF frequency at the base of a mixer transistor or IC terminal. Take critical voltage and resistance measurements on each IC or transistor to determine if the solid-state devices are defective. Test each transistor in circuit with a transistor tester or diode-junction test of the DMM. Remove and test once again.

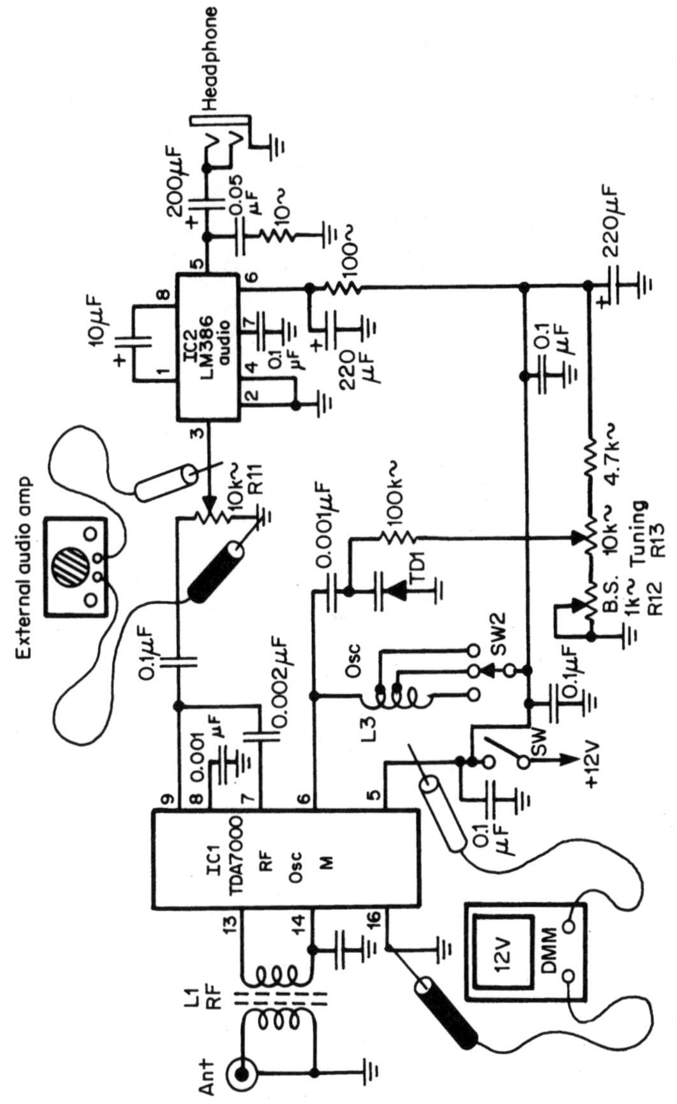

8-44 Check the shortwave receiver tests with the DMM and external audio amplifier.

9
CHAPTER

Testing electronic components R

(Rectifiers through relays)

Besides the diode, there are many different types in the rectifier family, such as boost rectifiers, damper diodes, diacs, fast-recovery rectifiers, high-voltage rectifiers, quadracs, Schottky diodes, sidacs, and triacs. The boost, high-voltage, and damper diodes are found in the TV chassis. A diac semiconductor might be used in control circuits, while the sidac is a surge protector for electronic equipment. The triac is found in the control circuits of the TV and microwave oven. The IC, transistor, and zener diode regulators are found in TV, radio, amplifier, and low-voltage sources (Fig. 9-1.)

Rectifiers

Boost rectifiers

What and where

In the early TV chassis, the damper tube served as a boost B+ voltage rectifier. Actually, the B+ boost is added to the low-voltage source. Later a selenium rectifier produced B+ boost voltage from a flyback winding. Today, the boost voltage might be produced from a boost silicon diode or a flyback winding. The boost voltage source within the TV chassis furnishes voltage to the color output transistors and cathode elements of the picture tube. Often, the boost voltage source is taken from a primary winding of the horizontal output transformer or flyback. This voltage might vary between 165 to 200 volts (Fig. 9-2). Locate the boost rectifier within one leg of the flyback connected to the screen control.

9-1 The electronics technician taking critical diode tests on rectifiers and transistors in a TV low-voltage power supply.

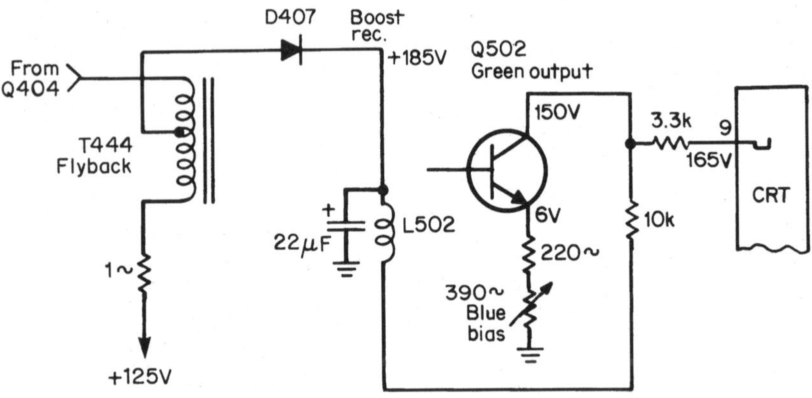

9-2 The boost rectifier (D407) provides +185 volts to the color output transistors.

When

The TV screen might be dark with a missing color, no raster, and only hum in the sound with a leaky or open boost rectifier. A leaky boost diode might cause the B+ fuse to open. A shorted boost rectifier might open a low-ohm isolation resistor. In the early TV chassis, an open or dried-up filter capacitor within the boost rectifier circuit might leave the left side of the picture with a dark shaded area.

Testing

Check for a missing boost voltage at the cathode terminals of the CRT or collector terminal of the color output transistor. Test the boost diode for leaky or shorted conditions with the diode test of the DMM (Fig. 9-3). Double-check for an open or burned isolation resistor between the boost diode and voltage source. Replace the boost rectifier with a 1000 or 1200-KV type. The NTE125, ECG125, and SK5010A universal replacements work well in boost diode circuits.

Bridge rectifiers

(See Diodes)

Diode rectifiers

(See Diodes)

Damper diodes (TV)

What and where

The early damper diode within the TV chassis was located alongside safety capacitors in the collector terminal of the horizontal output transistor. Today, the damper diode is part of the internal construction of the horizontal output transistor (Fig. 9-4). This diode is connected internally across the collector and emitter terminals. The damper diode prevents ringing and oscillations in the power supply of the TV chassis from the horizontal output circuits.

When

The damper diode can become leaky or shorted. Often, the B+ or main fuse is found open. Sometimes the horizontal output transistor is damaged. Always replace

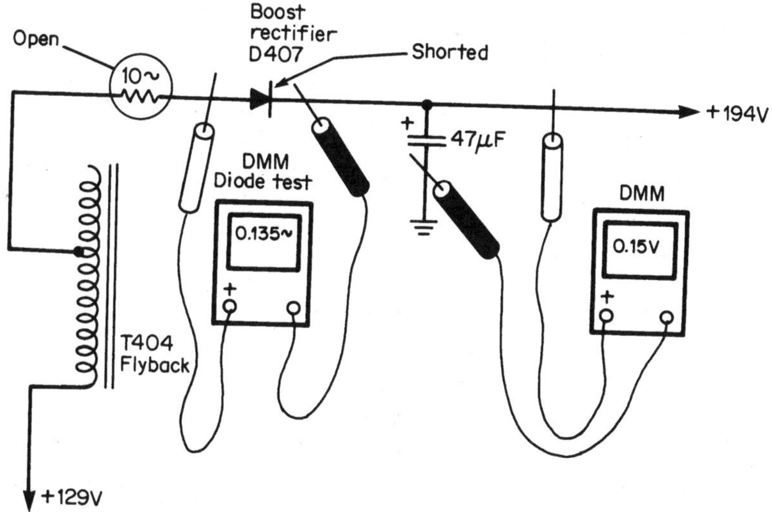

9-3 A shorted boost rectifier damaged the 10-ohm isolation resistor with 0.15 volts measured instead of +194 boost voltage.

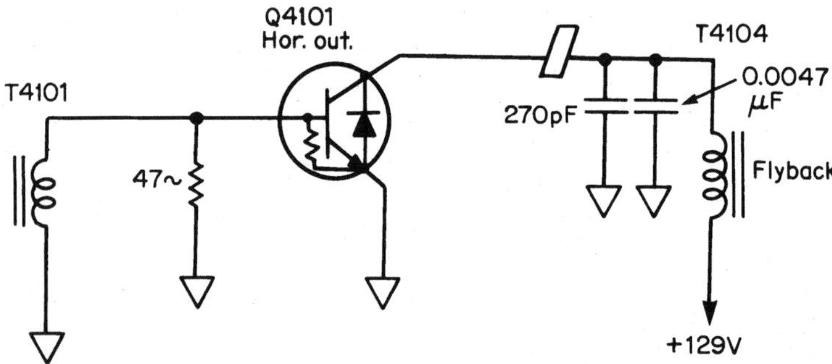

9-4 Today, the damper diode is mounted inside the horizontal output transistor.

the original output transistor that has an internal damper diode with another one. Replace the single damper diode that is outside the transistor with a 1500- or 2000-volt unit.

Testing

A quick damper diode test can be made from the metal shield of the output transistor to the chassis or common ground with a diode test of the DMM (Fig. 9-5). A normal damper diode will measure .500 to .575 ohms in one direction. The leaky diode will show a low-resistance measurement under 100 ohms with reversed test leads. A low-resistance measurement of 0.135 ohms indicates a shorted damper diode or leakage between the emitter and cathode of the output transistor. Replace the leaky output transistor in the latest TV chassis, which includes both damper diode and transistor in one component.

Diacs

What and where

A diac is a gateless triac. The diac has only two terminals, and the diac semiconductor might be a fullwave or bidirectional thyristor. Diacs are made of three-layer semiconductors with negative resistance. It is triggered from a blocking-to-conduction state for either polarity of the applied voltage. The diac conducts when the applied voltage exceeds the break-over voltage rating of the diac. Diacs are found in trigger-controlled circuits (Fig. 9-6).

Testing

The diac might look like a tiny switching diode. Check the diac with the diode test of the DMM or a low-resistance measurement. The normal diac will show an infinite measurement in both directions on the diode scale of the DMM. A low measurement indicates the diac is leaky or shorted. Infinite resistance measurement should be found on a normal diac with reversed test leads of the FET-VOM. On the highest resistance range of the FET-VOM or DMM (Fig. 9-7), any resistance measurement indicates that the diac is leaky.

Fast-recovery diodes or rectifiers

What and where

The standard silicon diode has a slow recovery time in low-frequency applications (60–120 Hz). The fast-recovery diode operates in high-frequency circuits with critical recovery time, and the diode must respond to very short duration spikes. A damper diode found in the flyback circuit is a special fast-recovery diode. The fast-recovery diode is found in computers, and might appear in DO-41, DL-41, DO-27, R-6, TO-220A, T0-220, and SMB packages. A fast-recovery diode might appear in a high-amperage stud rectifier.

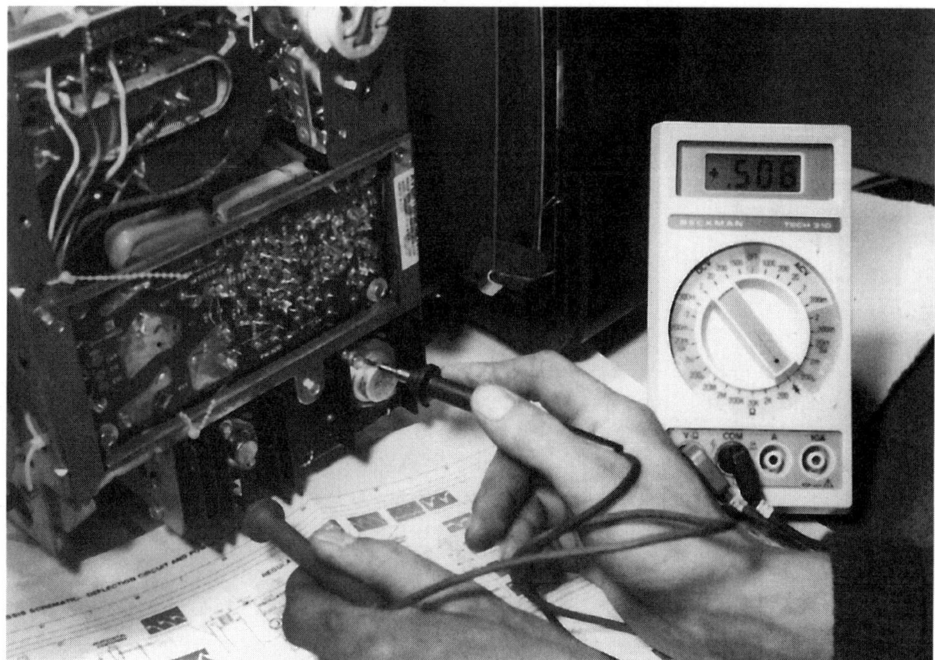

9-5 Check the damper diode with a diode test of the DMM to the metal case of the horizontal output transistor and common chassis ground.

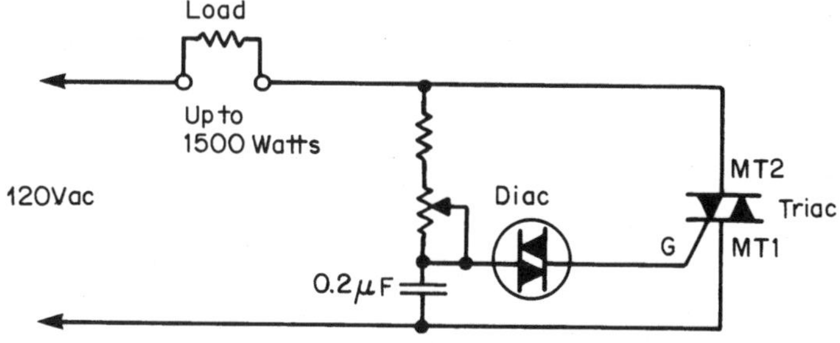

9-6 The diac might be found between the gate of the triac to the voltage control circuits.

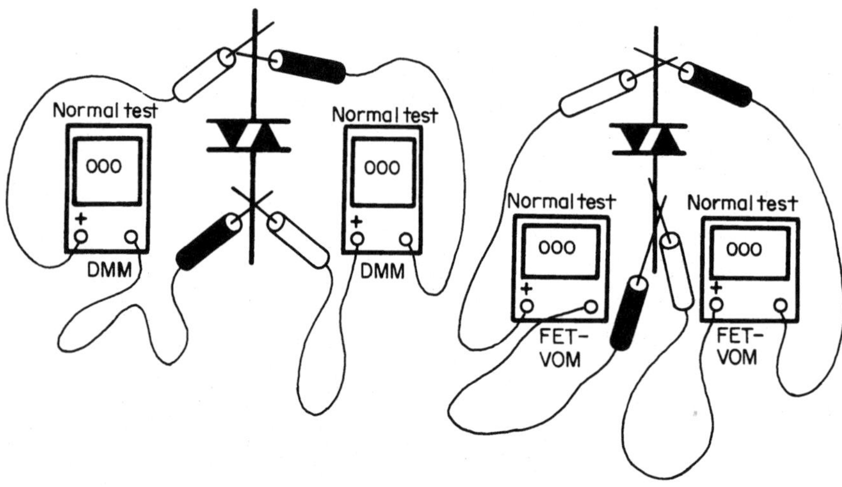

9-7 The normal resistance measurements of a gateless diac with the DMM and FET-VOM.

Testing

Check the recovery diode as any silicon diode. Test the diode with the diode-junction test of the DMM. Diodes can be checked rather quickly with the diode test of the digital multimeter. Test the suspected diode with a diode tester. A normal diode will provide a low-resistance measurement in one direction with the DMM (Fig. 9-8). The leaky diode indicates leakage in both directions or with reversed test leads. Remember, when taking diode tests with the VOM, place the positive probe to the positive (cathode) terminal of the diode and the black probe to the anode terminal. Reverse the probes when taking a diode measurement with the digital multimeter.

High-voltage rectifiers (TV)

What and where

The single high-voltage rectifier was found in the early black-and-white TV chassis. The silicon stick rectifier plugged into a fuse-type socket with a flexible socket lead for the other end. A high-voltage rectifier might have a 2-KV up to 20-KV rating. Today, the high-voltage diodes are found molded right inside the flyback transformer (IHVT). The resulting HV cable plugs into the anode socket of the picture tube.

The HV multiplier circuit consists of HV capacitors and diodes in an HV network found within the flyback and tripler units (Fig. 9-9). Often, the leaky HV diodes might cause a low arcing noise, hissing, firing, and arcover to the TV chassis. These HV diodes cannot be replaced separately within the HV multiplier circuits. Replace the defective flyback and tripler units when leakage occurs.

The HV diode or stick rectifier found in the black-and-white TV chassis might consist of many stacked selenium rectifiers within one fiber component. The high-voltage stick rectifies the RF voltage of the horizontal output transformer. If the rectifier becomes leaky, low, or no high voltage is measured at the CRT, you might smell a sweet odor with a shorted stick rectifier. Look for burned marks on the fiber tubing. These stick rectifiers should be replaced with the original part number. The defective HV rectifier might produce low or no high voltage to the picture tube.

Testing

Test the single HV silicon rectifier like any diode. Measure the high voltage at the anode terminal of the CRT with a high-voltage probe. Be very careful when taking high-voltage measurements around the picture tube.

High-voltage rectifier (microwave oven)

(See Diodes)

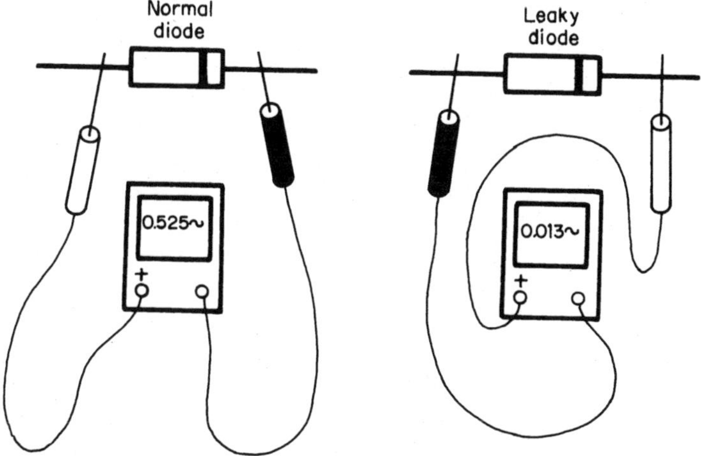

9-8 The normal and leakage diode test with the diode test of the DMM.

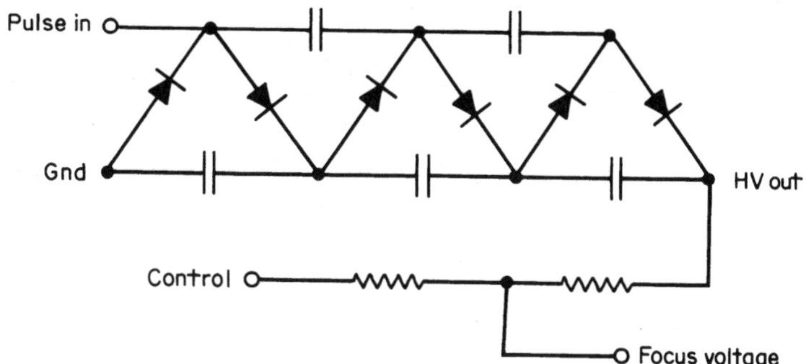

9-9 A stack of high-voltage rectifiers and capacitors within the flyback or tripler unit.

Quadrac

What and where

The *quadrac* combines both a triac and a diac in one package (Fig. 9-10). A quadrac is a bidirectional ac switch and is gate controlled by either polarity of the main terminal voltage. The quadrac is found in ac switching speed controls, temperature modulation controls, and lighting controls. The quadrac has a high surge capacity up to 150 volts and might operate in the 200- to 600-volt range. This component has a TO-220 package with MT1, MT2, and G terminals.

Testing

Check the quadrac with the triac tester. Take resistance measurements with the DMM, VOM, or FET-VOM and compare them with Fig. 9-11. A shorted quadrac will have a low-resistance measurement between MT1 and MT2.

Schottky barrier rectifier

What

The Schottky rectifier is a solid-state diode in which a metal and a semiconductor form the PN junction. This rectifier might be called a hot-carrier diode. Schottky diodes are very fast-switching diodes. A high-powered Schottky rectifier might have a stud mounting. The Schottky barrier rectifier might appear in DO-41, SMB, DO-201A, TO-220A, TO-220, and TO-247 packages.

Testing

Test the Schottky diode as any diode with the diode test of the DMM. A leaky or shorted diode will have a low resistance in both directions. An open diode has infinite resistance in either direction. The normal diode has a low resistance in one direction. Check the suspected diode with a diode tester or with the diode test of the DMM.

SCRs (silicon-controlled rectifiers)

What and where

The silicon-controlled rectifier might also be called a thyristor. An SCR also resembles the action of a thyratron tube. The anode, cathode, and gate elements of the SCR are about the same as the anode, cathode, and grid elements of the thyratron. SCRs are typically used to control ac voltages. The replacement SCRs might appear in 0.8- to 100-amp packages.

If a voltage is applied between cathode and anode terminals, with a zero gate voltage, no current will flow through the silicon-controlled rectifier. When voltage is applied to the gate terminal, current will flow from cathode to anode terminals. This current can still flow, even if the gate voltage is removed. Silicon-controlled rectifiers might be found within the ac power supply of the TV set, high-voltage regulation, microwave oven, and other ac-controlled devices (Fig. 9-12).

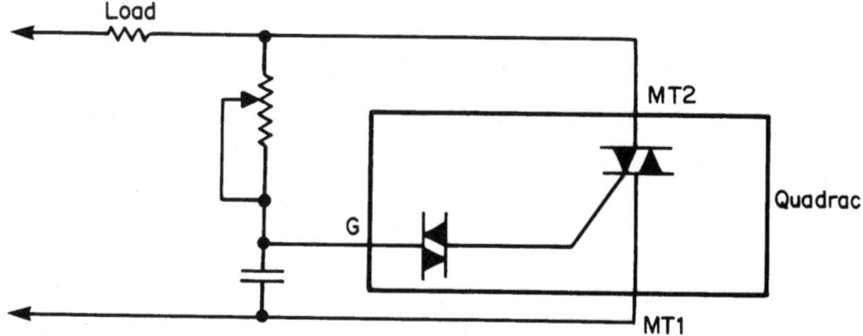

9-10 The quadrac component combines both a triac and diac in one package.

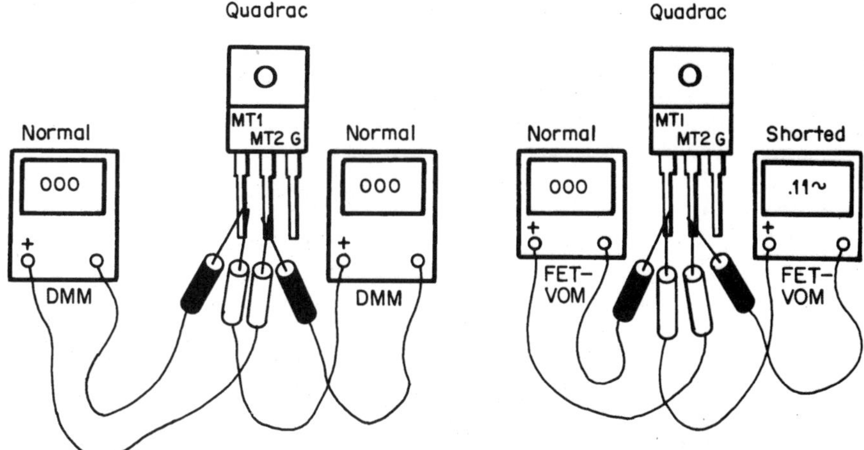

9-11 The normal quadrac resistance test of the DMM. A normal and shorted FET-VOM resistance test of a quadrac.

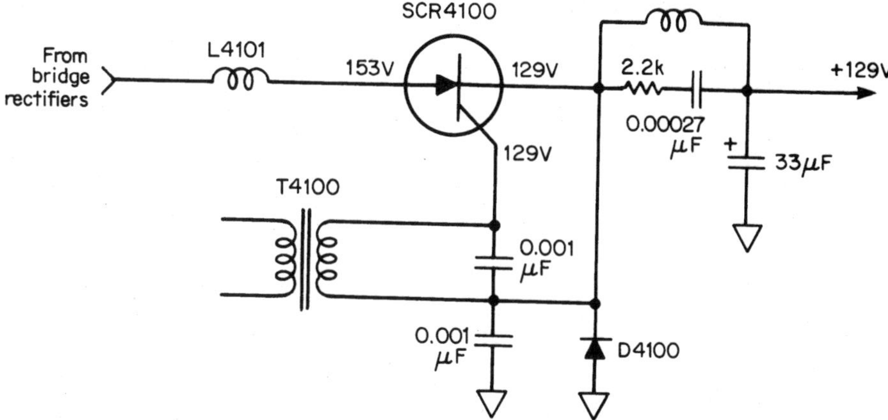

9-12 The silicon-controlled rectifier (SCR) found in the low-voltage power supply of a TV set.

Testing

Check the suspected SCR on the TRI-CHECK II tester. Disconnect the ac power cord. Disconnect all wires to the SCR. Turn on the TRI-CHECK, and the middle light will blink. Connect the black lead (T1) to the cathode terminal, clear (T2) to the anode, and red (G) to the gate terminal. Press the test button. The SCR light should light when the test button is pressed and turn off when the button is released. Both lights will light with a shorted SCR. No lights are on with an open SCR.

Test the SCR by taking critical voltage and resistance measurements. The SCR found in the low-voltage power supply and high-voltage rectifier circuits of the TV chassis can be checked with resistance measurements. A low-resistance measurement between the gate (G) and cathode (K) is normal (Fig. 9-13). Replace the SCR if the measurement is below 50 ohms. If any resistance measurement is found between the anode (A) and cathode (K) terminals, replace the leaky SCR. Like the transistor, the SCR can break down or become intermittent under load and should be replaced if suspected.

Sidacs

What and where

The sidac is a silicon bilateral voltage trigger switch with high-power current-handling capabilities. Upon exceeding the sidac break-over voltage point, the sidac switches through a negative resistance region to a low on-state voltage. Conduction will continue until the current is interrupted or drops below the maximum holding current of the sidac. The sidac might appear in a TO-92 or TO-202AB package.

The sidac is used in high-voltage lamp igniters, natural gas igniters, gas-oil igniters, high-voltage power supplies, xenon igniters, overvoltage protectors, pulse generators, and fluorescent lighting igniters (Fig. 9-14). A single sidac has MT1 and MT2 terminals. A dual sidac might have a MT1, center tap, and MT2 elements.

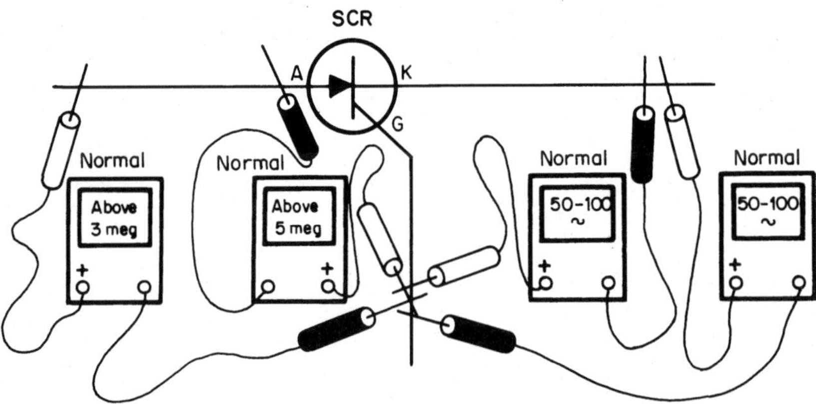

9-13 The normal resistance tests of an SCR with the diode test of a DMM.

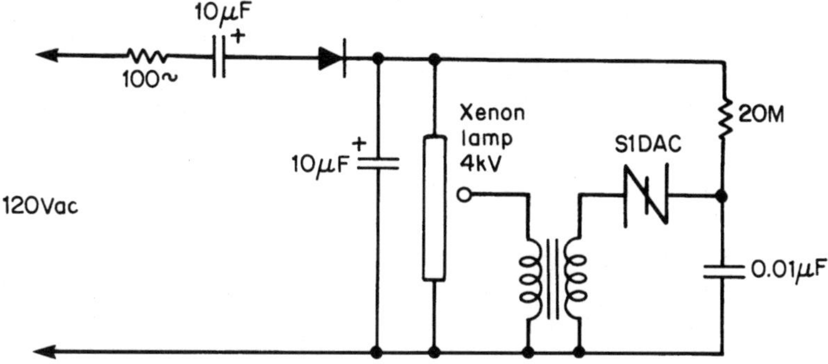

9-14 Here, a sidac component is found in a xenon lamp control circuit.

Testing

The normal sidac will have an infinite reading in both directions of any terminal. No resistance should be made with the diode on the 20-megohm scale of the DMM (Fig. 9-15). Likewise, no resistance measurement can be found on the highest RX100K range of the FET-VOM.

Sidactor

What and where

A *sidactor* is a nondegenerative crowbar fold-back device that protects telecom circuits from hazardous transient voltages. The sidactor is connected across the tip and ring to offer the fastest and most reliable overcharge position (Fig. 9-16). In standard mode, the sidactor is transparent to the circuit due to its high-state resistance. When a higher voltage is applied with break-over voltage, the sidactor will switch from its high-state resistance to a short, protecting the modem circuits.

Testing

The sidactor has an infinite resistance test on the diode test of the DMM in both directions (Fig. 9-17). Likewise, no resistance is measured between the two terminals with the RX100 scale of the FET-VOM. Any type of resistance measurement indicates that the sidactor is leaky and should be replaced.

Triac (general purpose)

What and where

The triac is a three-terminal, gate-controlled, semiconductor ac switching component. The triac resembles a dual SCR. It operates two separate SCRs connected in parallel and in opposite direction. Current of either polarity can flow through the component. When one side is conducting, the other side is off.

The triac switches on voltages within the microwave oven and on power supplies of a TV chassis (Fig. 9-18). Triacs come in TO-92, TO-22AB insulated, and nonisolated

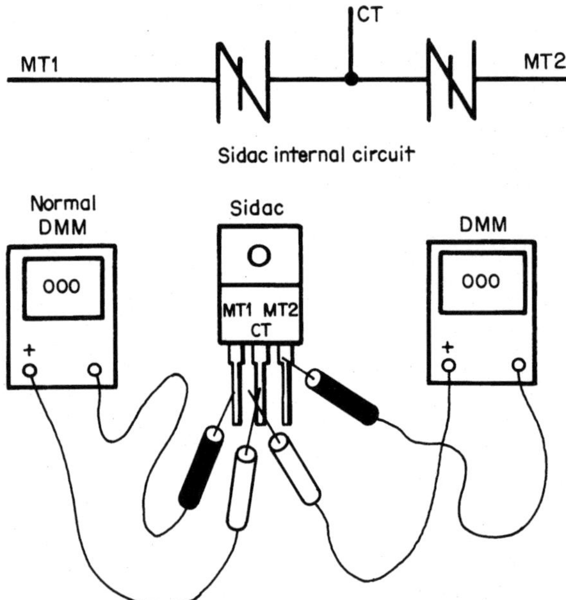

9-15 A normal resistance test of a sidac component with the diode test of the DMM.

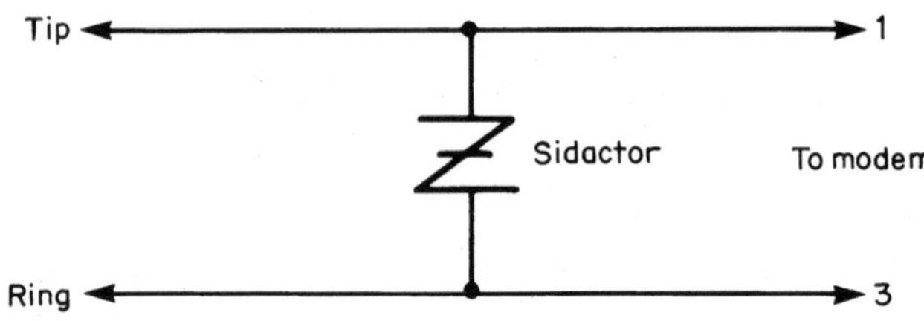

9-16 The sidactor is found in the tip and ring circuit of a modem.

packages. A gated triac might operate from 1 to 15 amps, while the logic triac operates from 1 to 8 amps. When a gate (G) voltage is applied, MT1 and MT2 conduct voltage.

Testing

Check the suspected triac on an SCR-Triac TRI-CHECK II tester. The defective triac usually becomes shorted between MT1 and MT2 terminals. When open, no voltage passes between MT1 and MT2. Take a quick resistance measurement across MT1 and MT2 with the DMM. A low resistance indicates a leakage between MT1 and MT2. The normal DMM measurement should be infinite ohms across MT1 and MT2 (Fig. 9-19). A normal FET-VOM measurement across MT1 and MT2 should have infinite resistance. The normal resistance measurement between MT1 and the gate on the DMM diode test should be around .447 ohms. The resistance measurement with the FET-VOM tester should be around 100 ohms.

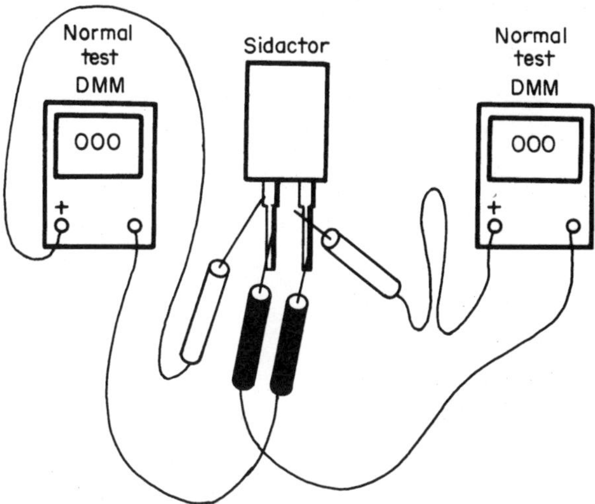

9-17 The normal resistance tests of a sidactor on the diode test of the DMM.

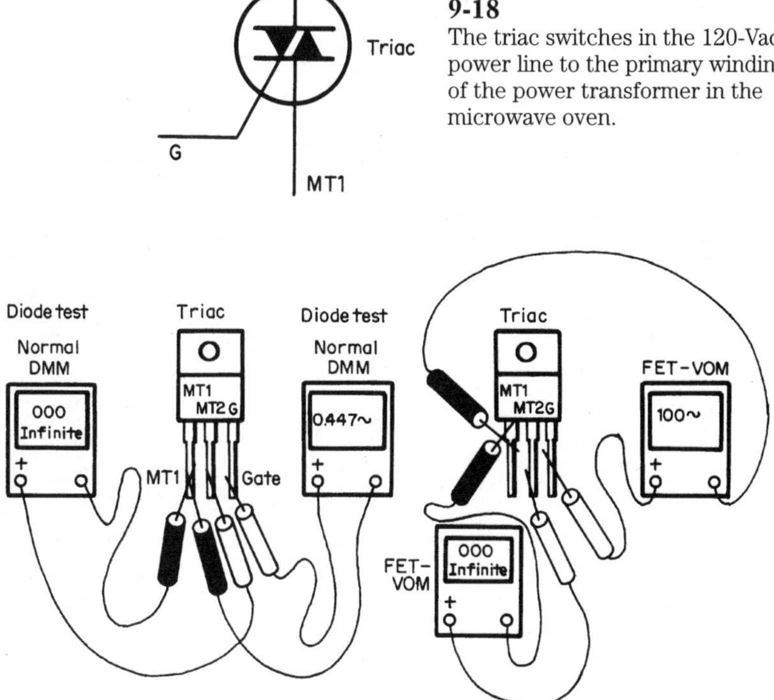

9-18
The triac switches in the 120-Vac power line to the primary winding of the power transformer in the microwave oven.

9-19 The normal resistance measurements with the diode tests of the DMM and FET-VOM resistance test of a triac component.

Triac (microwave oven)

What and where

The triac employed within the microwave oven is used in place of the oven relay (Fig. 9-20). The triac assembly applies 120 volts to the primary winding of the high-voltage transformer. Usually the triac module is controlled by the electronic controller unit with a gate voltage. When proper gate voltage is applied to the oven triac, the component will conduct applying 120 volts to the primary winding of the power transformer (Fig. 9-21).

When

The defective triac might become leaky, letting the oven run all the time, even when the oven is turned off. A defective triac might cause the oven to hum under load when the oven is turned on. Leakage occurs between MT1 and MT2. An open triac will not let ac voltage pass through the triac with no oven operation. Most oven triacs become leaky between MT1 and MT2 terminals. Improper or no gate voltage applied to the triac makes the triac act as an open component. Check the control board for missing voltage and triac control signal. Remove the triac from the oven circuit. But first, discharge the HV capacitor. Mark down each wire terminal. Check the color code of each wire.

Testing

Check the oven triac on the TRI-CHECK II tester. The tester was designed to test the suspected microwave triac assembly. This tester will test the triac for open, shorted, leaky, or normal conditions (Fig. 9-22). Remove all wires from the triac assembly.

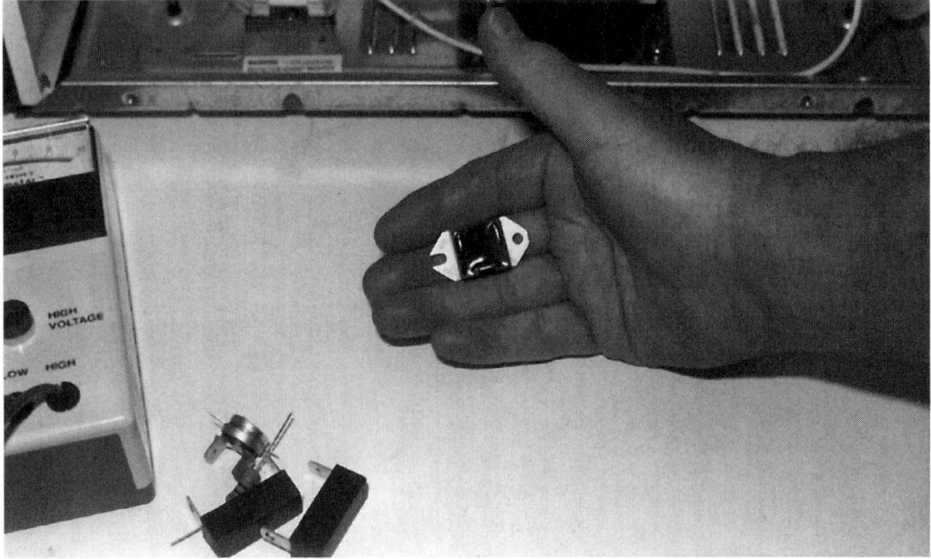

9-20 The present-day triac found in the microwave oven is very small compared to the early ones.

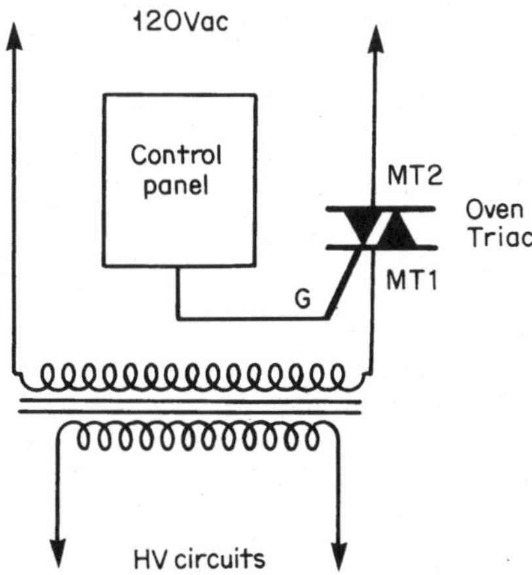

9-21 The control circuits provide gate voltage to the triac, which switches on the 120 Vac to the power transformer.

9-22 Test the microwave oven triac with a TRI-CHECK II tester. Discharge the HV capacitor before taking any tests.

Both lights will light when the test button is pressed and turn off when button is released on a normal triac. With an open triac, no lights will light up. Both lights will light with a shorted triac. Turn off the triac tester when finished testing, as two batteries are needed to power the TRI-CHECK II.

Check the continuity between any triac terminals with the DMM and FET-VOM. Switch the DMM to the diode test and read the resistance between G and MT1. This reading should be somewhere around 0.554 ohms in either direction (Fig. 9-23). Reverse the test leads and the reading should be the same for a normal triac. You will not read any resistance between other combinations of the two terminals for a normal triac. If a reading other than gate to MT1 is noted, the triac is leaky. A normal resistance of 15 to 20 megohms should be found between MT1 and MT2. A low resistance in both directions indicates the triac is shorted or leaky.

The low-ohm scale of the FET-VOM should measure around 35 ohms between gate and MT1, in both directions. A reading above 15 megohms from MT1 and MT2 is normal. Replace the triac assembly when a 5-megohm or lower measurement is noted between terminals other than MT1 and gate (G). Usually, the defective oven triac will appear shorted or leaky. Very seldom does the triac go open. Replace the triac with the original part number, and make sure the correct wires or cable go to the right triac terminals.

Zener rectifier

(See Zener diode)

Regulators

Voltage regulation might appear in IC, transistor, and zener diode low- and high-voltage circuits. Sometimes transistors and zener diodes form a low-voltage regulator circuit. IC regulators are found in lower-line voltage and low-voltage circuits of the TV set. Voltage regulators are found throughout the low voltage circuits of the cassette and CD player, VCR, and TV (Fig. 9-24).

IC voltage regulators

What and where

The IC fixed regulator found in the power sources might appear in just about every kind of electronic product, projects, or kits. IC regulators found in low-voltage sources of the CD player or deluxe receivers might have only three terminals: input, ground, and output terminals (Fig. 9-25). A higher dc voltage is applied to the input terminal with a regulated lower output voltage. IC regulators might appear in transistor, IC, T0-92, T0-220, IC-106, flat-line, and SMT packages.

A defective IC voltage-regulated circuit might cause excessive hum, a no- or low-voltage source, and intermittent chassis shutdown. A leaky IC regulator might produce a hum or lower or higher voltage source. The open IC might cause weak or no sound with low or no output voltage. Notice that large-filter capacitor preceded the input of the IC regulator and small electrolytic capacitors in the output voltage sources.

Testing

When a receiver or CD player appears with certain circuits or the whole chassis is dead, suspect an open voltage regulator. Test the input and output voltage. Usually, an open IC regulator will have a higher than normal input voltage and no output voltage. A leaky IC regulator might produce a lower dc voltage. Check for correct dc

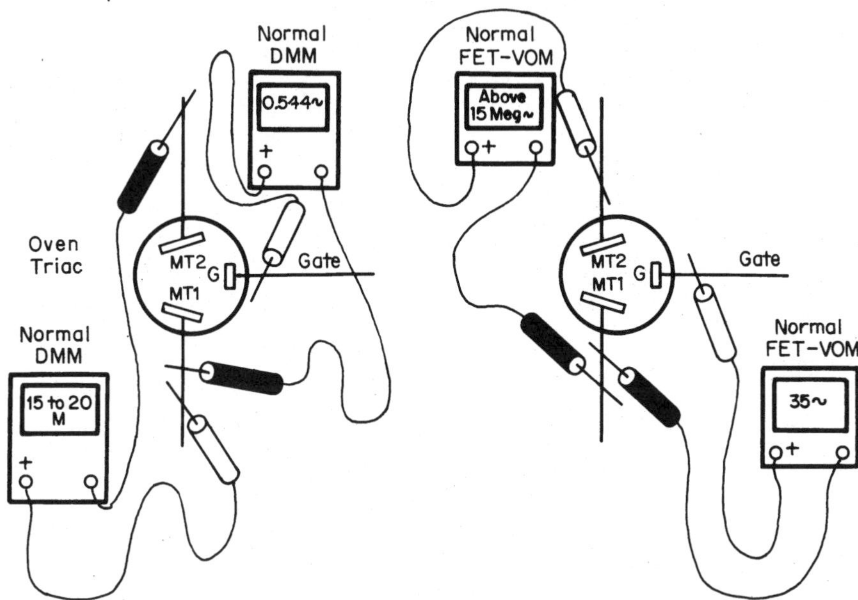

9-23 The normal oven triac tests with the diode test of the DMM and resistance tests of the FET-VOM.

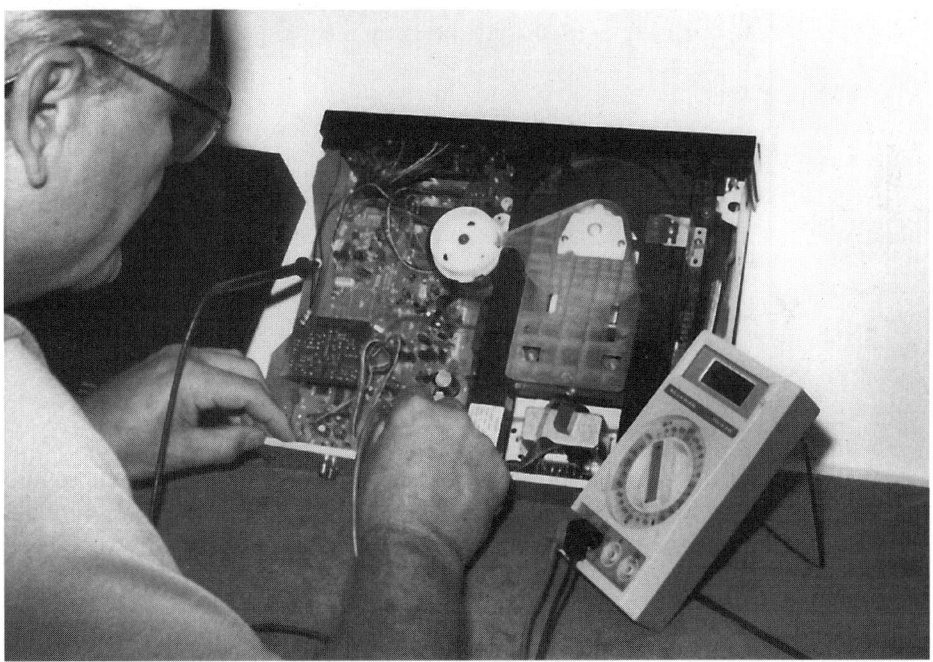

9-24 The electronics technician testing a regulator circuit in the low-voltage power supply of a CD player.

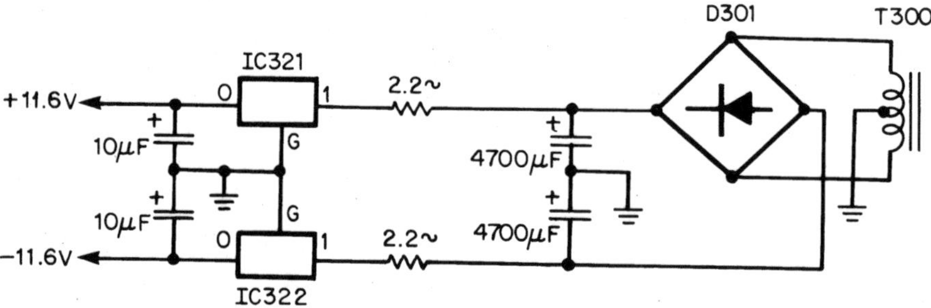

9-25 You might find IC voltage regulators in the low-voltage circuits of a cassette, CD player, and deluxe receiver.

voltage of another IC regulator in either the positive or negative source. You may find distortion in one or both audio channels with a leaky IC regulator.

Power line IC regulator (TV)

What and where

The power line IC regulator might appear as a power output transistor or a flat-type transistor. You can identify a power line regulator with four or five separate terminals (Fig. 9-26). The power output (IC-106) regulator IC has a common ground terminal in the middle with a base and output terminal on each side. The metal case is the input terminal and must be insulated from a metal heat sink. Locate the power line IC regulator after the main filter capacitor or bridge rectifier circuits.

The power line IC regulator operates in the power line voltage source with a fixed output voltage. The early transistor and flat IC regulators were large and bulky compared with the recent power line regulators. These flat transistor-type regulators are mounted on a separate heat sink. Notice that this transistor-looking power line regulator has five terminals (Fig. 9-27). You can identify the IC power line regulator from a vertical output IC by the terminals, letters, and numbers stamped on the component.

Notice that the last three numbers are the fixed output voltage of the regulator. Usually, the flat line regulators start with a STR lettering with output numbers as voltage. An STR30135 IC regulator has a regulated 135-voltage source.

When

The defective fixed IC regulator might become shorted, leaky, or appear open. Very low or no voltage output is noted with an open IC regulator. Very high output voltage might occur with a shorted IC. When lightning damage hits the TV, you may find damaged PC wiring, silicon diodes, and line voltage regulators. Sometimes the IC regulator might have blown sections with power outage or lightning damage. The chassis might come on and then shut down at once with a defective line regulator IC. A shorted or leaky IC regulator might blow the main power line fuse.

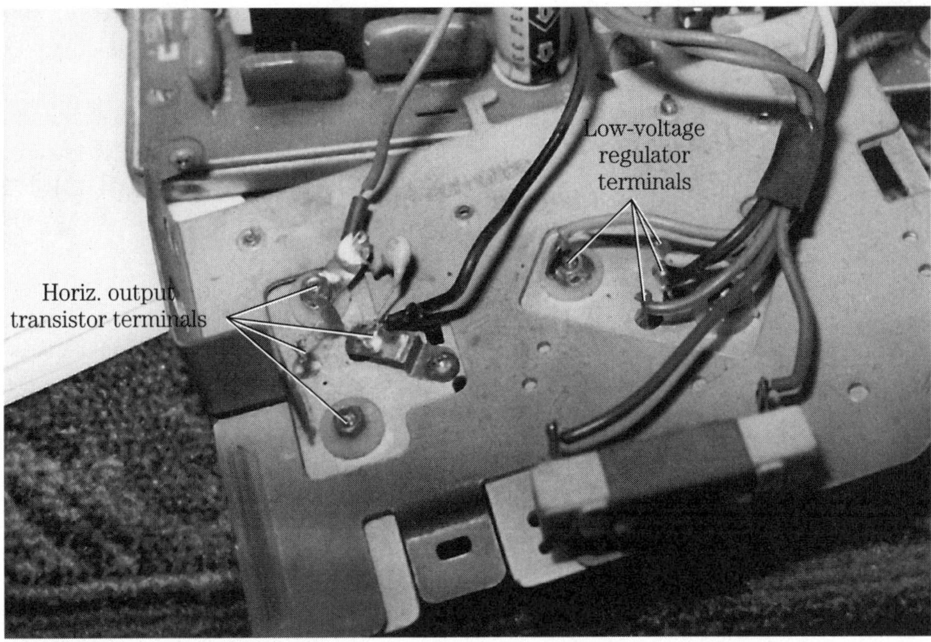

9-26 The early power line regulator resembled a power output transistor, except the regulator had four terminals.

9-27 The last three numbers on today's power line regulator STR30135 equals 135 volts.

Testing

Remove the horizontal output transistor or B+ fuse when the fuse keeps blowing or the chassis shuts down. Measure the dc line output voltage. Check the input voltage applied to the power line regulator (Fig. 9-28). Suspect a defective fixed IC regulator when the voltage is higher than normal, very low, or there is no dc voltage. Inspect all components tied to the power line regulator terminals. Check all resistors and diodes within the regulator circuits. Sometimes a separate fuse is found at the input terminal. This fuse should open when the IC regulator becomes leaky or an overload occurs in the dc power line source.

Transistor regulators

What and where

A voltage regulator holds the voltage constant during variations of load or input voltage. The transistor regulators are found in low-voltage sources of the low-voltage power supply and flyback secondary circuits of the TV chassis. One or two T0-220 low-voltage regulators are found in radio receivers, CDs, VCRs, and cassette players. This flat-mounted voltage regulator is often mounted on a small heat sink (Fig. 9-29).

You may find several transistor voltage regulators within the flyback secondary circuits of the TV chassis. These regulated voltage sources feed critical TV circuits such as the tuner, control microprocessor, and standby circuits. The transistor regulator might have a combination of transistor and zener diode within the regulation circuits (Fig. 9-30).

Q4012 provides a regulated 12-volt source, while a regulated 5-volt source is found at the emitter terminal of Q4011. Zener diode D402 provides 5.6 volts at the

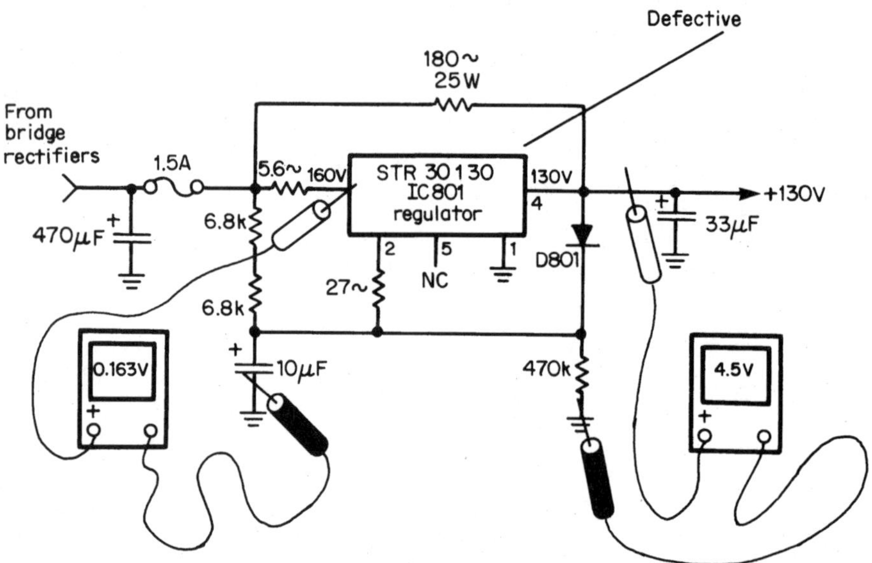

9-28 Test the power line regulator with correct input and output voltage tests.

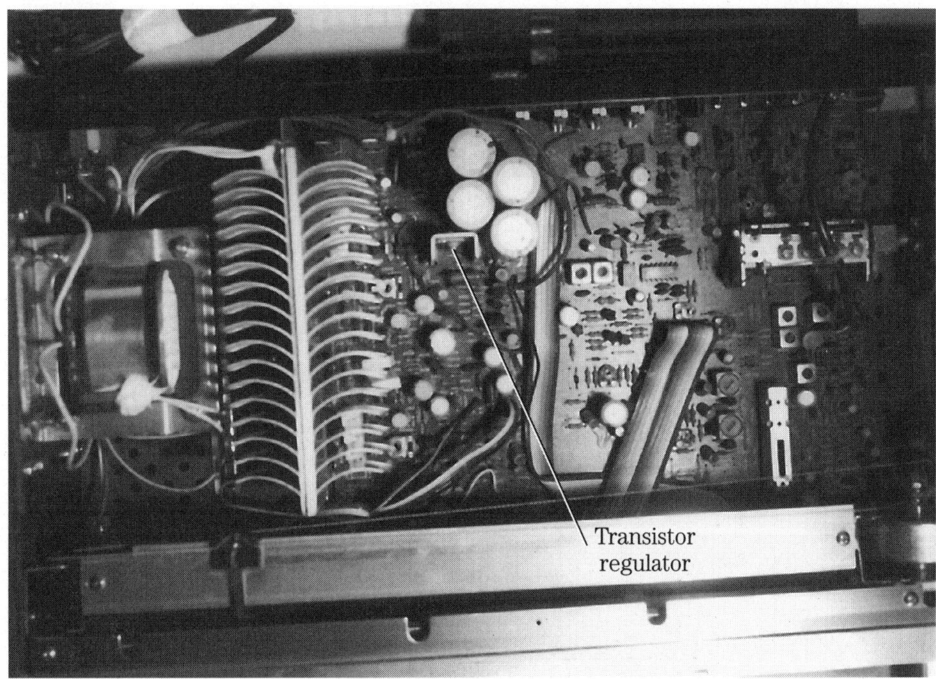

9-29 The transistor voltage regulator within the large receiver is mounted on a small heat sink.

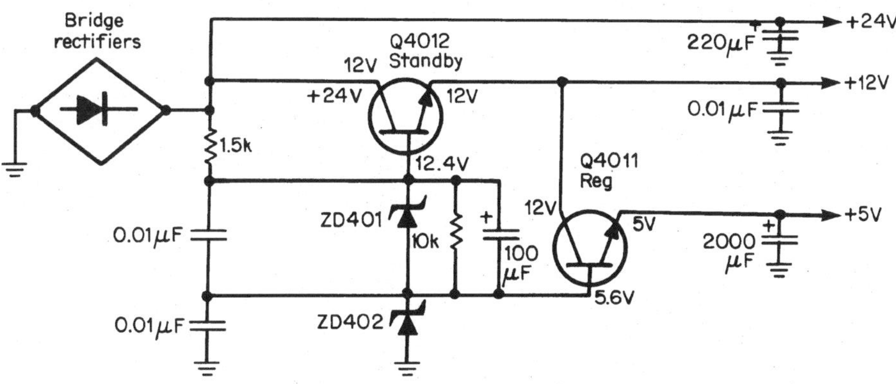

9-30 The scan-driven secondary voltage of the flyback might contain a transistor and zener-diode regulators.

base of Q4011. The 12.4 volts found at the base terminal of Q4012 is the combination of D401 and D402. A change of value in the 1.5K-ohm resistor, leaky Q4012, and shorted or leaky D401 and D402 might produce a dead chassis. A leaky D402 might result in cycling off and on of the TV chassis. An open Q4012 and Q4011 regulator transistor can result in no 12- or 5-volt sources. Intermittent shutdown might result from an intermittent D402. Locate the defective transistor regulator mounted on a large heat sink within the TV chassis (Fig. 9-31).

Testing

After locating the suspected voltage regulator transistor, take critical voltage measurements on each terminal. An open regulator transistor will have no or very little output voltage. The shorted or leaky transistor might have a lower dc voltage and quite similar voltage on all three terminals. When the chassis becomes intermittent, monitor the emitter terminal with the voltage range of the DMM.

Check the regulator transistor and zener diode with in-circuit transistor tests with the diode-junction test of the digital multimeter or transistor tester. The intermittent transistor might restore to a normal measurement when touched by test prods. Double-check all components within the regulator circuits when a regulator transistor has a high-leakage or shorted condition. Carefully remove the transistor and take another out-of-circuit test. Replace any regulator transistor that shows erratic or intermittent measurements.

Zener diode regulators

What and where

A zener-diode voltage regulator is a simple regulator whose output is a constant voltage drop across the zener diode in reverse breakdown. Zener diode regulators might be found within the base circuits of a transistor voltage regulator or in a separate regulator circuit. Within the combination transistor and diode regulator circuit, the zener diode provides a fixed voltage. Zener-diode regulators are found through the dc low-voltage power supplies in many different consumer electronic products (Fig. 9-32).

Often the defective zener diode will become shorted or leaky. Burned or charred signs might be found on the body of the zener regulator. Check for a damaged zener or a change of resistance if found in the same circuit of a regulator transistor. White marks around the diode might indicate overheating.

Testing

Check the zener diode with in-circuit tests of the DMM. Check the zener diode as a fixed diode with the diode test of the DMM. Remove one lead of the zener diode for accurate tests. Critical voltage measurement across the diode can indicate if the diode is shorted or leaky. When found in the base circuit of a voltage transistor regulator, take critical voltage measurements and test the transistor with a transistor tester. Both components can run warm when either one is found leaky (Fig. 9-33).

Relays

There are many different relays found in the consumer electronics field. The microwave oven has a power and oven relay. The new TV chassis might have a degaussing

9-31 The low-voltage regulator is found on a large heat sink within the TV chassis.

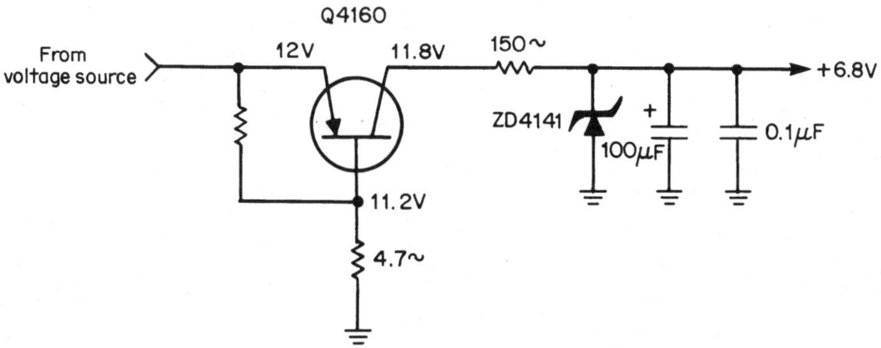

9-32 The zener diode regulates and reduces the dc voltage from 11.8 volts to +6.8 volts.

and power relay. The 12-volt SPDT relay might be used in car stereos and alarm systems, while general-purpose relays might be used in telecommunications, security, and power switching. Usually, only one relay is located within the TV chassis. The relay is a signal-actuated switching component. The defective relay might hang up and not turn off or on, contain dirty switching contacts, and produce intermittent switching. A defective relay amplifier might cause the relay to come on without a signal or keep the relay in off position.

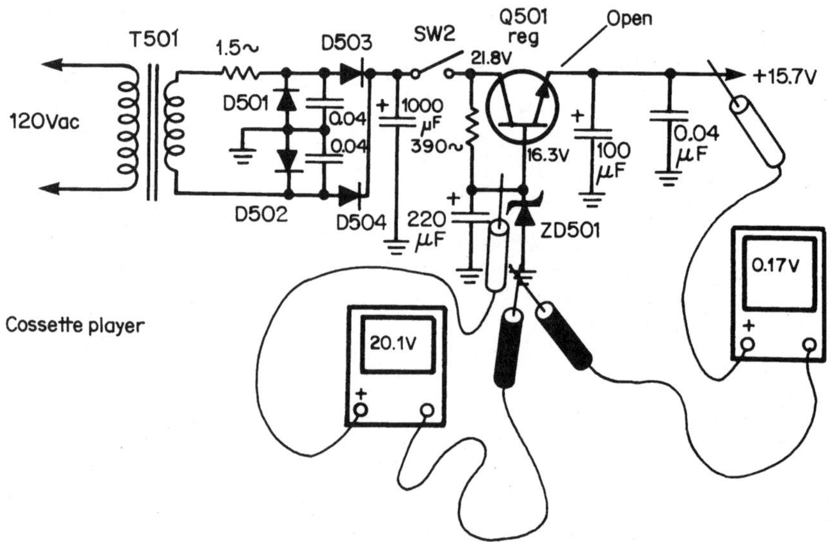

9-33 An open transistor regulator (Q501) has practically no output voltage and higher-than-normal voltage on the collector and emitter terminals.

Ac-dc relay

What and where

The ac-dc relay operates in either an alternating or direct-current circuit. These relays often turn power off and on to the various electronic products. The ac-dc relay might operate in SPST or DPDT switching circuits. The power relays might have a nominal voltage of 12 Vac–12Vdc, 24 Vac–24 Vdc, 120 Vac–12 Vdc, and 240 Vac. These power relays might appear in IC, PCB, and standard mounted relays.

Testing

Suspect dirty contact points with intermittent turn-on-and-off operations. Check the ac or dc voltage applied to the relay solenoid winding (Fig. 9-34). If signal or power voltage is found at the solenoid terminals and no relay points are energized, suspect an open solenoid winding. Check the solenoid resistance with the DMM. Since relays have different resistance in different switching circuits, check the coil winding given on the schematic. Often, a continuity resistance test of the coil winding might indicate the relay should close contacts. Place a screwdriver blade next to the solenoid winding and notice the pull of the metal blade when the relay is turned off and on.

Degaussing relay (TV)

What and where

The degaussing relay within the TV chassis turns the degaussing coil on and shuts the relay down after turn on. The degaussing relay is controlled by a transistor degauss switch or amplifier from a controlled signal of the main control microprocessor (Fig. 9-35). Sometimes the degaussing switch might also operate in conjunction with the kill switch of the vertical sawtooth signal in other TVs.

When the remote control turns on the micom or control processor, the degaussing switch provides power line voltage to the degaussing coil. The 120 Vac is applied across a low-resistance thermistor that limits the ac voltage to the degaussing coil. The thermistor increases in resistance as it heats up and shuts off the ac voltage to the degaussing coil. Locate the degaussing coil relay contacts within the degaussing circuits.

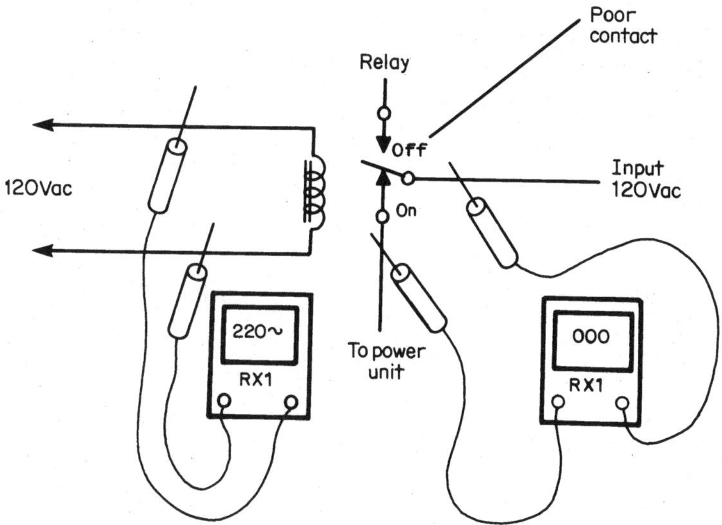

9-34 Check the resistance of the solenoid and switch contacts when the relay fails to function.

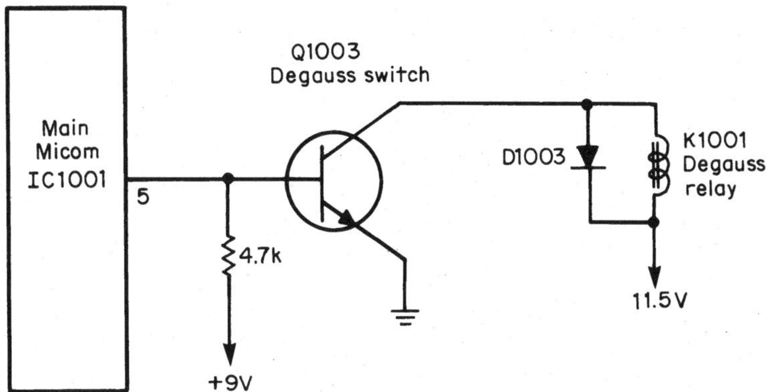

9-35 In the latest TV chassis, the main control IC turns on the degaussing circuits with a transistor amplifier and relay.

Testing

Sometimes one can hear the relay click off and on. A defective relay or no voltage applied to the degaussing coil can leave impurities or different colors on the CRT. Intermittent degaussing can be caused by an intermittent degauss transistor switch, dirty relay controls, and intermittent signal from the main control IC.

A quick ac voltage test across the degaussing coil can tell if the degaussing coil, thermistor, or relay points are defective (Fig. 9-36). When the power line ac voltage is measured across the degauss switch contacts or thermistor, that component is open and defective. Suspect an open degaussing coil when 120 Vac can be measured across the coil winding and there is no action. Take a quick voltage test across the solenoid winding of the relay with open relay contacts. If voltage is present and there is no action, suspect an open solenoid winding. Check for broken or dirty contacts when the relay is energized and no ac voltage is applied to the degaussing coil.

Power relay (TV)

What and where

The power relay within the TV chassis turns on the TV with a remote control standby system. The remote transmitter turns on a decoder or control IC, which in turn applies voltage and signal to a power transistor amplifier and relay (Fig. 9-37). The relay switching points are found in one side of the ac power line to apply voltage to low-voltage rectifiers or a bridge circuit.

When

Suspect a defective control IC, relay transistor, relay, and improper standby voltage when the remote transmitter will not turn on the TV. It has been known for these power relay circuits to come on without anyone operating the remote transmitter. Peek at the relay and notice if the points are energizing. If not, check the standby voltage power supply circuits. Often, the standby voltage is supplied to the control IC and relay circuits.

Testing

Measure the voltage applied to the control IC and relay. Remove one terminal of the fixed diode across the solenoid winding. Check continuity across the coil winding. Measure the voltage applied across the solenoid. Test the fixed diode with the diode test of the DMM. Sometimes these diodes become leaky and will not let the relay energize. Check for the power line voltage across the switch terminals with the DMM or VOM (Fig. 9-38).

General-purpose relay

What and where

The general-purpose relay might have SPST, DPDT, and 4PDT switching. These relays might plug into a socket, or be surface-mounted or PC-mounted relays. The solenoid may operate from a 5- to 120-volt source. The solenoid resistance is from 40 to 4500 ohms.

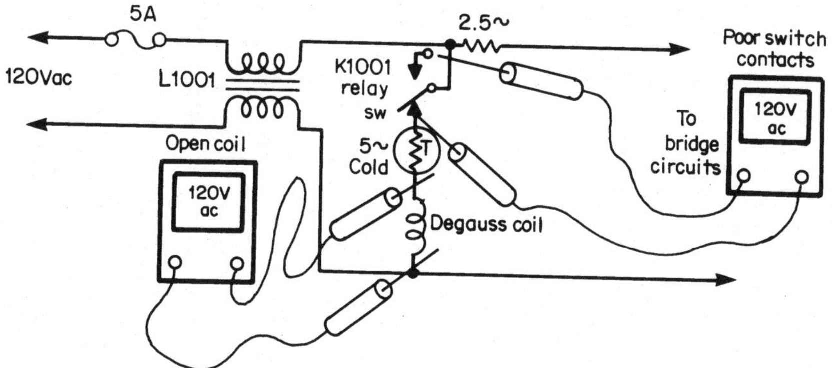

9-36 Take a quick ac voltage test across the degaussing coil and relay to determine what components are defective.

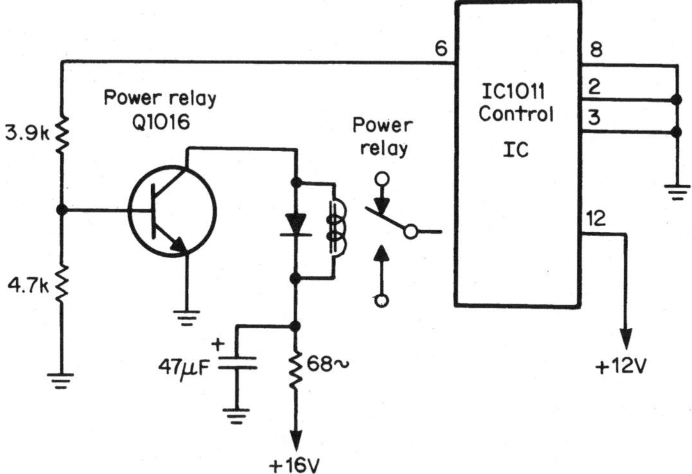

9-37 The power relay turns on the TV with control from IC1011 and the remote transmitter.

Testing

Sometimes it is difficult to find the solenoid and switching terminals of a relay on a PC board. It's best to remove the relay from the PC wiring with solder wick and a heavy-duty soldering iron. The solenoid wires can be traced to each pin contact terminal. Take the continuity of the relay solenoid winding with the RX10-ohm scale of the DMM or VOM. You may assume the winding is normal if a continuity measurement is obtained. Remove the plastic top container and depress the switching contacts. Take a low-ohm measurement between a set of switching contacts to determine if the points are burned or dirty.

Some relays have a diagram layout with two spade terminals at different directions than the others, indicating the solenoid winding. The next two bottom-spade

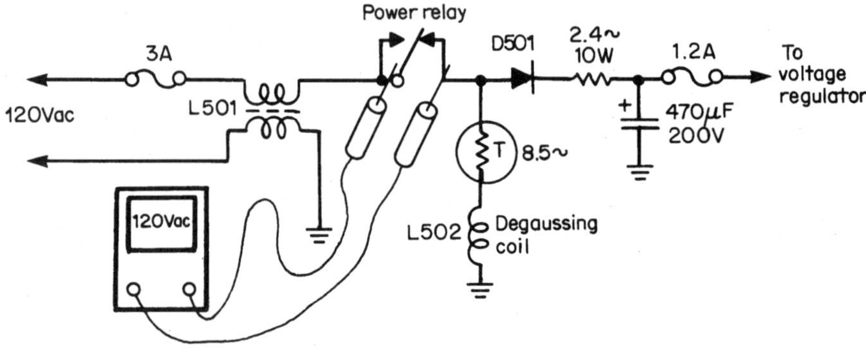

9-38 Measure the ac voltage across the dirty relay points to check for a poor set of contacts.

terminals equal the switching terminals to a DPDT pair of contacts (Fig. 9-39). By measuring the voltage applied across the solenoid winding and taking the continuity of the solenoid, you can locate a defective relay.

Microwave oven relay

What and where

The deluxe microwave oven might have a cooking or oven relay, variable power switching relay, auxiliary relay, browner relay, main relay, damper relay, and convection relay (Fig. 9-40). Although the average ovens do not have more than three different relays, several relays listed might do the same job. For instance, within the GE JEBC200 oven, there are five different relays: The damper relay closes and opens the damper door during convection cooking (Fig. 9-41). A heater relay provides power to the heater coils during convection cooking. The main relay provides power to the high-voltage section, while the power relay provides 120 Vac to the main heater and power transformer. The convection relay turns on the convection motor during convection cooking. A stirrer motor rotates and stirs the RF energy at the bottom of the oven to provide even cooking with no dead spots.

In some ovens, the cook, main, or power relay might serve as only one relay to provide 120 Vac to the power transformer and high-voltage circuits. The browning and heater relays serve the same purpose—to turn on the cooking heater circuit for convection cooking. The oven or power relay is found in place of a triac in other ovens (Fig. 9-42). A variable-power switching relay is located with the HV section to control the on/off time of the magnetron in the secondary circuit.

The advantage of this system is that it results in less stress to the components in the HV circuit. It switches the circuit when no high voltage is applied, so no transient currents are generated. The switching device, however, requires a higher grade of insulation because the device is located in the HV circuit.

When

When no high voltage is found at the magnetron, there is no cooking action, no heater cooking, and no browning of food, suspect a defective relay. Line voltage is applied to the primary winding of the HV transformer by a power, main, or oven relay. Some ovens might call the relay a power relay, as power is applied to the power transformer. Often, you can hear the oven relay energize. When 120-volt ac is not applied to the heating element in convection cooking or browning of food, suspect a heater or browning relay. Remember, the control panel provides voltage to each solenoid winding.

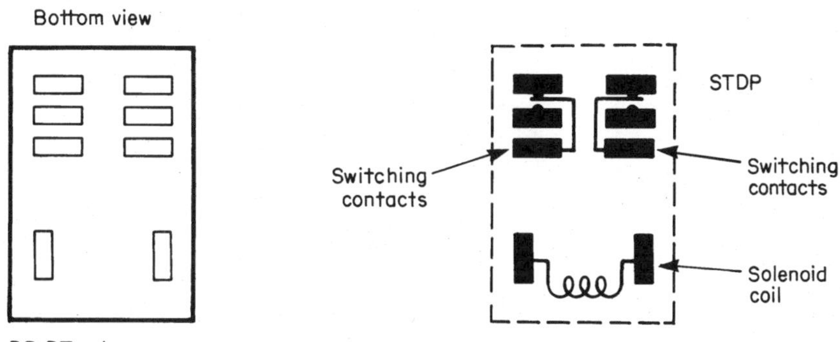

9-39 The bottom view and relay contact points of a general-purpose relay with the solenoid coil.

9-40 You might find one or more relays mounted in today's microwave oven.

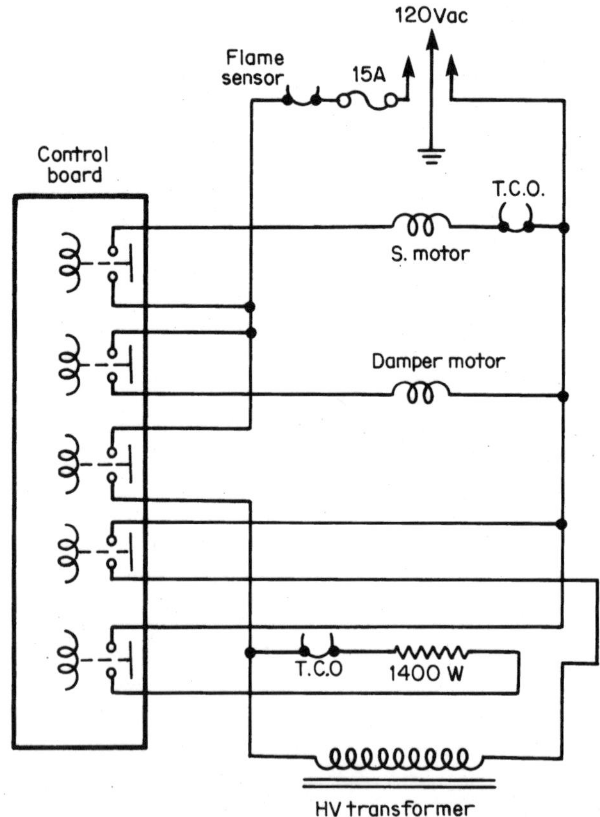

9-41 Five different relays are found in the GE JEBC200 microwave oven.

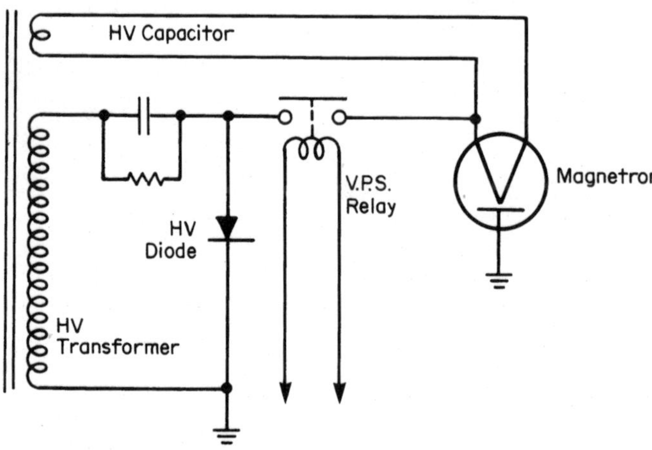

9-42 A variable-power switching relay switches the HV off and on during the cooking process.

Testing

The cook or power relay can be checked with the ohmmeter range of the VOM or DMM. Discharge the HV capacitor before taking any kind of tests. With the power off, locate the solenoid coil-winding terminal and check for continuity (Fig. 9-43). Most oven relays have a resistance between 150 and 300 ohms. If in doubt, remove the external cable leads from the solenoid terminals. Then take another resistance measurement. Of course, if the coil is open, the relay must be replaced.

A continuity check between the switch terminals should indicate an infinite resistance. Only if the coil is energized should you measure a short across the switch terminals. A continuity check between each switch terminal and ground should indicate an infinite resistance. The switch terminals can be checked by removing the top cover and using an insulating tool to push the solenoid together. Very little resistance is measured between normal closed-switch contact.

In case voltage is not reaching the solenoid terminals, the oven relay will not operate. Some oven relays operate from the ac power line, while those with a control-board assembly operate from 10 to 24 Vdc. The voltage can be checked at the solenoid terminals when the oven will not turn on. Always clip the meter to these terminals. First determine if the voltage is ac or dc. Low or no voltage at the relay indicates a defective control-board circuit. Correct voltage at the solenoid terminal indicates an open or defective relay solenoid.

When everything else has been checked and the relay will not energize, remove the solenoid terminal wires. Mark where each wire goes before removing them. Now apply 5 to 14 Vdc from a bench power supply to the low-voltage relay solenoid. You should hear the relay energize. Suspect a defective solenoid when the relay will not

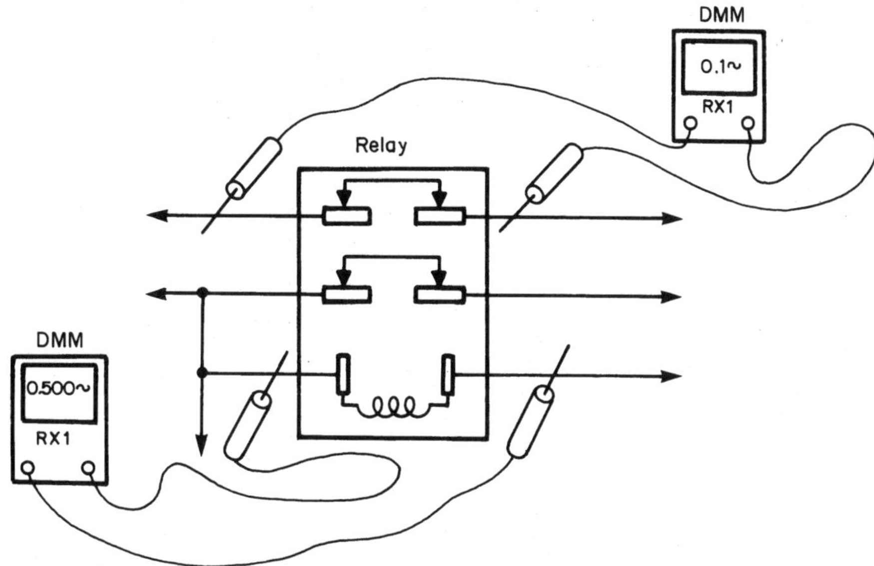

9-43 Check the continuity of the power or cook relay and switching contacts with accurate resistance tests.

close contacts. Inspect controls for burned areas. Replace the relay if it has burned or pitted contacts.

The main or power relay might have a resistance of 150 to 650 ohms. An auxiliary relay solenoid might have a resistance of 1.2K ohms. You can assume the solenoid is good when continuity is measured across the coil terminals.

PC-mounted relays

What and where

The PC relay mounts on top of the printed circuit board and is soldered directly to the PC wiring. PC relays come in micro-miniature and miniature. These PC relays operate in communications, networking in security, power switching, instrumentation, and control interface equipment. The printed circuit relays are now found directly on the control board of the microwave ovens. You may find a SPDT, DPDT, or a deluxe PC relay with a 4PDT, 6PDT, and 8PDT switching contacts (Fig. 9-44).

The micro-miniature PC relay solenoid might operate from 5 to 24 volts with a 75-, 440-, or 1550-ohm coil resistance. The gold-clad silver palladium contacts might be rated at 2 amps. The miniature printed circuit relay might operate from a 5- , 6-, 12-, and 24-volt dc source with a coil resistance of 47, 68, 155, 275, 660, and 1100 ohms. The switch contacts are made up of silver alloy and silver cadmium oxide.

Testing

Measure the voltage applied across the solenoid coil (5 to 24 volts). Suspect the voltage control panel with no or low voltage. Check the continuity resistance of the coil (47 to 1100 ohms). Measure the resistance across each switching contact for poor contacts. If the resistance is high or erratic, suspect a burned or dirty switch contact.

Reed-type relays

What and where

The reed switch is a frequency-sensitive type in which the movable contact on a thin metal strip is activated by a coil. You might find an experimental reed switch that operates inside a glass tube with a magnet moved up to the switch that closes the switch contacts. An early-type reed switch was found in the remote on/off circuits in the early TV chassis (Fig. 9-45). The reed-type switch appears in a PC board, SIP, DIP, and SMD mounting.

The reed-type switch might have SPST and DPDT switching contacts. The normal coil-applied voltage is 5, 12, and 24 volts. The solenoid resistance is from 200 to 3000 ohms, depending on what circuit the relay operates in.

The magnetic reed switch or sensor consists of two ferromagnetic reeds hermetically sealed in a gas-filled capsule or glass tubing. When these reeds are introduced to a magnetic field, they attract each other and close the switch contacts. When a magnet is pulled away from the switch contacts, they will open up (Fig. 9-46). These reed switches are used in electronic projects where moisture is a problem. They operate freely in airy, watery, oily, and dusty environments.

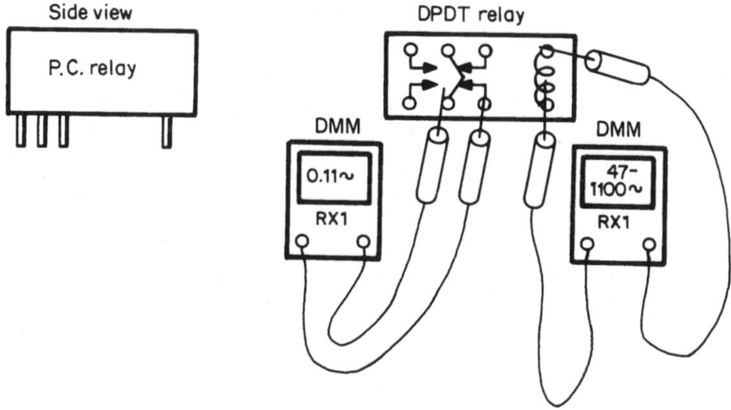

9-44 Check the resistance of the solenoid and switching contacts on the PC-mounted relay.

9-45 The pen points at a reed-type relay found in the early RCA TV chassis of the low-voltage power supply.

Testing

Check the switch contacts with the RX1 scale of the digital multimeter. Measure the ac or dc voltage applied across the solenoid terminals. Take a continuity resistance measurement of the solenoid coil winding. Replace the magnetic reed switch when the contacts will not close or open with a stronger magnet applied.

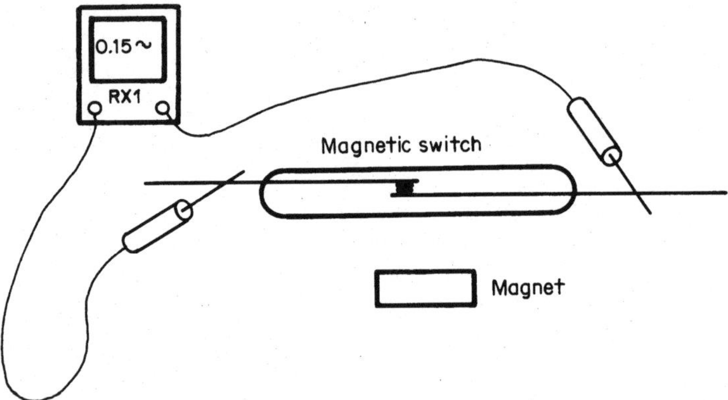

9-46 A reed-type switch might consist of a magnetic thermal switch contact and external magnet.

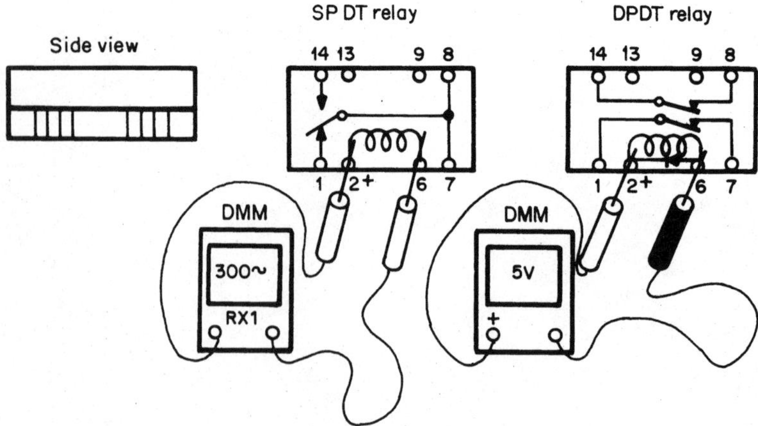

9-47 The surface-mounted relay mounts on the PC wiring. Notice the fixed diode across the DPDT solenoid winding.

Solid-state relay

What and where

A solid-state relay might have a complete electronic relay without any moving parts. The relay might consist of switching transistors, thyristor, triac, and SCRs. The solid-state relay might provide high voltage and high-current switching for motors, transformers, and heating elements, etc. The high-current relays might have a heavy-duty heat sink with ac or dc input voltage.

Testing

Replace the solid-state relay when input voltage is found and there is no relay switching action. Most solid-state relays are sealed, and some are molded within a PC board mounting.

Surface-mounted relays (SMD)

What and where

Surface-mounted relays might appear in DIP reed packages. The SPDT or DPST switches form a normally open or latching action. The normal solenoid voltage might be from 5 to 12 volts, with a coil resistance from 150 to 200 ohms. You may find in some SMD relays a fixed diode across the solenoid winding (Fig. 9-47).

Testing

Check the voltage applied to the solenoid. Measure the resistance of the coil winding. Test the contact point resistance with the low-ohm scale of a DMM. Replace the SMD relay if it has pitted points and an open solenoid.

10

CHAPTER

Testing electronic components R

(Resistors)

The symbol R, or ohm, represents resistance. A resistor in a circuit opposes the flow of current. Resistors are made up of carbon, carbon film, flameproof metal film, metal oxide, nonflammable metal, thick film, and vitreous enamel. Resistors come in high-wattage, wire-wound, and surface-mounted types. Resistors might appear in circuits with a fixed value, semifixed value, or variable resistance (known as potentiometers). (See Fig. 10-1).

Resistors and capacitors are the most common electronic components found in electronic products. Several resistors can be placed in series for added resistance or placed in parallel for less resistance. The defective resistor can be replaced with a higher wattage if the correct wattage is not handy or available. Replacement resistors can be purchased separately or in tape-and-reel packaging.

Auto dummy load resistors

What and where

The dummy load resistor is used instead of a speaker when making tests and adjustments on the auto receiver or amplifier. A dummy load consists of a wire-wound resistor that equals the output impedance of the amplifier. Most electronic technicians use high-wattage 4-, 8-, or 10-ohm, 10–150-watt wire-wound resistors. These noninductive audio dummy load resistors might be fixed or have a variable lug to vary the resistance (Fig. 10-2).

10-1 The latest TV chassis might contain carbon, metal film, wire-wound, and power resistors spread over the entire chassis.

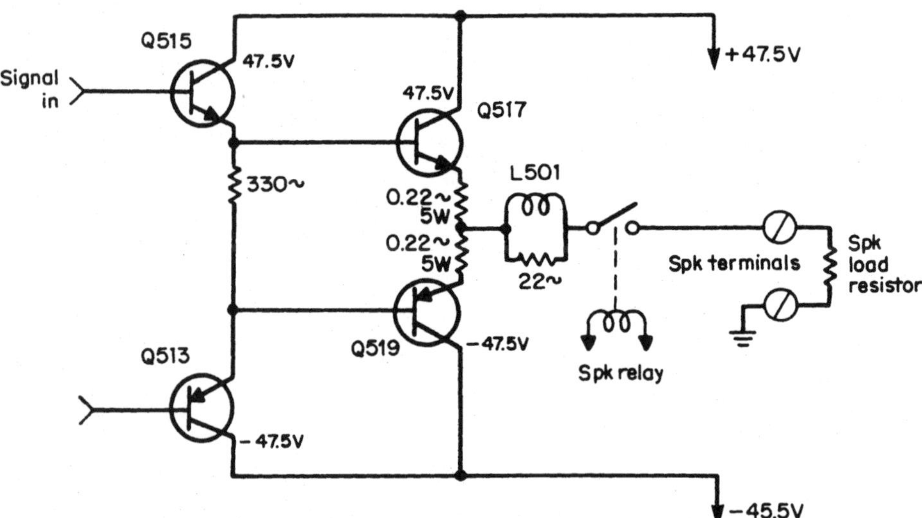

10-2 A load resistor is connected to the speaker output terminal instead of the speaker in alignment and troubleshooting procedures.

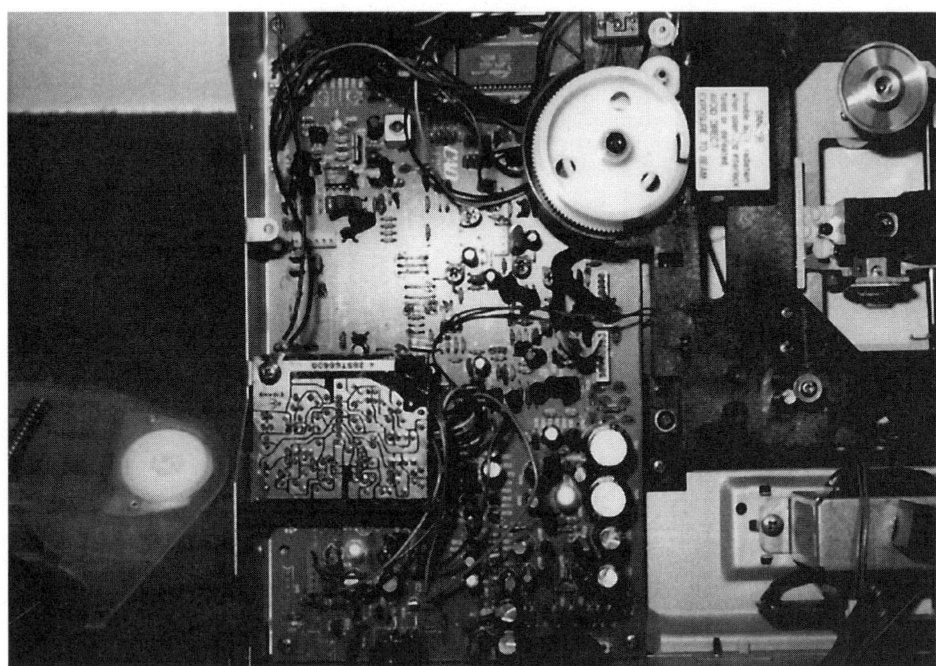

10-3 Many carbon resistors are mounted on the PCB chassis within the CD player.

Testing

Very little problems are found with high-wattage dummy load resistors. Check for correct resistance on the RX1 scale of the DMM. An open resistor has no resistance measurement. Look for poorly soldered connections.

Carbon resistors

What and where

Carbon resistors are found in just about every electronic product. The carbon resistor is made from carbon, graphite, or a composition of carbon material (Fig. 10-3). The early carbon resistor is now being replaced with carbon film resistors. Carbon resistors appear in $\frac{1}{8}$-, $\frac{1}{4}$-, $\frac{1}{2}$-, and 1-watt sizes at 5% and 10% tolerance.

Testing

The defective resistor might appear open, increased, or decreased in resistance and become charred or burned. The charred or burned resistor always decreases in resistance or might be burned to an open resistor. The cracked resistor might increase in value. Usually, resistors that have a higher ohm rating also have a tendency to increase in value. Check the suspected low-ohm resistors with the RX1 scale of the DMM. Remove one end of the resistor to test accurately with in-circuit tests (Fig. 10-4).

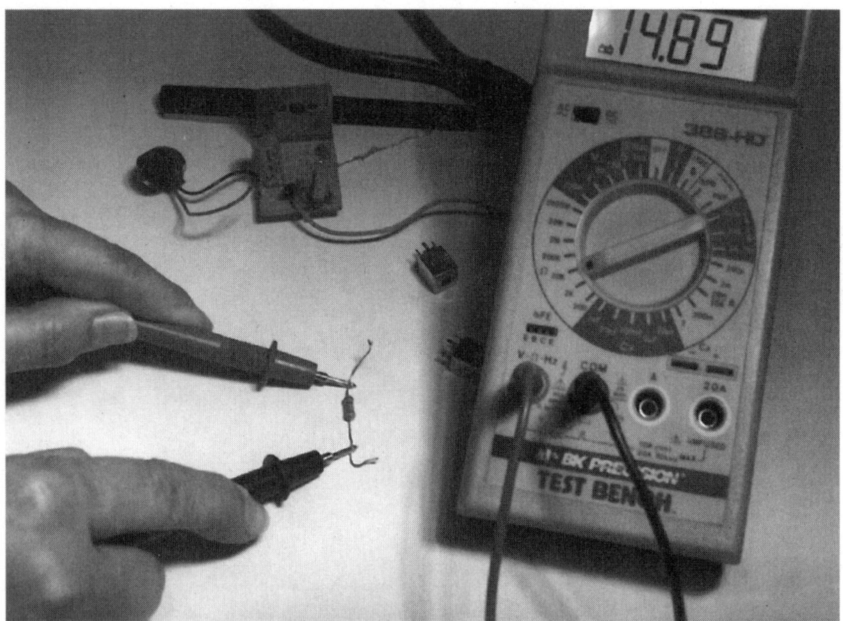

10-4 After locating a defective resistor, remove the suspected resistor out of the circuit and test with the DMM for correct resistance.

Carbon film resistor

What and where

The carbon film resistor has excellent reliability, solderability, low-noise, moisture-stable, and heat-stable characteristics. A carbon film resistor has the resistance film element deposited on a ceramic body or substrate. The carbon film resistors appear in $\frac{1}{8}$-, $\frac{1}{4}$-, and 1-watt sizes. Replacement of carbon film resistors begins at 1 ohm up to 100K ohms with a 5% tolerance. You will find carbon film resistors in most electronic products in the consumer electronics field.

Testing

Test all low-ohm resistors on the RX1 scale of a DMM for accurate measurements. Remove the suspected resistor from the circuit for accurate tests. Replace the defective resistor with exact ohm resistance and wattage. Never replace a 1-watt resistor with a $\frac{1}{8}$- or $\frac{1}{4}$-watt size. Most resistors are identified by a color-coded chart (Table 10-1).

Cermet power resistors

What and where

Cermet power resistors are sealed with a special cement, have excellent flame and moisture resistance and high-temperature stability, and appear in a ceramic flame-retardant package (Fig. 10-5). They might be found in high-powered amps

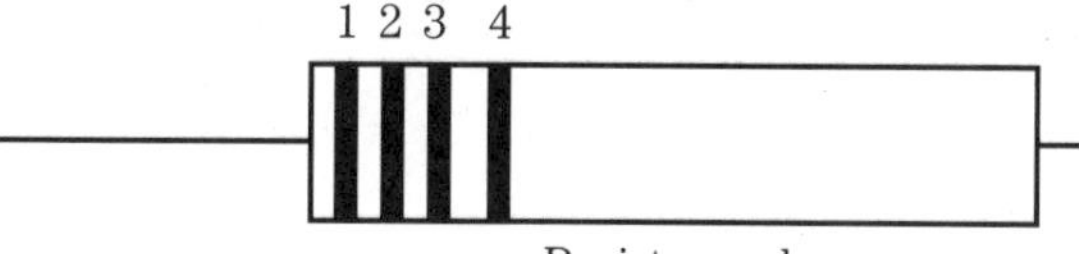

Table 10-1. The resistor color code chart

Color	Band 1	Band 2	Band 3	Band 4
Black	0	0	1	....
Brown	1	1	1	1%
Red	2	2	100	2%
Orange	3	3	1000	3%
Yellow	4	4	10000	4%
Green	5	5	100000	....
Blue	6	6	1000000	....
Violet	7	7	1000000	....
Gray	8	8	1000000000	....
White	9	9		
Gold				5%
Silver				10%
No color				20%

10-5 The cermet power-emitter resistors are found in the high-powered amps and receivers.

and TV sets. These cermet power replacement resistors might appear in 3-, 5-, and 10-watt resistors with a resistance from 0.1 to 25K ohms.

Testing

The cermet power resistors might be found burned open, blown apart, cracked, and with a change in value. Check for an open wire-wound resistor with a shorted high-powered output audio transistor. Remove one end for a correct resistance test on the low-ohm scale of the DMM. You might find that these resistors are less than one ohm when found in the emitter circuit of the audio output transistor or bias circuit of an IC.

Cermet resistor

(See Potentiometer)

Chip resistors (SMD)

What and where

Chip or surface-mounted resistors are thick film types that mount directly on the PC wiring (Fig. 10-6). The features of the chip resistor might include high-purity alumina substrate, wave or flow solderability, wrap-around termination, excellent high-frequency characteristics, tight temperature coefficient resistance, inner electrode protection, high-quality thick film element, excellent electrical stability, and reduced lead inductance. The chip resistor might appear in thick film chip, ultra low ohm chip, and metal film chip types.

The thick film resistors appear in the 0603/0805 and 1206 case sizes. The resistance might be from 1 to 10 megohms at $\frac{1}{10}$, $\frac{1}{8}$, and $\frac{1}{16}$ watts. You can purchase SMD replacements of 1% to 5% tolerance. These small chip SMD resistors are found in small electronic products such as cassette and CD players, AM-FM-MPX radios, small and large screen TV sets, and small electronic products. Replacement chip (SMD) resistors can be purchased in a 5- or 10-lot sack or on cardboard (Fig. 10-7).

Testing

Check the SMD resistance like any resistor. Test with the low-ohm scale of the DMM across the end terminals. Remember, a look-alike resistor might be a solid connecting chip with no resistance. Try to take accurate resistance measurement within the circuit. When the SMD resistor is removed, it should be replaced like all SMD parts. Discard the removed SMD resistor. Notice that the chip resistor and capacitor look somewhat alike. Check for a layout part location within the TV, radio, or sound system for the correct part number.

Fixed resistance

What and where

A fixed resistor is a nonadjustable resistor with a fixed value of resistance (Fig. 10-8). Resistances that are variable load, trim pots, and potentiometers are variable-type resistors. The fixed replaceable resistor might have a resistance 0.1 to 22 megohms. Fixed resistors are found in practically every type of electronic product in the consumer electronics field.

10-6 Chip or surface-mounted resistors are mounted on the PC wiring within portable CD players

10-7 The replacement surface-mounted (SMD) resistors might be mounted on a card or packed in plastic bags.

10-8 The fixed resistor has a set value and can be mounted through the PCB or on top of it within special constructed projects.

Testing

Check all fixed resistors within the circuit with the correct scale of the DMM. You will find that the digital multimeter and FET-VOM take accurate resistance measurements compared to the small VOM. The VOM takes quick continuity measurements, but not as accurate as the DMM on low-ohm resistors.

Flameproof resistors

What and where

The flameproof resistor is quite accurate in value and is placed in circuits where higher-than-normal current and voltage is found, causing the resistor to overheat. Flameproof resistor replacements might have a $\frac{1}{4}$-, $\frac{1}{2}$-, or 1-watt rating with a 1% or 5% tolerance. Flameproof resistors have a high stability and reliability and are coated with a flameproof covering. They are found in critical circuits of the TV chassis (Fig. 10-9).

You might find more flameproof resistance in the American TV chassis than foreign models. For instance, in a Goldstar CMT-2612 model, flameproof resistors are found in the emitter and bias circuits of the color output transistor. These resistors are 1% variety. Critical flameproof resistors are also found within the 120-volt power source at 5%.

In the RCA CTC166 chassis, flameproof resistors are found in the sound output voltage source, run voltage, B+ source, flyback voltage sources, yoke return resistors, vertical output voltage source, voltage regulator source, color output transistor circuits, and high-voltage shutdown circuits (Fig. 10-10). Many of these resistors have 1%, 2%, and 5% tolerances. Most of these flameproof resistors are listed as safety components. Replace with only flameproof resistors.

10-9 Several flameproof resistors are found on the TV chassis.

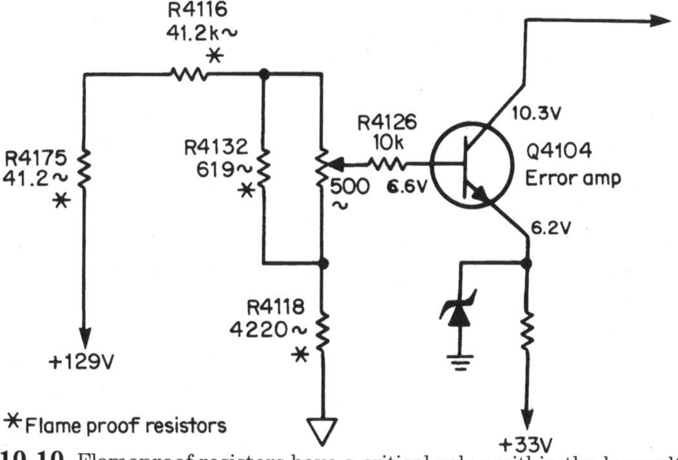

10-10 Flameproof resistors have a critical value within the low-voltage error amp circuits of the TV power supply.

Testing

Check for correct resistance with the DMM. Remove one end for accurate measurement when combined with other resistors or coils in a circuit. For accuracy and avoiding overheating and fire, replace with flameproof resistors and not carbon resistors.

Fusible resistors

What and where

Fusible resistors are placed in circuits for fuse protection and have a very low ohm rating. The fusible resistor might be called a surge resistor. These resistors are found within the low-voltage power line as isolation or fuse resistors of the black-and-white and color TV chassis (Fig. 10-11). Fusible resistors are also found in appliances, the low-voltage source feeding the horizontal output transistor, low-voltage sources, and within high-powered amplifier circuits. The early fusible resistor is covered with a fine-sand coating.

Testing

The defective fusible resistor runs very hot and sometimes cracks open the sand-coated component wound with nichrome wire. Fusible resistors might be wound on a ceramic form. Check the fusible resistor for accurate resistance with the RX1 range of the DMM. A R311 (1-ohm) fusible resistor protects the vertical circuits within the flyback secondary voltage source of the TV chassis (Fig. 10-12).

Large-wattage resistors

What and where

Usually, large-wattage resistors are wire-wound type resistors. These replacement resistors come in 2-, 3-, 5-, 7-, 10-, 15-, and 25-watt components. A wire-wound resistor is made up of nichrome or manganin wire and is wound on a round, ceramic, or card form. Some large-wattage resistors are silicone and vitreous enamel coated, while others are sealed with a special cement within a ceramic package (Fig. 10-13).

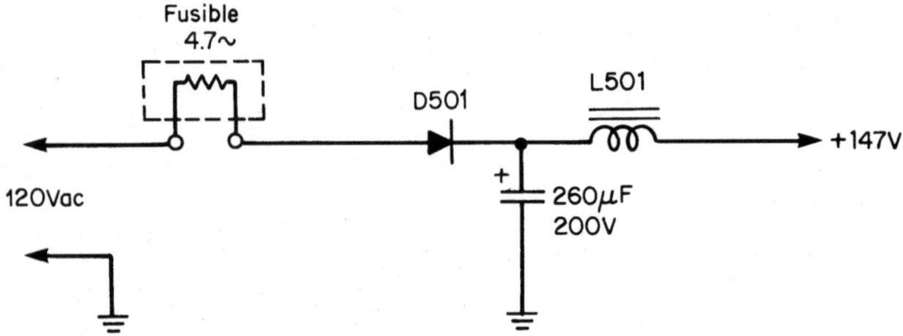

10-11 Fusible resistors were used in place of a fuse in the early black-and-white TV chassis.

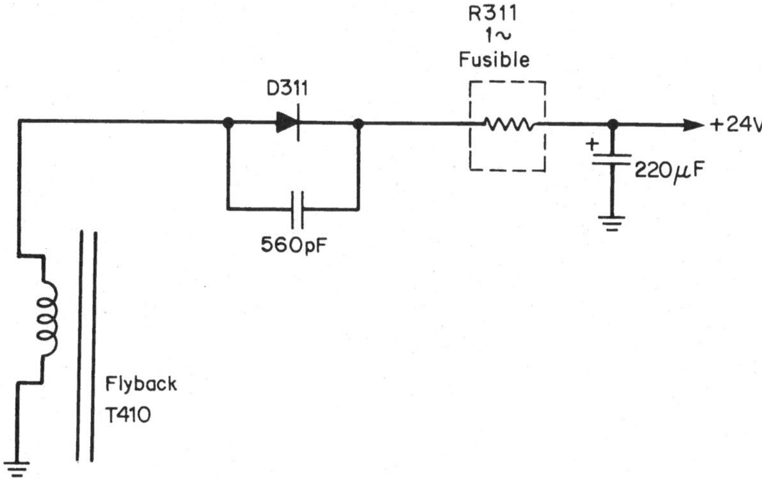

10-12 A R311 (1-ohm) resistor is a fusible resistor protecting the vertical components within the 24-volt source.

10-13 Large-wattage resistors are found in the low-voltage power supply and horizontal circuits of the TV.

You might find some high-wattage resistors (75 to 300 watts) placed within the anodized aluminum housing. These are designed for chassis or heat-sink mounting for maximum heat dissipation. The resistant element is made from copper/nickel or nickel/chrome alloy.

Large-wattage resistors up to 25 watts are found in the low-voltage power supplies in isolation and voltage regulation circuits of the TV chassis. One or more high-wattage resistors might serve as audio dummy loads. The defective high-wattage resistor might be found cracked, broken, or burned open. One or more shorted silicon diodes, transistor voltage regulators, and leaky horizontal output transistors might destroy or open up the wire-wound resistors (Fig. 10-14). These resistors run quite warm and should be mounted at least 1 inch above the PCB TV chassis.

Testing

Inspect the ceramic cover for cracks or broken resistors and burn marks. Peek around the terminal leads for PC wiring, burn marks, or signs of overheating. Sometimes removing excess solder, cleaning up solder tabs, and resoldering resistor terminal leads might solve the dead chassis. Check the resistor for correct resistance with the RX1 scale of the DMM.

Metal film resistors

What and where

A metal film resistor contains a resistance made of thin or thick film metal alloy placed on a plastic or ceramic body (Fig. 10-15). Metal film resistors are quite accurate in value, with a 1% and 5% tolerance. These resistors are less sensitive to temperature change and less noisy. The metal film resistors are manufactured using a vacuum sputtering system to deposit multiple layers of mixed metals into a carefully treated ceramic substrate. The metal oxide resistors have a high surge and overload capability, and are noninductive, with a special lacquer coating to resist trichlorethelyne, freon,

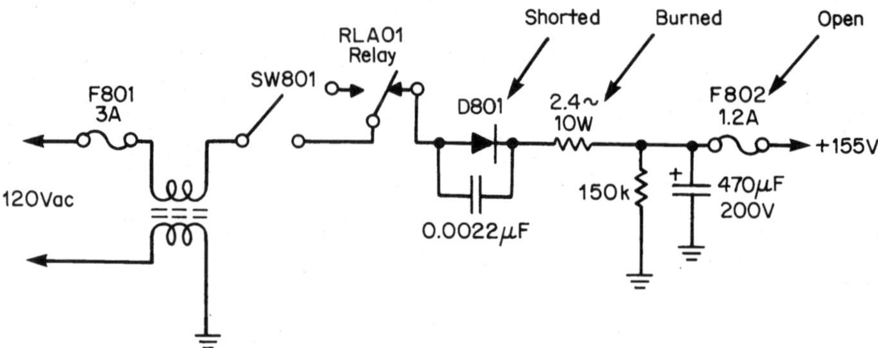

10-14 A shorted horizontal output transistor opened up the 1.2-amp fuse, 2.4-ohm 10-watt power resistor, and shorted the half-wave rectifier (D801).

and other cleaning agents. The metal film and metal oxide resistors are from the same family.

These flameproof oxide resistors are found in the low-voltage, voltage regulator sources, transistor color output, and special scan-derived voltage circuits of the TV chassis (Fig. 10-16). You might find some form of metal film resistors in high-powered auto and home receiver-amplifiers.

10-15 Metal film resistors are found mounted on the CRT board within the TV color output transistor circuits.

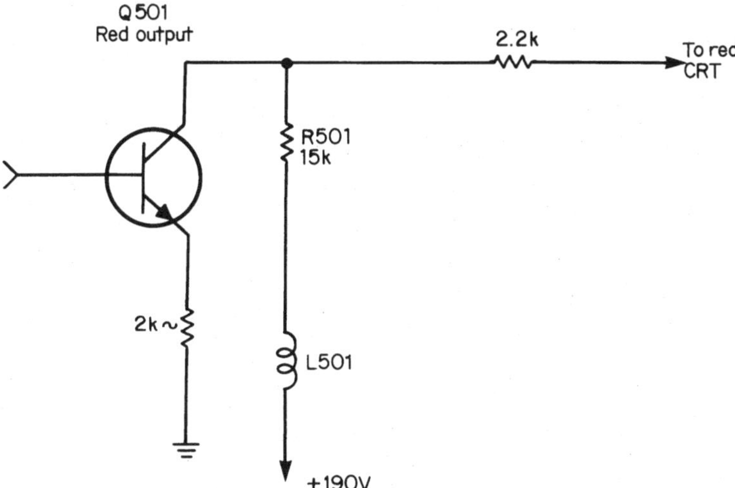

10-16 R501, a (15K-ohm) resistor, is a metal film resistor serving as the plate load in the red output transistor circuit.

Testing

A burned or cracked body will really show up in the metal oxide resistor. Test each transistor within the circuit with the ohmmeter range of the DMM for accurate resistance measurements. Take a closer peek with the magnifying light at the terminal connections of suspected resistors. Brown or burned marks around the terminal board connections indicate the resistor has been operating quite warm. Clean up the terminal pads for a good soldered connection. If the terminals are corroded, extend the resistor lead to the next component-soldered connection.

Metal oxide resistors

What and where

Metal oxide film resistors cost less than carbon and general-purpose metal film resistors. Metal oxide film resistors can replace many low-power, general-purpose wire-wound resistors. These metal oxide resistors can meet overloaded tests without producing a fire hazard. The metal oxide resistor has a high surge overload capability.

The metal oxide resistors are small in size and have a higher power rating at the same size as a regular resistor. The main feature of a metal oxide resistor is that it is made of flame-retardant, insulated material and will never flame or arcover (Fig. 10-17). Metal oxide resistors are found in electronic circuits that might overheat, such as the TV chassis. These resistors are also found in emitter circuits of high-powered audio output transistors. They are replaceable in 1- and 2-watt types from 0.47 to 1 megohm.

Testing

Check for correct resistance with the DMM. Remove one end from the PC board for correct measurement. Check for cracked areas.

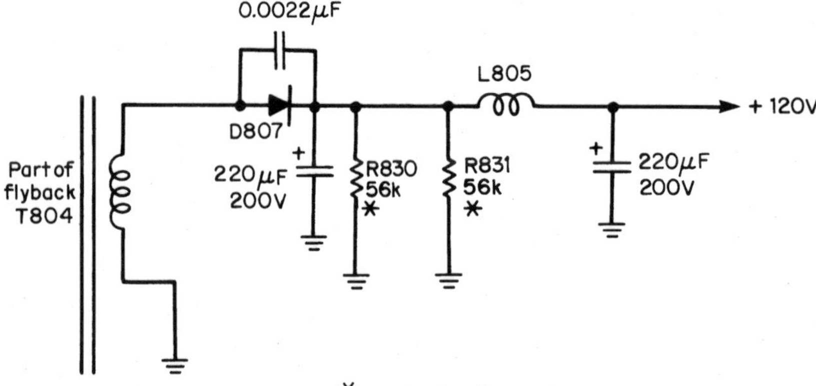

10-17 R830 and R831 are metal oxide film resistors found within the +120-volt source of the secondary flyback winding in the TV chassis.

Power resistors

What and where

Power resistors might have a fixed resistance, adjustable sliding tab, lug and fixed lead, and appear in high-wattage types. These replaceable power resistors might appear in 25- to 225-watt components. Power resistors might have a vitreous-enamel, metal-oxide, silicone-coated, and wire-wound coating. The resistance might vary between 0.5 and 100K ohms. The power resistors are used in circuits that provide isolation or voltage drop in high-current circuits such as the low-voltage power supply and transistor voltage-regulated circuits (Fig. 10-18). The 2.2-ohm 7-W, 5.6-ohm 5-W, and 180-ohm 25-W are listed as power resistors.

When

Power resistors have a tendency to overheat, crack, and go open. A shorted or leaky bridge rectifier (D801) might open up the 2.2-ohm resistor. If regulator IC801 goes open, the 180-ohm 25-watt resistor will run excessively warm and may open up. A leaky or shorted IC801 might destroy the 5.6-ohm 5-watt power resistor.

Testing

Take a critical low-ohm measurement with the DMM to determine if the power resistor is open or contains the correct resistance. Replace all cracked or broken resistors. Check for open power resistors on the emitter circuit of the power output transistors of the audio amplifier section.

Resistor networks

What and where

The resistor networks consist of more than one resistor in a single component. Most resistor networks consists of 3 to 9 resistors in one body. These resistor networks

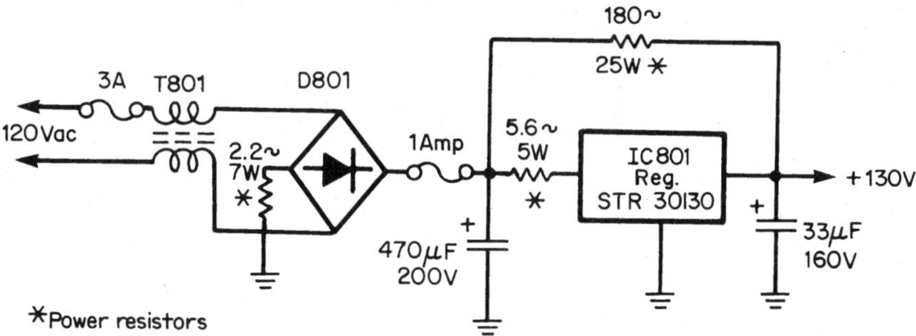

10-18 The 2.2-ohm 7-W, 5.6-ohm 5-W, and 180-ohm 25-W resistors are found in the TV 130-volt low-voltage power source.

might be surface mount (SIP), thick film, surface-mounted chip, dual-in-line, and thick film chip resistor arrays. A surface-mount-SIP resistor network might have the internal resistors connected with a bussed terminal or isolated terminal (Fig. 10-19).

The thick film resistor networks contain a low-profile single-in-line package that is highly stable with a resistance range from 22 ohms to 1 megohm. These thick film networks might appear in a common terminal, isolated terminator, and line terminator. Pin 1 is the common terminal to the other resistors (Fig. 10-20). Each resistor is isolated from the others with each terminal connection. Resistors are connected in series and parallel in the line terminator network. Resistor networks are found in auto receivers, TV, and computers.

The conformal coated SIP resistor networks might appear in a common bussed or isolated element component. The common terminal is marked by a dot and the rest in numbers. The lead terminals contain 60/40 tin-lead plating, alumina substrate, cermet resistance, and body material of conformal epoxy resin (Fig. 10-21). The isolated element network might have only three resistors with six terminals. Replaceable conformal coated SIP resistor networks might have from 5 resistors in a 6-pin terminal bussed network, while 5 resistors might have 10 terminals in the isolated resistor network.

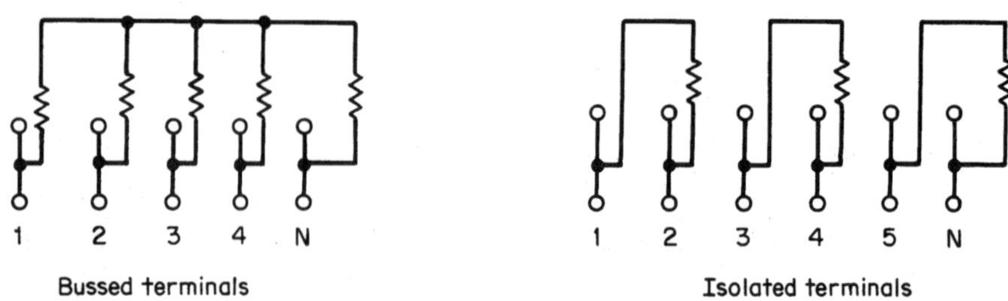

10-19 The resistor networks might consist of bussed or isolated terminals in the surface-mounted SIP resistor network.

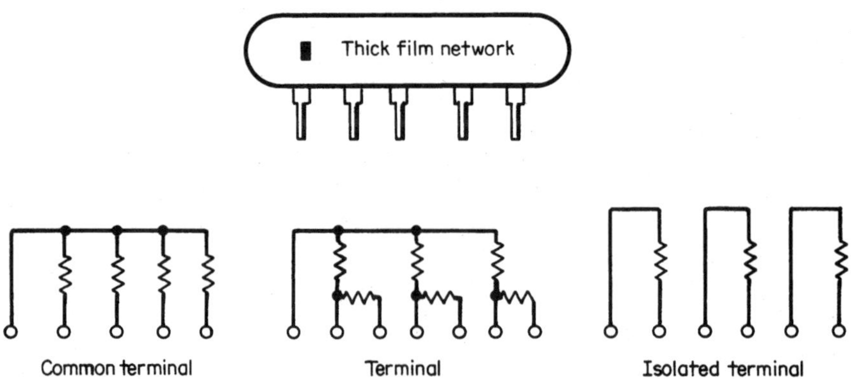

10-20 A common isolator and line-terminated resistor might be found in the thick film resistor networks.

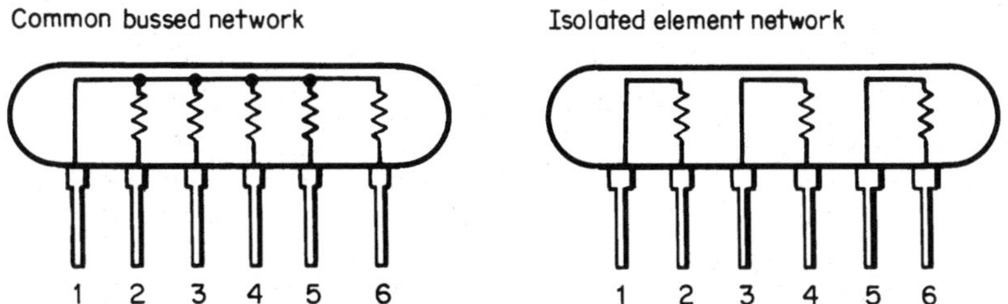

10-21 Pin 1 is the common bussed terminal within a resistor network.

Testing

Determine if the resistor network has a common or a separate connection of each resistor. Start at terminal 1 (a dot) and take critical resistance measurements. Compare the resistance tests to the schematic. The whole component must be replaced when one resistor is found defective. Often, if one resistor has been burned open or there is a change in resistance due to a leaky transistor, then replace the whole resistor network.

Silicone-coated resistors

What and where

The silicone-coated resistor has high stability and reliability in circuit performance. The silicon resistor is a temperature-sensitive resistor made of silicon material. It has an all-welded construction and excellent insulation resistance. The silicon coated wire-wound resistor has a resistance tolerance of 1% or 5%. The replaceable resistance values might range from 0.1 ohms to 25K ohms. These resistors might appear in 1-, 3-, 5-, and 10-watt components. The silicon resistor is found in amplifiers, TVs, VCRs, and CD players.

Testing

Check the suspected resistor with the lowest range of the DMM. Replace with either a silicon or vitreous enamel-coated resistor.

Semifixed resistors

What and where

Semifixed or adjustable power resistors have a sliding tab with a nut and bolt that can preset the required resistance. Often they have lug-type terminal connections. These replaceable semifixed resistors might appear in 10- to 250-watt components with a 10% tolerance. They may have an outside vitreous enamel coating. These adjustable power resistors might be found in the early TV's high cermet power supplies, speaker load, amplifiers, and appliances.

Testing

Check and adjust the semifixed resistors with the low-ohm scale of the DMM. Readjust the sliding-bar lug for correct resistance. Check for burned or overheated connections. Sometimes the resistor has become so hot, the solder is melted at the terminal connections or has burned wiring.

Standoff resistors

What and where

The standoff type of resistors have metal stands, clips, or legs to hold the large-wattage resistor up from the PCB chassis. The metal standoff supports are molded right in the ceramic, silicone, or cement resistor. The metal standoffs are actually the power resistor terminals. These replaceable standoff resistors appear in 5-, 7-, 10-, and 15-watt components. Standoff resistors are mounted on the large amplifier, VCR, high-powered auto amp, and TV chassis (Fig. 10-22).

These replaceable wire-wound resistors might have a resistance from 0.33 to 10 ohms with a 10% tolerance. You might also find several standoff resistors on power board within the low-voltage power supply of a TV chassis. In the latest TV chassis, these power resistors are mounted vertically to dissipate more heat away from the PCB and other components (Fig. 10-23).

Testing

Inspect the wire-wound standoff resistors for possible cracks and burned marks. Measure the correct resistance with the DMM across the resistor terminals. Inspect for cracked or broken areas.

Surface-mounted resistors

(See Chip resistors)

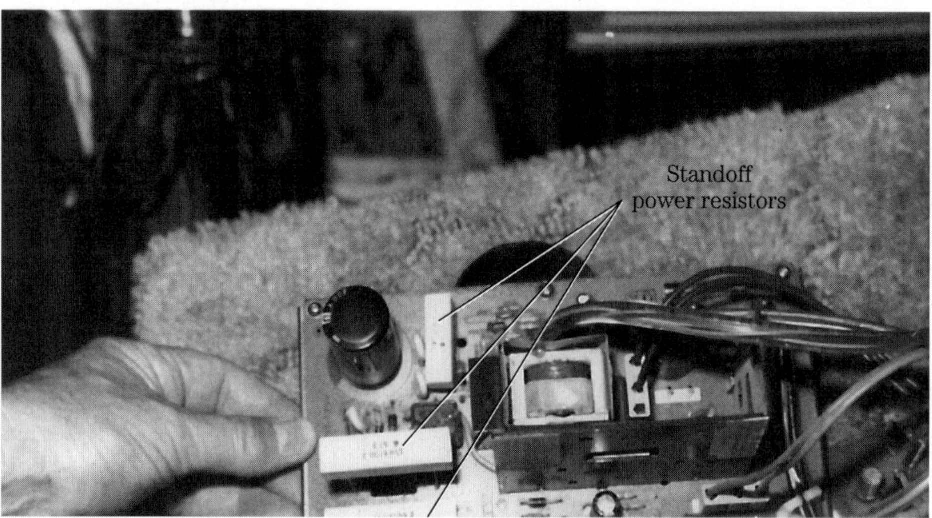

10-22 Several standoff power resistors are found in the power supply of a 13-inch TV and VCR chassis.

10-23 Notice the three vertical mounted wire-wound standoff resistors in the latest 13-inch RCA TV chassis.

Trim-pot resistors

(See Potentiometers)

Variable resistors

(See Potentiometers)

Vitreous-enamel resistors

What and where

The vitreous enamel coating on power-type resistors has a glasslike finish material. Often it is constructed on a tubular component that provides excellent heat dissipation. Most have welded terminals for positive electrical connections. Vitreous enamel resistors have a long-term stability fired at temperatures that exceed 1000 degrees Fahrenheit. Uniform winding and casting provides uniform heat dissipation and has a sharp appearance. These resistors might have lug, welded-lug, or adjustable bar terminals.

Vitreous enamel resistors are wire-wound power-type 5-, 10-, 25-, 50-, 100-, and 200-watt components. Replaceable resistor range from 0.1 to 5K ohm types with 5% and 10% tolerances. You might find them in the large power amps, auto high-powered amps, TV sets, and appliances.

Testing

Check for correct resistance with the low-ohm scale of the DMM or FET-VOM. Inspect for cracked or broken areas.

Wire-wound resistors

What and where

Wire-wound resistors are found in cermet, power silicone-coated, vitreous-enamel, adjustable-power, and resistor networks. The resistor wire-wound material might be manganin or nichrome. Wire-wound resistors are used in high-current, low-ohm circuits. Often these wire-wound resistors are mounted at least 1 inch above the PCB. They are found in low-voltage power supply, high-powered emitter amp circuits, auto receivers, VCRs, CD and cassette players, amplifiers, and large-wattage receivers.

Testing

Check the power transistor terminals for correct resistance with the low-ohm scale of the DMM. Always replace the wire-wound resistor with another of the same resistance and wattage. Do not replace with a carbon, flame, or metal oxide resistor.

Conclusion

Resistors have a tendency to decrease or increase in resistance. Very high ohm values seem to increase in value. Low resistance or emitter resistors can become overloaded, burned, and decrease in resistance. Visually inspect the resistor for cracked or burned areas. If in doubt, remove one end of the resistor from the PC wiring with a sol-

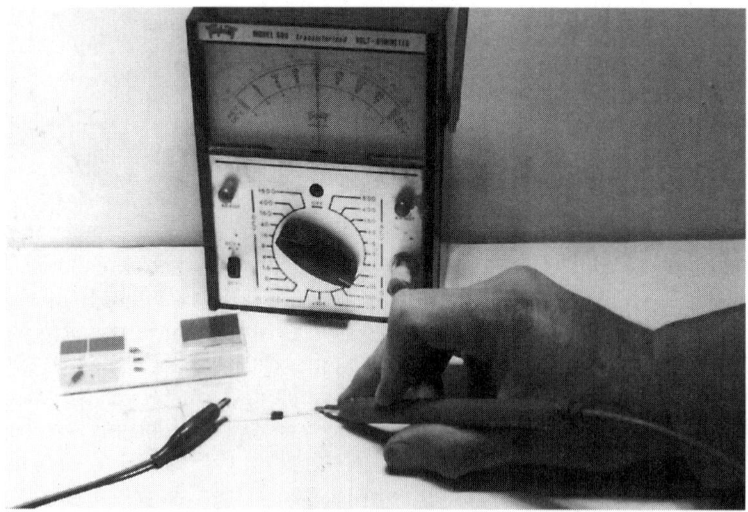

10-24 For correct resistance measurement, double-check the suspected resistor out of the circuit on the DMM or FET-VOM.

dering iron and solder wick. Take another critical resistance measurement with the ohmmeter range of digital multimeter (DMM) or FET-VOM (Fig. 10-24).

Remember, the DMM is more accurate when taking resistance and voltage measurements. Replace the defective resistor with the same wattage and resistance. Choose a higher wattage resistor when in doubt or if the same kind of resistor cannot be located. If a 5-watt, 25-ohm resistor is burned into, and you do not have a 5-watt resistor, replace it with a 7- or 10-watt resistor of 25 ohms. Replace high-wattage resistors at least 1 inch above the PCB. Double-check for poor PCB terminal connections.

11
CHAPTER

Testing electronic components S

(Saw filters through semiconductors)

Semiconductor devices—such as diodes, rectifiers, transistors, or ICs—are found in today's electronic products in the consumer electronics entertainment field. A solid-state diode might be constructed from germanium, selenium, and silicon. The transistor is a three-terminal device that can amplify, oscillate, or contain switching action. An integrated circuit (IC) (internal connection) is made up of wires, semiconductor material classified as monolithic, thin-film, and hybrid. The IC might be referred to as a chip of silicon material. Locating a defective diode, transistor, or IC component might repair 85% of the problems within any electronic product.

Saw filter (TV)

What and where

Today a new circuit found in the TV-IF stages is called the *surface acoustic wave* (SAW) network. The surface wave filter is made up of a piezoelectric material with two pairs of transducer electrodes. The ac input signal applied to one electrode sets up acoustic waves that travel along the surface of the plate or electrode to the other electrode, which generates an ac output voltage. One electrode is the input and the other the output of the transducer. The voltage applied across the positive and negative terminals causes distortion and mechanical waves. The SAW filter establishes the proper IF frequency, which in turn eliminates IF alignment.

Often the SAW filter in the TV chassis is located between the IF amplifier and IF-IC or microprocessor (Fig. 11-1).

Testing

Although the SAW filter network does not cause too much trouble, you can check it with the ohmmeter and crystal checker. It's best to remove the SAW filter from the circuit because low-ohm resistors are found in the input and output circuits. The primary leakage can be checked with ohmmeter probes across terminals 3 and 4. Likewise, the output circuit can be checked across terminals 1 and 2. No resistance measurements should be read across these terminals. Remove the SAW filter from the circuit for accurate tests.

A resistance check does not necessarily indicate that the SAW filter is functioning. Check the SAW filter in or out of the circuit with a crystal checker. The crystal checker will determine if the SAW filter is oscillating (Fig. 11-2). Check the input and output terminals in the same manner. The crystal meter will provide a high reading when tested out of the circuit and a lower reading in the circuit. Remove the SAW filter for a correct test on the crystal checker. If a crystal tester is not handy, sub another SAW filter component.

Semiconductors

Bridge rectifiers

(See Rectifiers)

CMOS integrated circuits

What and where

The COS/MOS integrated circuit (IC) might be an operational amplifier, having metal-oxide silicon components in a complementary circuit. CMOS stands for com-

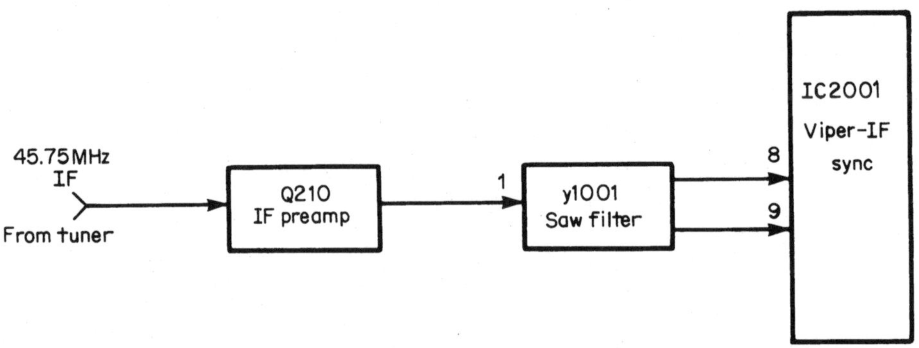

11-1 The SAW filter is located between the IF amp and video IC (IC2001) in the TV chassis.

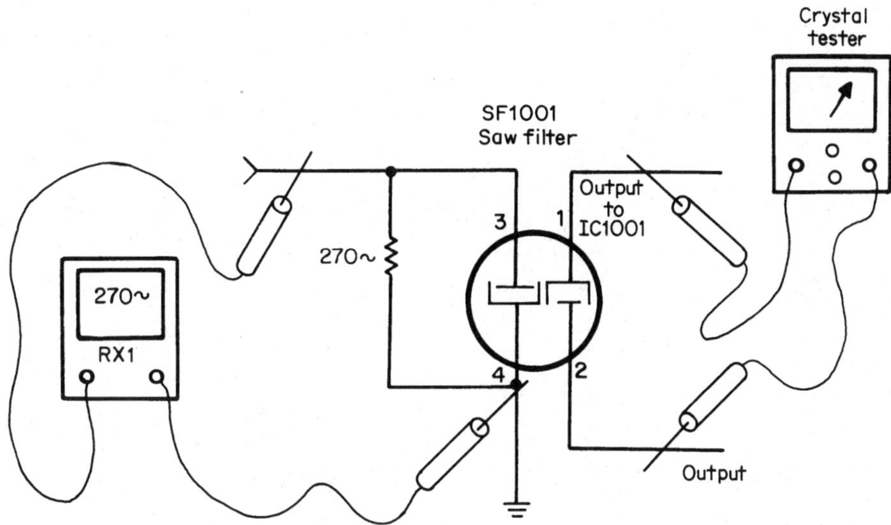

11-2 Check the SAW filter for leakage with the ohmmeter and a crystal tester for correct operation.

plementary metal-oxide silicon. These integrated ICs operate at a very high speed with very low voltage. They are more stable than the TTL units. The operating voltage might be 3 to 12 volts. CMOSs are found in computers, TVs, and fast-switching circuits.

CMOS chips are easily damaged from static electricity. Just touching the chip without the hand or body grounded can damage the IC. Keep the CMOS chip inside the protective cover until ready for mounting. Make sure the body and electronic product is grounded on a grounding plate or mat.

A CMOS integrated circuit contains two common numbering systems: TTL and 4000 series. The TTL and CMOS ICs begin with 74HC/74HCT/74LS/74S/4000/80C/ and 22000 series.

Testing

Take critical voltage measurements on each pin terminal and compare with the schematic. Sub a new IC.

Integrated circuits (ICs)

Audio TV-radio

What and where

Integrated circuits (ICs) within the receiver audio section might include one large IC component. Audio circuits within the TV chassis might all be included in one large IC or combined with other TV circuits within one component (Fig. 11-3). You might find transistors and IC components within the TV stereo channels. The TV stereo sound

11-3 The CMOS IC is found in the shielded remote control area of a TV.

channels begin with the VIF/SIF IC fed to an MPX stereo IC, audio video control IC, stereo IC amp, and dual-stereo output IC amplifier (Fig. 11-4).

The clock radio receiver circuits are quite simple compared to the TV stereo circuits. In clock radio-cassette audio circuits, IC1 is a preamp, with another IC as an audio output amplifier. Locate the audio output IC coupled to the pm speaker. The radio or tape input is switched into IC1 where it is amplified to the volume control (Fig. 11-5). C11 couples the audio signal from the preamp circuits to the audio output IC3. C13 couples the power output audio to the speaker. IC3 has a supply voltage of 11.72 volts, while the preamp IC1 is powered with a 5.9 voltage source.

Signal tracing

All audio circuits can be signal traced with an external audio amp, radio signal, scope, signal injection, tape signal, and pm speaker. Check the audio signal by the number. Start at number 1 and proceed until the weak, distorted, or dead signal is located (Fig. 11-6). Likewise, inject a sine or square wave at terminal 1 and proceed by the numbers using the speaker as indicator. The scope, external amp, and internal speaker can be used as indicators. If the signal is normal at the volume control, proceed to the preamp audio stages until the signal cannot be heard. Test the transistor and ICs and take critical voltage measurements when the signal becomes weak, dead, or intermittent.

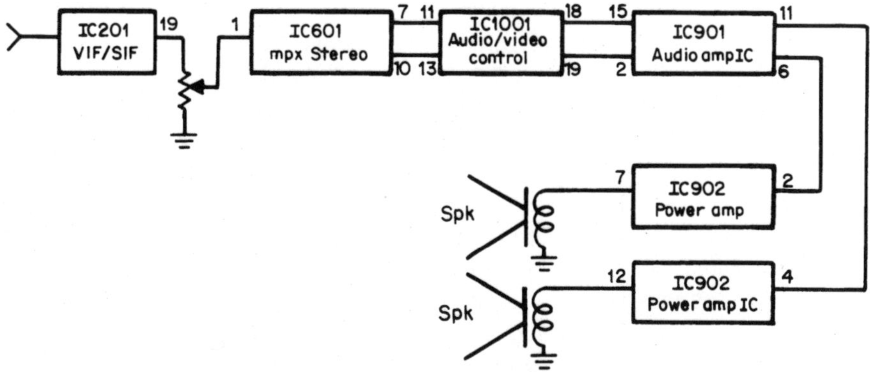

11-4 The stereo audio circuits might consist of four separate ICs in the latest TV stereo audio circuits.

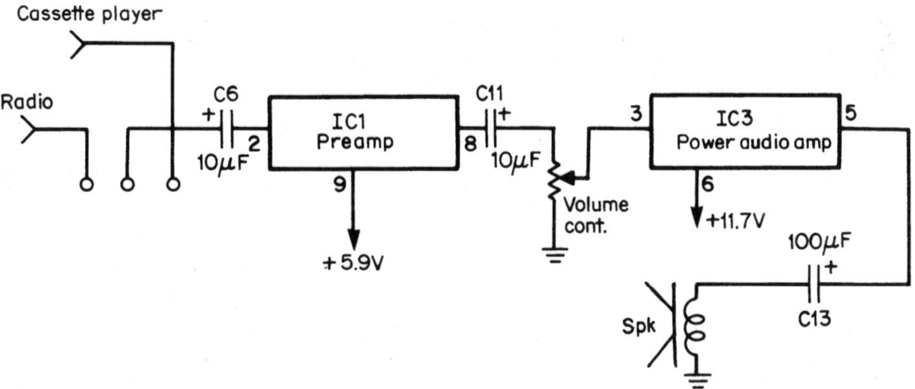

11-5 Another preamp IC might drive the audio output IC in the clock radio audio circuits.

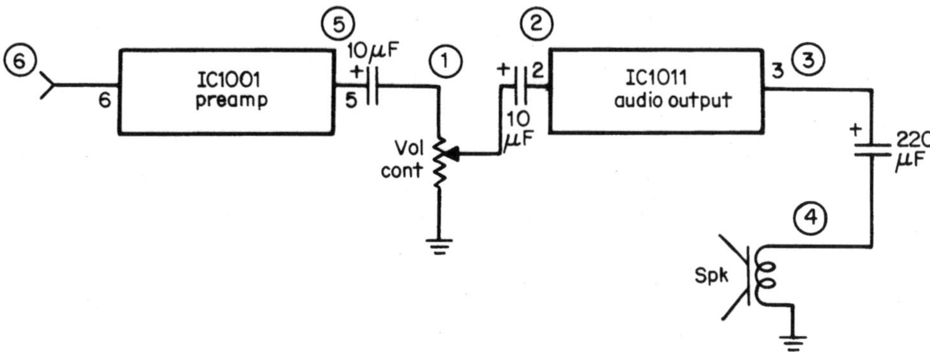

11-6 Start at the volume control and signal trace the audio circuits by the number.

Testing

Locate the defective circuit by signal-tracing the audio circuits. Suspect a defective IC or voltage source when a signal is applied to the input and there is no signal at the output IC terminal. First take a voltage source measurement. Check the voltage on each IC pin terminal. Mark the voltages on the IC and compare them to the schematic. Take a resistance measurement between each pin and common ground.

If the supply voltage is low, suspect a leaky IC or improper voltage source. Very low voltage on several terminals of the IC can indicate a leaky component (Fig. 11-7). Remove the terminal pin from the circuit and take a resistance measurement between the pin and chassis ground. Pick up the excess solder around the supply voltage pin with solder wick and iron. A low-ohm measurement, below 1K ohms, indicates a leaky IC. Now check the supply voltage at the disconnected pin terminal to chassis ground.

Open internal connections inside the IC can cause a dead or intermittent symptom. If the signal is applied to the input and there is no or a low signal at the output terminal, suspect a defective IC. Sometimes voltage and resistance measurements are normal compared to the schematic, and replacing the IC is the only way to solve the problem.

Remove the suspected IC and replace with the original part number. If the original IC is not handy, universal audio transistors and ICs work well in audio circuits. Look the part number up in the universal semiconductor replacement manual. Often, the part number is stamped on the top of the IC (Fig. 11-8). Apply silicon grease to the back of high-power amp ICs when mounted on a heat sink.

Deflection IC (TV)

What and where

In the early deflection circuits, a tube and then the transistor were used as the horizontal oscillator circuits. The horizontal deflection IC, countdown, or deflection IC provide a drive waveform for the horizontal and vertical circuits (Fig. 11-9). The countdown or deflection horizontal circuit provides a drive voltage or waveform to the horizontal driver transistor. The horizontal driver transistor provides a drive signal to the horizontal output transistor. Today the deflection IC might be located in one large IC that includes the IF/luma/chroma/deflection processor (Fig. 11-10).

Integrated circuit IC1001 performs the functions of horizontal synchronization (horizontal oscillator) and vertical countdown. This combined circuit supplies both horizontal drive for the horizontal output stages and vertical drive for the vertical deflection amplifier (Fig. 11-11).

Horizontal sync is applied to pin 1 of IC1001. Inside the IC, horizontal sync is applied to the input of the first phase loop, and vertical sync is applied to a vertical sync comparator for further processing. IC1001 is powered from an internal 10-V shunt regulator that is biased through the deflection startup circuit or a 24-V supply at pin 11.

The horizontal deflection synchronization system uses two phase-lock loops and an oscillator operating at 250 kHz exactly 16 times the horizontal rate. The frequency of the oscillator is determined by the LC network between IC1001 (pins 4 & 5).

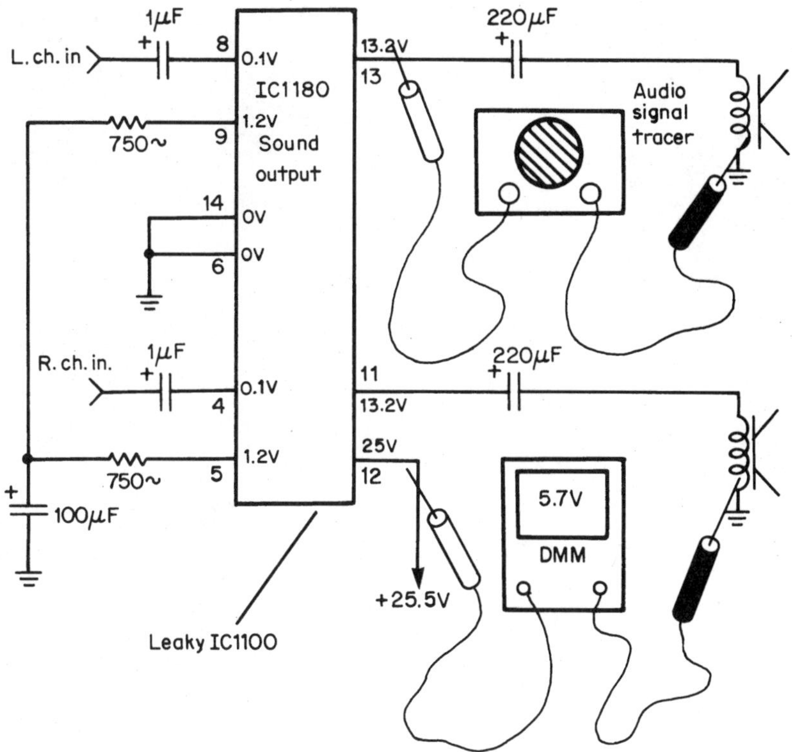

11-7 Remove the supply pin terminal 12 and take a resistance measurement to common ground to determine if IC1100 is leaky.

A horizontal output pulse is coupled from IC1001 (pin 16) to the horizontal driver (Fig. 11-12). The drive stage drives the horizontal output transistor. IC1001 can be checked with a pulse waveform at pin 16.

Testing

Scope the horizontal waveform at the output terminal of the deflection IC. Check the vertical output deflection waveform. If there is no waveform, suspect a defective IC, improper supply voltage source, or defective components tied to the pin terminals. Measure the voltage applied to the supply source. Often, this voltage is the highest dc voltage applied to the deflection IC. Suspect a leaky IC if the voltage is low.

First determine if the voltage source is supplied to the IC from the low-voltage power supply or flyback secondary voltage sources (Fig. 11-13). When the voltage is low or improper and supplied by the low-voltage power supply, simply troubleshoot the low-voltage circuits. The horizontal circuits must operate before a voltage is generated from the flyback circuits. If the supply voltage is generated within the flyback circuits, an injection voltage must be applied to the horizontal deflection IC to make cause the chassis to shut down.

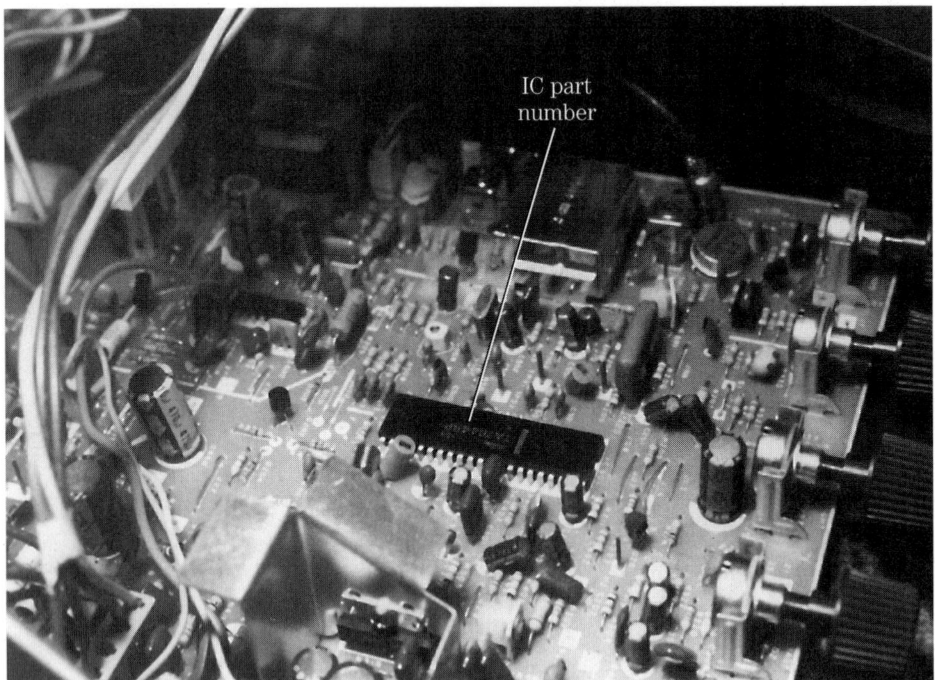

11-8 Check for a part number on top of IC (TA7644BP), and look it up in the universal semiconductor manual.

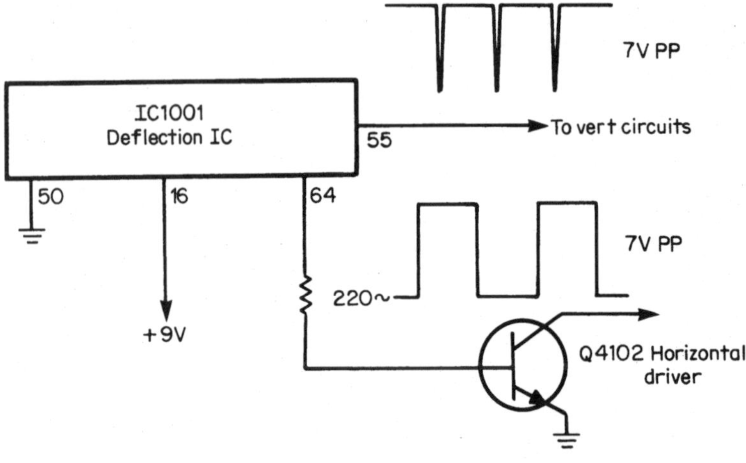

11-9 The deflection IC1001 provides a vertical and horizontal drive sweep waveform for the vertical and horizontal circuits.

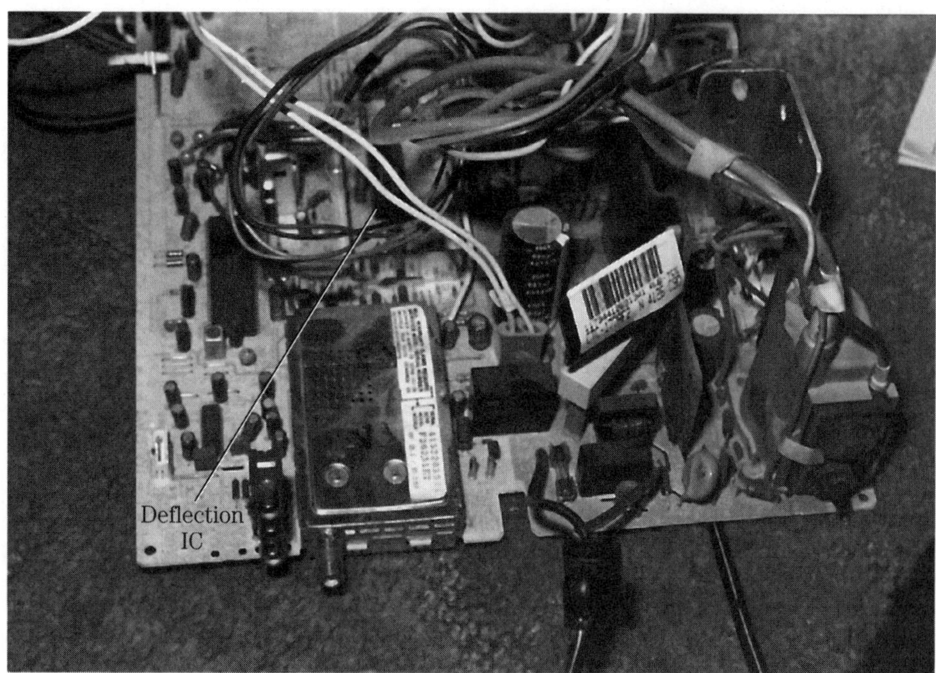

11-10 The deflection IC is found with other circuits in an RCA CTC177 chassis.

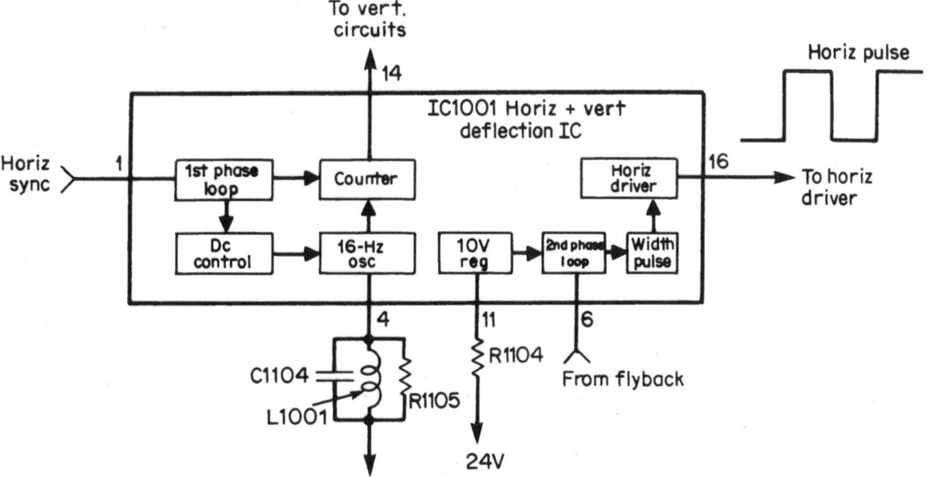

11-11 The internal horizontal circuits within the horizontal IC1001 deflection circuits.

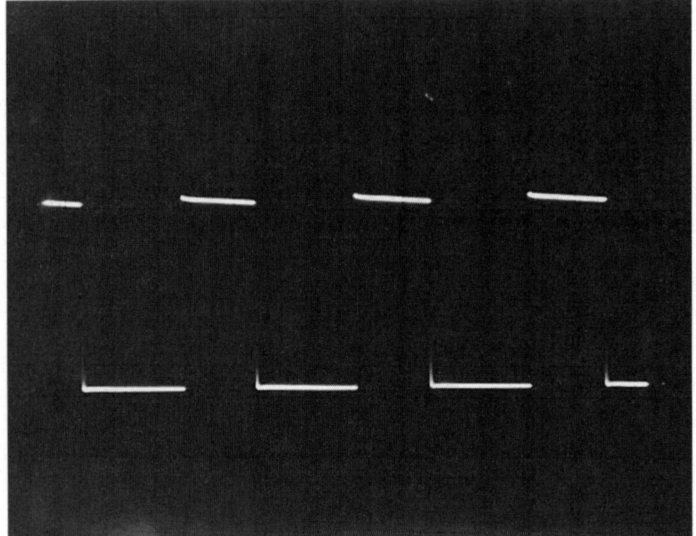

11-12 A horizontal square wave is scoped at pin 16 of the deflection IC to the horizontal driver transistor.

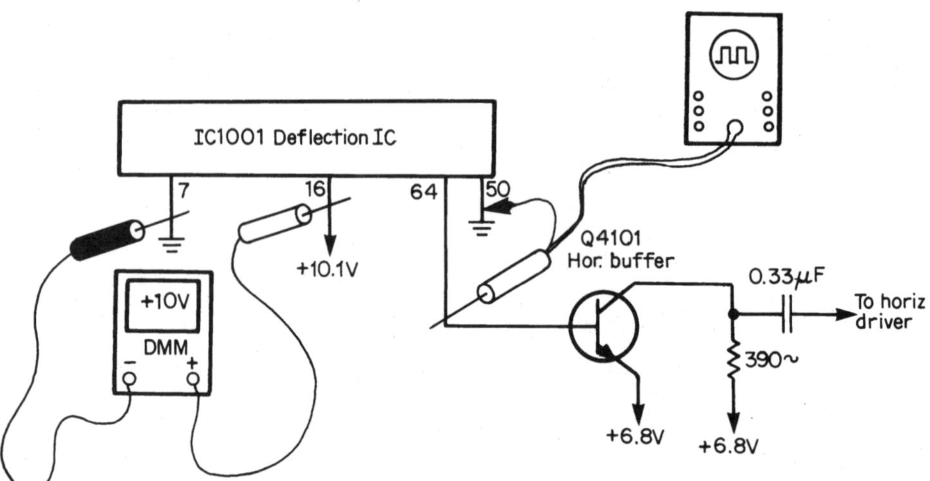

11-13 Check the voltage source at pin 16 and notice if it comes from the low-voltage power supply or flyback secondary circuits.

Locate the correct supply voltage pin of the deflection IC on the schematic. If no schematic is handy, look up the IC part number in the semiconductor manual and locate the correct supply voltage terminal. Most deflection IC supply voltage sources are below 18 volts, with the average around 10–12 volts. Inject a 10-volt supply source from an external power supply to the voltage pin and common ground (Fig. 11-14). Scope the horizontal output drive waveform that feeds to the horizontal

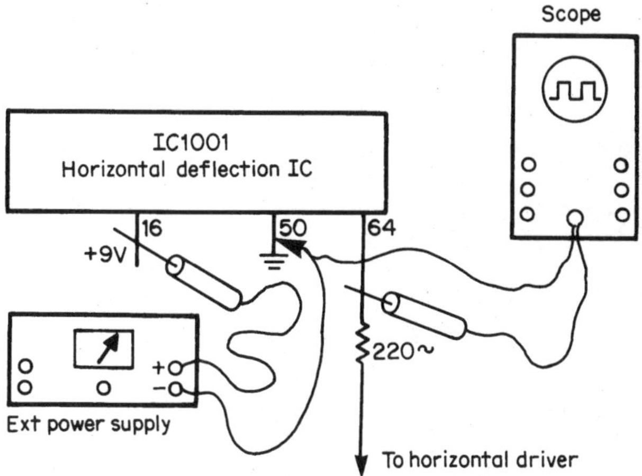

11-14 Inject an external power supply source to the supply pin (VCC), and scope the horizontal drive waveform at pin 64 to determine if the IC and circuits are defective.

driver transistor. If a square waveform is found, you can assume the horizontal deflection circuits are operating and the horizontal IC is normal. If not, take critical voltage and resistance measurements on the IC terminals and compare to the schematic. Remove the ac power cord when making these voltage injection tests. If in doubt, replace the deflection IC.

Croma/IF/AGC/video/luminance IC (TV)

What and where

The large IC might have five or six different sections of the TV set within one large IC. When one section becomes leaky or open, this does not mean all circuits are not functioning. You can have one or more sections not operating and the others normal in a defective multiple-operating IC. Sometimes all circuits are dead with a leaky IC or improper voltage supply source. At other times, all voltages and resistance measurements are fairly normal, and replacing the suspected IC is the only answer.

With a no-color or one-color-missing symptom, check the color 3.58-MHz crystal, color supply voltage, and waveforms to the red, green, and blue color output transistors (Fig. 11-15). Suspect a no-red-color-output waveform of IC1001 on pin 9. Check the voltage at pin 9. No waveform and improper voltage at pin 9 indicate a defective IC1001. Take critical voltage measurements on all pin terminals of the color circuits and resistance measurements to common ground. The voltage measured at the color output of IC1001 should be quite close on all three terminals (5.4 V) (Fig. 11-16). Suspect a defective color output transistor or CRT when the red color is missing in the picture and the color output waveform is normal at pin 9.

Check the 3.58-MHz oscillator circuits when no color is found in the picture. Take a color oscillator waveform at pins 4 and 5 of IC1001 (Fig. 11-17). Measure the voltage on all color pin terminals of IC1001. Scope the color input and output terminals.

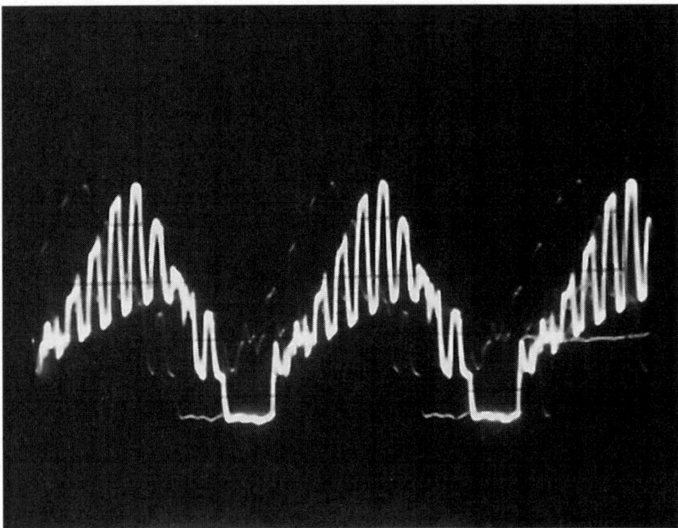

11-15 Check the waveform at the three color output pins of the IC to determine if more than one color is missing.

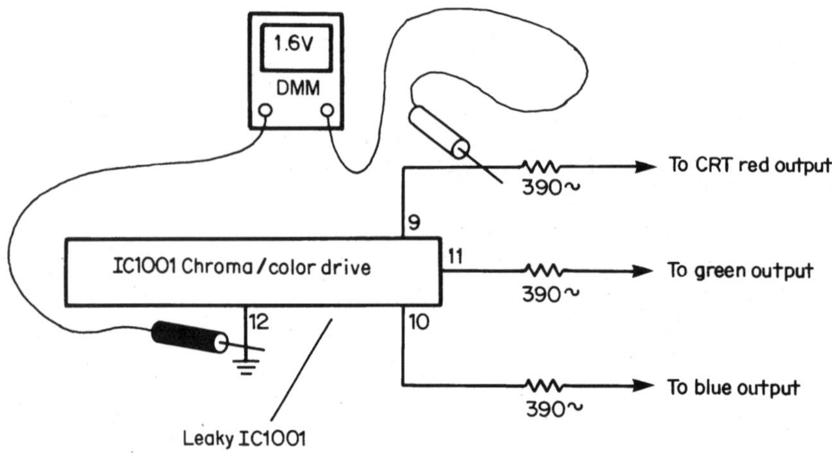

11-16 The voltage and scope waveforms on pins 9, 10, and 11 should be quite close and similar.

Take a resistance measurement of each IC pin terminal to common ground. Sub another 3.58-MHz crystal. Replace IC1001 when all other tests are normal (Fig. 11-18).

Vertical output IC (TV)

What and where

The deflection IC provides a drive waveform to a transistor or IC vertical output circuits. Today, most TV chassis have a vertical output IC that provides sweep to the vertical yoke winding. The vertical IC is often found mounted on a heat sink in the TV chassis (Fig. 11-19). The vertical drive signal is fed directly from the vertical deflection IC or to a buffer transistor before the vertical power output IC. Locate the vertical output IC mounted on a separate heat sink.

When

The defective vertical symptom might be a horizontal white line, improper height, intermittent sweep, vertical foldover, rolling, vertical crawling, black at the

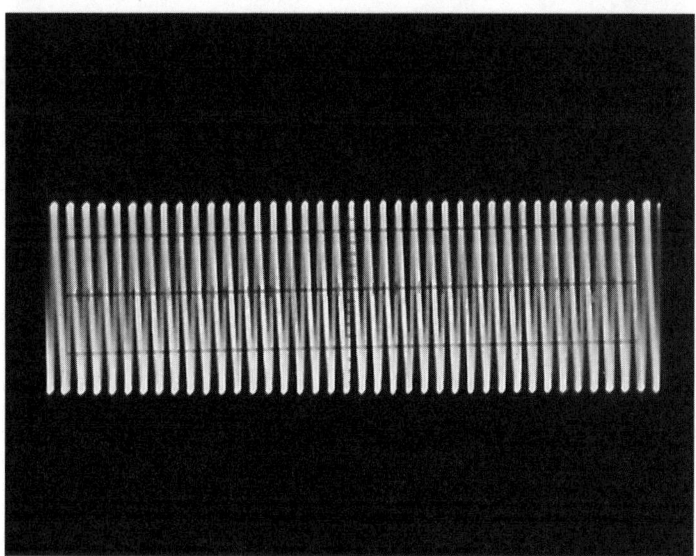

11-17 The 3.58-MHz crystal oscillator waveform taken from pin 4 or 6 of IC1001.

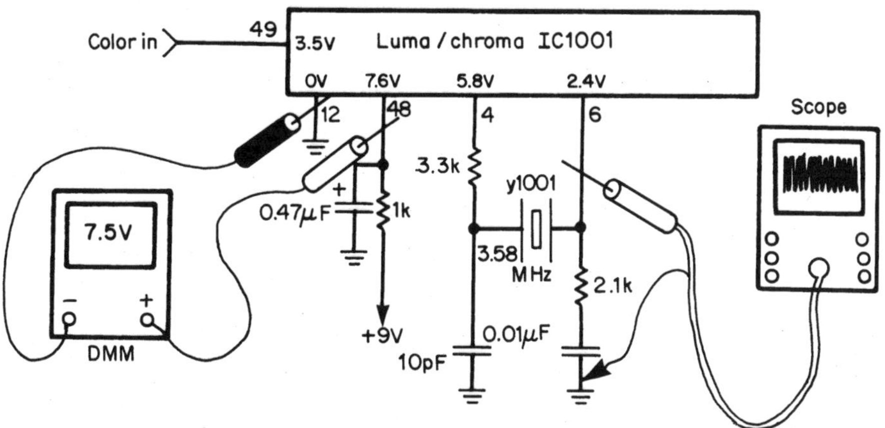

11-18 Check the voltage and waveforms on pins 4 and 6 to determine if the color oscillator circuits are functioning.

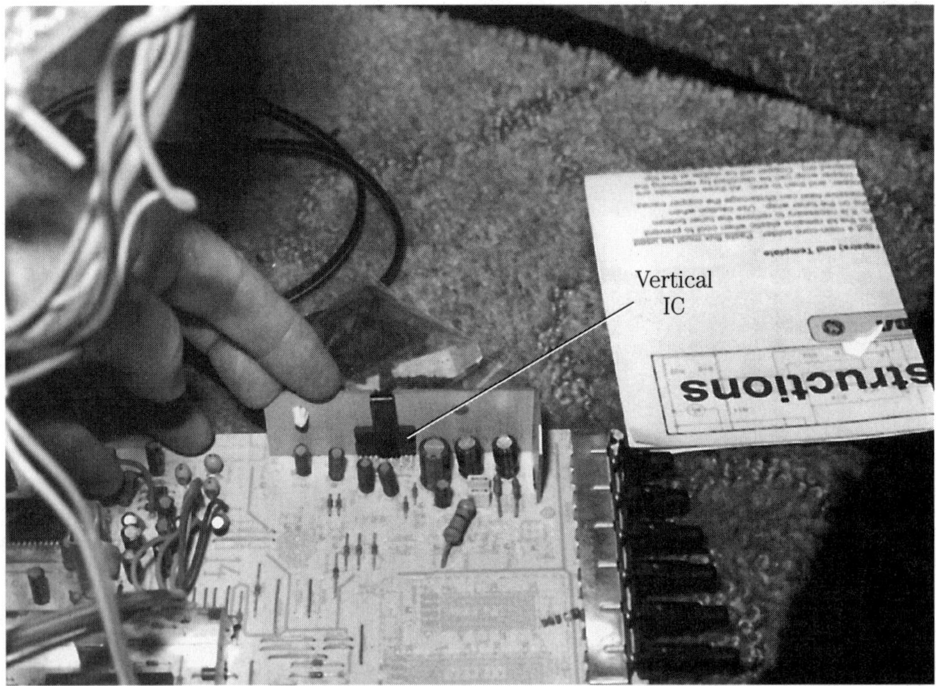

11-19 The vertical IC might be clipped or bolted to a separate heat sink in the latest TV chassis.

top, or bunching vertical lines. Just about any component within the vertical circuits can cause a horizontal white line or no vertical sweep. Insufficient vertical sweep can appear with only 1 to 8 inches of raster (Fig. 11-20).

Improper height is when the top and bottom of the raster cannot be adjusted with either the vertical height or linearity controls to fill out the screen. Improper vertical sweep can be caused by a leaky vertical output circuit. Intermittent sweep might result from transistors, ICs, or poor board-wiring connections. Check for leaky output transistors with improper bias resistance for vertical foldover. Also, vertical foldover can be caused by leaky or open electrolytic capacitors in the vertical feedback circuits. Vertical crawling and rolling can be caused by defective filter capacitors in the power source and improper sync signal.

Testing

Servicing the vertical output circuits is fairly easy with only one output IC. Check the vertical drive waveform at the deflection IC with a scope and voltage tests. Check the vertical sawtooth waveform fed to the vertical driver or output transistor. Improper or no waveform signal can indicate a defective IC, defective surrounding components, or no voltage source. Measure the B+ voltage source at pin 8 of IC4301 (Fig. 11-21). If the voltage is low, suspect a defective IC, components, or voltage supply source. Remove the voltage source at the supply pin 8 with solder wick and iron. Suspect a leaky IC if the correct voltage returns.

Scope the input signal at pin 6 and the output at pin 4 of IC4301. If input signal is normal and there is no output sweep, take voltage and resistance measurements on each pin terminal. Check for a defective 220-uF electrolytic capacitor, D4303, D4305, C4303, and open 3.3-ohm resistor. Replace IC4301 when other components test normal. All vertical ICs and transistor parts can be replaced with universal components when the original part is not available.

For no vertical sweep, suspect an open sweep return resistor or capacitor in the grounded leg of the vertical yoke winding. An open yoke winding, capacitor, or low-ohm

11-20 Only 2 or 3 inches of vertical sweep on the TV screen indicates improper vertical sweep in the TV chassis.

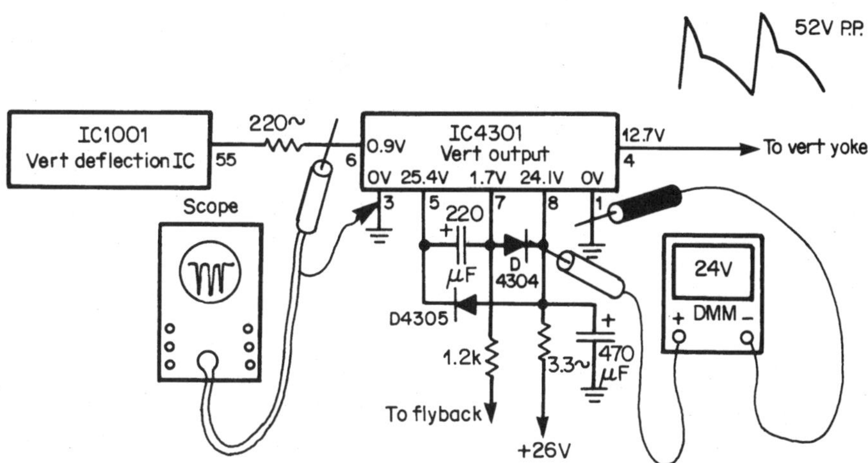

11-21 Measure the supply voltage at pin 8 and scope the input and output terminals (6 and 4) of IC4301 to check the vertical output IC.

resistor can cause a horizontal white line or no vertical sweep (Fig. 11-22). The defective large coupling capacitor C4307 can cause no vertical sweep, bunching of lines, and a jumping picture. Check for an open or dried-up feedback electrolytic capacitor for vertical foldover symptoms.

TTL IC

What and where

The transistor-logic (TTL) IC is a high-speed device. A low-power version of a TTL is a 7400 quad NAND gate or 74L00. A high-speed TTL IC begins with a 7400 or 74H00. The 74H00 operates twice as fast as a regular TTL and consumes more power. You may find that the high-speed TTL integrated circuits are used less, as Schottky devices are superior and produce less noise problems. TTL ICs are found in computers and switching circuits. These devices are very cheap compared to other ICs.

Testing

Remember, TTL ICs are protected in a static-sensitive package. Keep the TTL IC in the package until ready to be mounted. Take critical voltage and resistance measurements on each pin terminal. Scope in and out signals, and sub a new component.

Microprocessor

What and where

A microprocessor is a single chip or component that controls a unit or product and contains control processing, logic, and complicated circuits. The microprocessor might be called a central processing unit (CPU). Sometimes the microprocessor might be called a computer in a chip. The microprocessor might control the tuning in present-day receivers and control computer functions, TVs, camcorders, CDs, and VCR operations (Fig. 11-23). The microprocessor is located with many chip or gull-wing terminals.

The microprocessor or control micro in an RCA CTC167 TV chassis actually controls most functions of the chassis. U3101 controls the power on and off, contrast, degaussing, brightness, color, tint, sharpness, tuning, shutdown, logic, volume, mute, tone, stereo sense, mono/stereo, oscillator, LED switching, data, clock, sync, vertical, horizontal, and IR remote control operations (Fig. 11-24).

The microprocessor (IC601) in the Samsung 8mm camcorder controls the still position adjustment; head-switching adjustment; battery detection; clog detector; S reel and T reel sensor; end sense; top sense; dew in; record proof; cassette in; T/E LED; mode switch A, B, and C; LM forward; and LM reverse. You might find that the MICOM in a camcorder is of the surface-mounted type (SMD). Often, pin 1 is noted with a dot on the body of the microprocessor or stamped on the PCB (Fig. 11-25). You might find more than one microprocessor within the camcorder and VCR units.

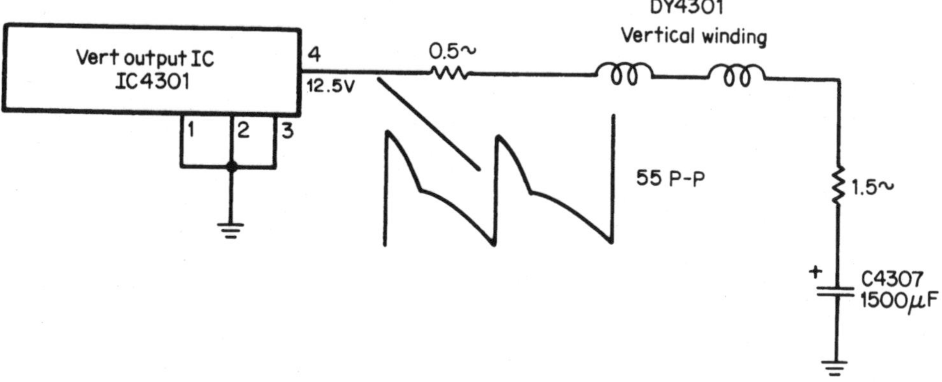

11-22 For a horizontal-white-line symptom, check C4307, the 1.5-ohm resistor, and the vertical yoke winding.

11-23 A microprocessor with many terminals controls the operational functions in the TV chassis.

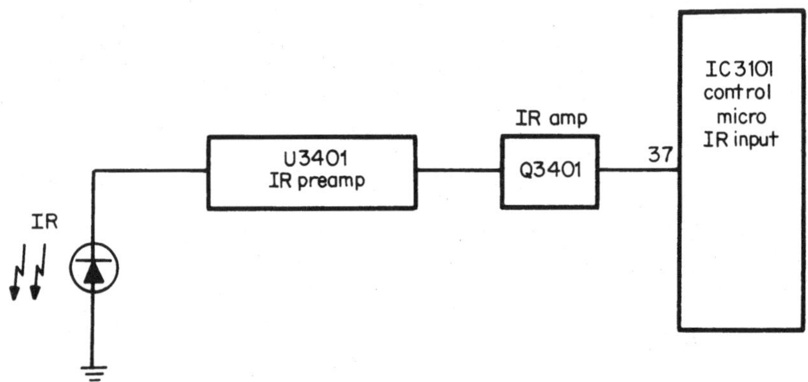

11-24 The infrared (IR) components of remote control receivers are fed to pin 37 of the microcontrol IC3101.

11-25 Pin 1 of the surface-mounted microprocessor or a notched corner might identify the start of pin numbers in a camcorder.

Testing

Always use a pair of test probes that have sharp points to take voltage and resistance measurements on the CPU. Be careful not to short out the closely mounted terminals. Check for the supply voltage pin on the schematic of the microprocessor. Take critical voltages and resistance measurements on each terminal. Scope in and out signals. These microprocessors might have more than 80 pin terminals (Fig. 11-26).

Special ICs

Hall effect IC

What and where

The Hall effect transistor or IC is placed in a magnetic field with either a voltage or magnet operation. A Hall field effect is when the electric field of a conductor carrying current operates in a magnetic field. Within the camcorder, the auto focus MICOM might control the Hall detect block, in and out. The Hall effect IC might be found within a VCR, cassette player, or camcorder (Fig. 11-27). Locate the Hall IC next to a rotating magnet in the cassette player.

The electronic automatic shutoff circuit might cause the cassette player to shut off as soon as it is turned on, after a few minutes, or not at all. The automatic shutoff circuit may be controlled with a round magnet rotating close to a sealed magnetic switch or IC component. In some models, a Hall IC component is mounted close to the rotating magnet (Fig. 11-28).

When the magnet stops rotating, the IC triggers the electronic control circuit, which in turn energizes a control relay, shutting off the player. If the magnet stops,

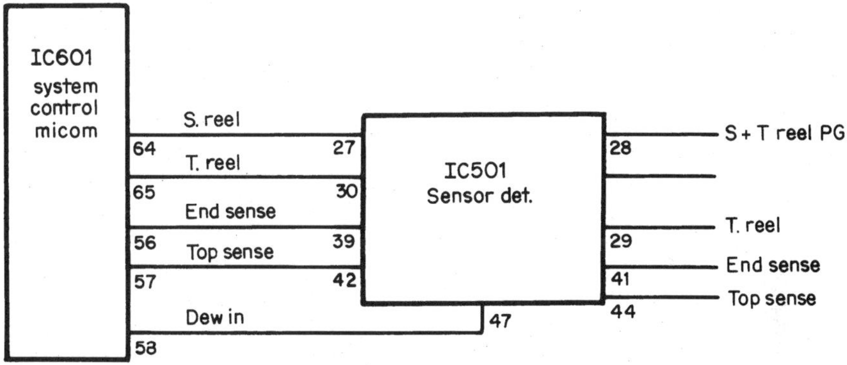

11-26 The microprocessor of a VCR or camcorder might have up to 80 pin terminals.

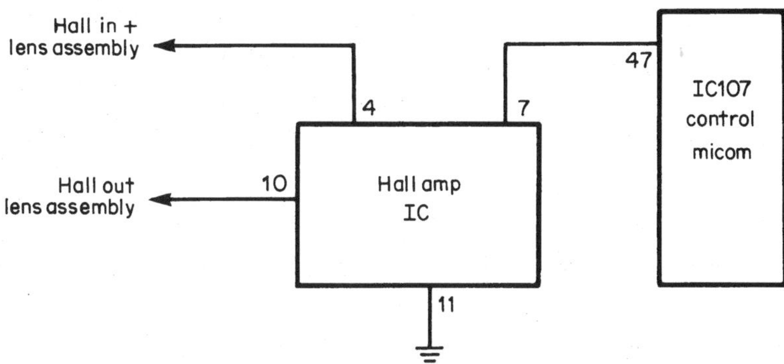

11-27 The Hall IC controls the lens assembly with signal from the control MICOM (IC107) of a camcorder.

11-28 A magnet is rotated in front of a Hall IC to shut off the tape rotation in a cassette player.

notice if the belt is broken or off the pulley. The rotating magnet is driven with a rubber belt from the take-up reel.

Testing

First determine if the automatic shutoff is electronic or mechanical. Note if the relay energizes or shuts down. Replace the Hall IC if the magnet is rotating. Take critical voltage measurements on the suspected IC and compare to the schematic. Replace the Hall IC if it has normal voltage measurements.

LED flasher IC

What and where

The most common flasher LED has a built-in flashing circuit. The flashing rate or blinking frequency might be from 1.5 to 2.5 Hz. They may operate from 2.1 to 10 volts dc. A small black speck inside the plastic case might be a tiny oscillator IC. The flashing IC makes eye-catching displays or indicator devices. LED flashers are found in many electronic projects as indicators.

The most common LED IC flasher component is an LM3909 chip. You will find many different flashing circuits built around this flashing IC. The LM3909 IC might drive four or more LEDs from one low-voltage battery. A very simple LED flasher IC circuit is shown in Fig. 11-29. Batteries powering the flashing LED circuits will last for a long timc.

Testing

Check for correct voltage across the battery terminals. Check for correct voltages on each pin terminal. Sub another LED. Then replace the defective IC component.

LSI processor

What and where

LSI is a large-scale integrated circuit. There are many integrated circuits such as the small-scale integration (SSI), medium-scale integration (MSI), large-scale integration (LSI), and very large-scale integration (XLSI). The LSI device might be found in pocket calculators, CDs or cassette players, camcorders, VCRs, computers, and TV chassis. Locate the LSI processor by looking for up to 80 terminals.

The large-scale integration IC is a complicated internal circuitry with many silicon wafers made up of transistors, diodes, resistors, and capacitors (Fig. 11-30). LSI chips are designed to operate complicated circuitry and special functions. An LSI device might be called a CPU or control system IC.

Testing

Scope the input and output signals. Take control voltage on the voltage supply pin terminal. Compare all pin terminal voltages and resistance measurements to common ground with those of the schematic. Replace the suspected LSI device. Keep the new replacement inside the protective envelope until ready to mount into the circuit.

Memory ICs

What and where

The memory IC might be called a ROM, RAM, PROM, or EPROM. A ROM is a read-only memory. Data that is ROM stored is permanent data. The data entered by

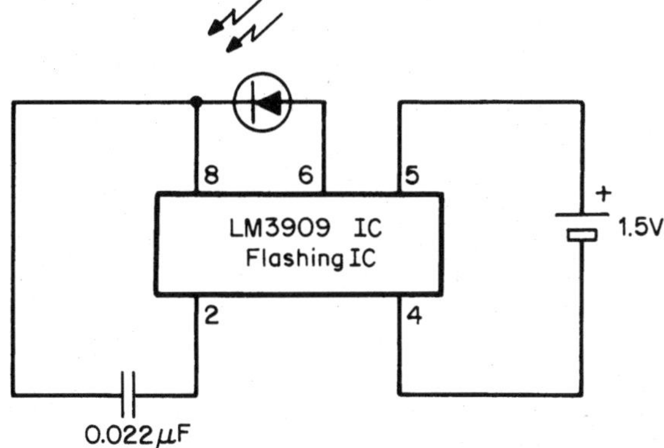

11-29 A simple LED flasher IC circuit powered by a flashlight battery.

11-30 LSI or CPU microprocessors control operation of the tabletop CD player.

a programmer of a computer might be called PROM (programmable read-only memory). Once the data is entered into a ROM or PROM device, the program cannot be altered. The ROM data cannot be lost in the computer if the power is removed. Memory chips are found in computers, VCRs, and TV sets.

Another memory IC that can be programmed like a PROM is the EPROM (erasable programmable read-only memory). The message or memory can be erased with a strong violet light.

The RAM (random access memory) can be written into a read-out memory by a control microprocessor or a CPU. The data in a RAM can be lost in a power outage, lightning, or when power is interrupted. RAM ICs are typically used to store individual programs and data, while ROMs are used to store language translation programs on the computer.

Testing

Check for the operation that is missing or not functioning within the computer. Take critical voltage measurements on each terminal pin, and sub another memory IC.

Motor control IC

What and where

The IC might control the speed and start and stop of a camcorder, VCR, CD player, and receiver control motor. The motor control IC is often controlled by a control system microprocessor or MI-COM. Then, the motor IC controls the operation of the motor (Fig. 11-31). Locate the motor IC within the motor speed circuits.

The motor control (IC502) in a CD player might control both the slide and spindle motors. The IC501 servo control IC provides signal to the motor control IC (Fig. 11-32).

The loading motor in a Samsung SCX854 camcorder is controlled by the loading motor IC504 at pin terminal 14. A loading motor voltage to unload the cassette is

11-31 A motor control IC powers the motors within the CD player.

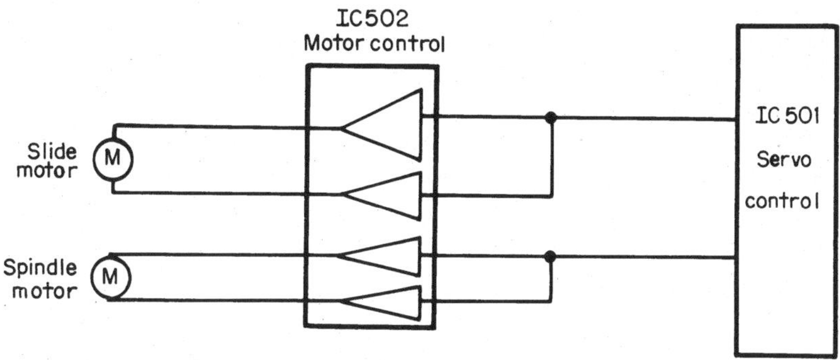

11-32 IC501 and IC502 provide motor control for the slide and spindle motors in the CD player.

applied to the loading motor from IC504. The LM load voltage from the loading motor is fed from LM REV of pin 74 of IC601 MI-COM (Fig. 11-33). A LM FWD load is fed to pin 6 of the loading motor IC. Check the dc voltage applied to the motor terminals when loading mode is entered. Reverse dc voltage is fed from pin 14 of IC504 to unload the cassette.

Testing

Check the voltage across the motor terminals when the circuit is activated. Take a continuity test across the motor terminals to determine if the motor is open. Measure the supply voltage found on the Vcc pin of the motor control IC. Scope the signal from the MI-COM system control to the motor control IC and onto the motor terminals.

Surface-mounted IC

What and where

Electrostatic-sensitive devices can be damaged by static electricity. These sensitive devices include ICs, field-effect transistors, and semiconductor components. Carefully handle all sensitive devices in their original electrostatic packet; do not unwrap them until they are ready to be mounted.

Keep your body drained of electrostatic charges by keeping all clothing at ground potential. Wear a conductive ground wrist strap. Make sure the camcorder, CD player, or VCR and test equipment are grounded (Fig. 11-34). A grounded mat is ideal to place the test equipment and electronic product on while servicing the unit. Use a grounded-tip soldering iron and solder removal tool to prevent antistatic buildup. Keep freon-propelled chemical spray cans away from electronic products with electrostatic-sensitive devices. Prevent extra body motion that might create static electricity from a carpeted floor. Store the sensitive device in the black foam packet it came in.

Most of the surface-mounted chips have gull-type wing terminals, mounting the IC right on the PC wiring. After locating a defective SMD IC with voltage and resistance measurements, locate the position of terminal 1. Mark the position on the PC wiring, if the starting dot is missing on the PCB. The excess solder can be removed with a special iron tip that covers all terminals, or lift each terminal separately. When the solder melts, carefully twist the iron. Raise up and remove the IC terminal with iron and a small screwdriver blade or pick. Remove each terminal in the same manner.

Some technicians use a sharp knife or tool and cut the gull-wing tip terminals close to the body of the IC. Then remove the gull-wing tips from the PC wiring. After removing the defective IC, remove the excess solder with solder wick and iron. Throw away the removed chip; do not try to use it again.

Testing

Scope the signal in and out of a suspected SMD IC. Take critical voltage resistance measurements on each terminal and compare to the schematic (Fig. 11-35). If the resistance measurement is quite low, remove the pin terminal from the PC board with solder wick and iron. Then take another measurement. Often, a low-voltage

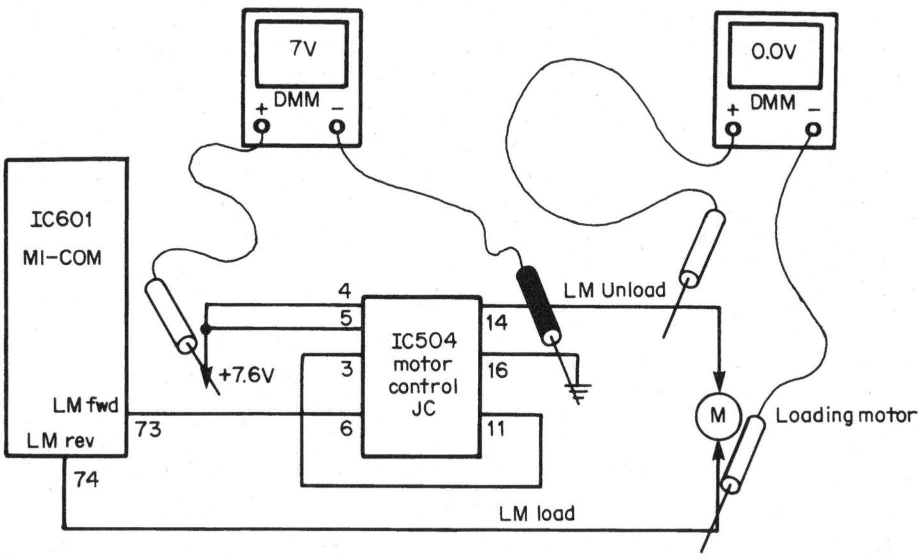

11-33 Check the voltage at the motor terminals to load and unload the cassette or CD in the VCR or CD player.

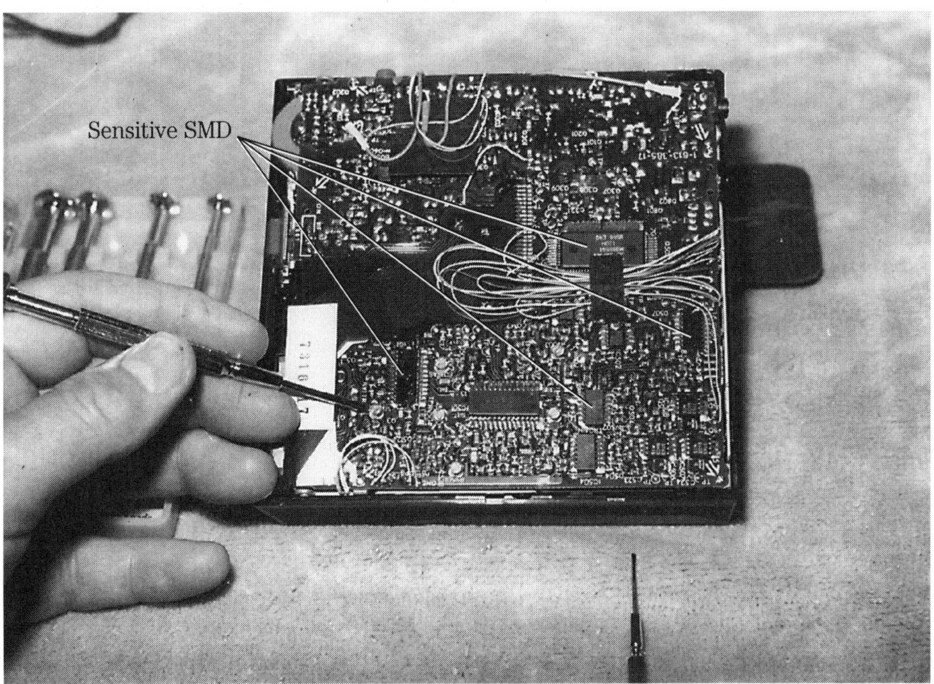

11-34 Make sure your body is grounded when working around electrostatic-sensitive SMD components.

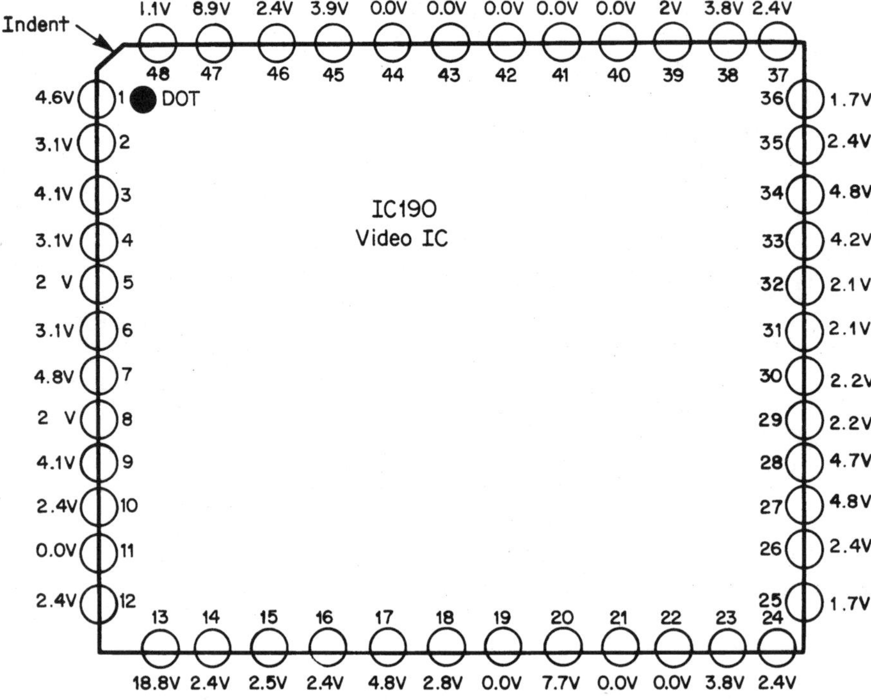

11-35 Take critical voltage and resistance measurements on the many surface-mounted pin terminals to common ground to determine if the IC is defective.

measurement on the supply voltage pin (Vcc) indicates a shorted or leaky IC. Test each component tied to each terminal pin before removing the chip.

Laser diode

(See Diode)

LED

(See LED)

Transistors

There are many different types of transistors found in radio receivers, cassette and CD players, amplifiers, camcorders, VCRs, TVs, and electronic projects. They may be used as amplifiers, rectifiers, regulators, detectors, and oscillators. The plastic or metal-case transistors can be found in electronic projects, radios, and amplifiers. High-powered transistors are located in amplifiers and high-powered auto receivers. Transistors can be soldered directly into the PCB or plugged into sockets. SMD transistors are soldered directly on the PC wiring. High-powered transistors might be mounted on heat sinks. The transistor has an emitter (E), collector (C), and base terminal (B) elements (Fig. 11-36).

General-purpose transistor

What and where

The general-purpose transistor might be used in audio amplification, detection, and oscillation circuits (Fig. 11-37). The common-emitter amplifier might be used as a common reference to both the input and output with the emitter at common-ground potential. In a common-base transistor amplifier, the base is positive with respect to the emitter and negative to the collector terminal. The common-collector amplifier is at ground with all other voltage negative in respect to the collector terminal. The emitter is the most negative, and the base is less negative (more positive).

The general-purpose transistor might be a PNP or NPN type. Today most of the electronic products have NPN transistors. The bias voltage between the base and emitter terminals of a PNP transistor is 0.3 volts, while 0.6 volts is found in the NPN transistor. Simply measuring the bias forward voltage on a transistor with in-circuit voltage tests can indicate a defective transistor. Locate the suspected transistor with these terminal connections and a schematic diagram.

The unknown or unmarked transistor terminals mounted within the chassis can be checked with the transistor tester or junction-diode test of the DMM (Fig. 11-38). Sometimes the transistor leads are marked E, B, and C on the transistor of PCB, and

11-36 Four different power transistors are mounted on a large heat sink.

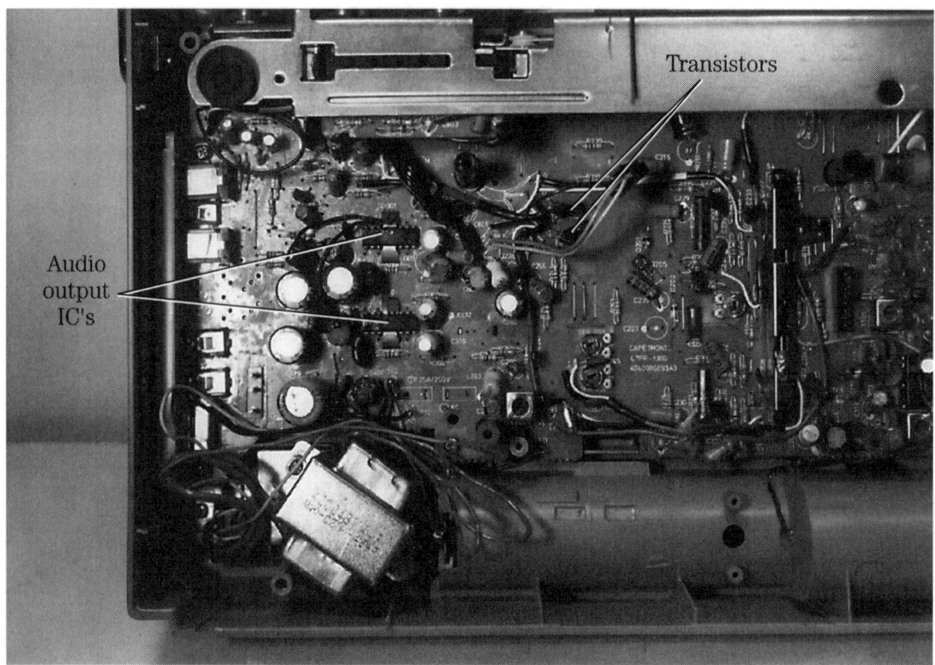

11-37 Separate audio preamp transistors are found within the cassette-radio audio circuits.

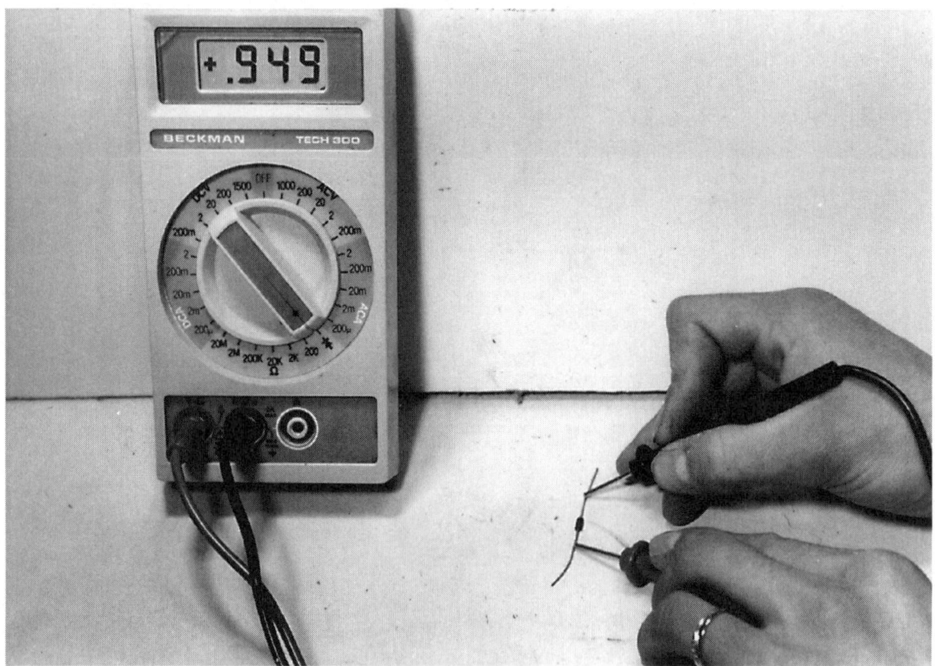

11-38 The junction-diode test of a DMM checks the condition of a normal switching diode. Remember, the simple transistor has diodes back to back.

other times not. Remember, the base terminal measurements are common to the collector and emitter terminals with the diode test of the digital multimeter (DMM). Likewise, the resistance measurement is lower from base to emitter than from base to the collector terminal. The positive terminal from the DMM is placed on the base terminal with NPN measurements. Place the negative probe of the DMM on the base terminal for a PNP transistor test.

When

The defective transistor might be open, leaky, shorted, intermittent, or have a high-resistance junction. For instance, an open transistor might not show any measurement between the base and collector or the base and emitter terminals (Fig. 11-39). A leaky or shorted transistor will show a low resistance between any three elements. Most transistors become leaky or shorted between the emitter and collector terminal. It's possible to find a transistor leaky between all three terminals.

The intermittent transistor might have a poor junction between any two elements. Sometimes the transistor will test normal when a test probe touches the terminal or when tested out of the circuit. Just touching the terminal might make the transistor restore back to normal. If the intermittent transistor tests open or intermittent while being tested, replace it. Replace the general-purpose transistor that has a high-resistance junction test compared to the other elements (Fig. 11-40).

Testing

Locate the defective stage with scope, signal, and external audio amplifier tests. Check the suspected transistor with in-circuit transistor tests or DMM diode tests. A

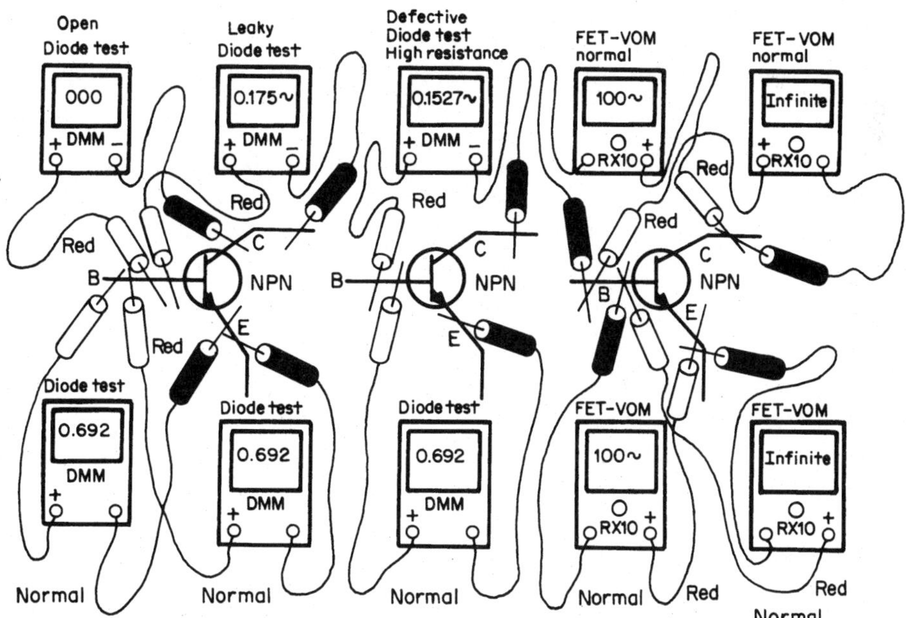

11-39 The different resistance measurements of an open, leaky, and high-resistance junction with DMM diode tests.

11-40 Check the transistor condition in circuit with the DMM diode test before removing from the chassis.

quick diode-junction test can determine the condition of all transistors in a chassis within minutes. Take a forward bias voltage test between the emitter and base terminals. The normal NPN bias voltage should be 0.6 volts on the DMM, or within a half of a volt. By making quick bias voltage tests between the base and emitter terminals, you can quickly locate the defective stage. Some electronics technicians quickly test each transistor within the circuits to locate the defective stage.

Measure the voltage on each transistor terminal and compare to the schematic. A low-voltage measurement on the collector terminal might indicate a leaky transistor or improper voltage source. A close voltage measurement between all terminals might indicate a leaky transistor (Fig. 11-41). The collector terminal voltage is higher in an NPN transistor than the emitter or base. No voltage measurement on the emitter terminal, with an emitter resistor in the circuit, might indicate an open transistor or emitter resistor. To prevent damage to the transistor, be real careful not to short the transistor leads while taking transistor tests. Remember, the forward bias voltage of an NPN transistor between the base and emitter is around 0.6 volts. A forward bias on a PNP transistor is 0.3 volts.

Bipolar transistor

What and where

A bipolar transistor is a very commonly used transistor with two PN junctions. The base (P) terminal is found between the N junctions, emitter, and collector terminals (Fig. 11-42). A two-junction transistor is current operated, while a field-effect transistor is voltage-operated. The bipolar transistor might be enclosed in a plastic or metal case. Sometimes the collector terminal or ground is connected to the tab or metal case.

The TO-72, TO-220, and TO-40 series are the most commonly used bipolar transistors. Power TO-220 or TO-40 series transistors might be mounted on heat sinks. The metal heat sink helps dissipate the heat created by the power output transistor. The center terminal of the TO-220 transistor might contain the collector terminal, which has a positive voltage with the NPN type and should be insulated from the metal heat sink. A piece of mica insulation prevents the voltage from shorting to the ground

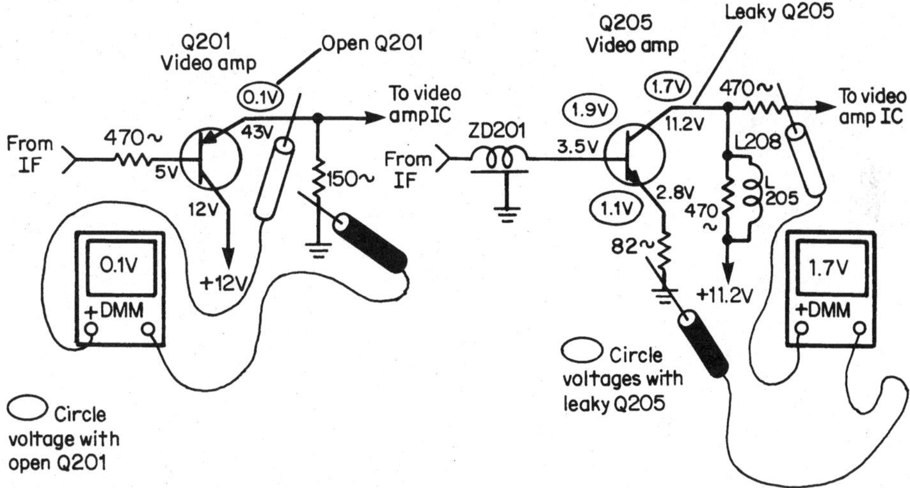

11-41 The different voltage measurements of a PNP and NPN video amp transistor with correct DMM tests.

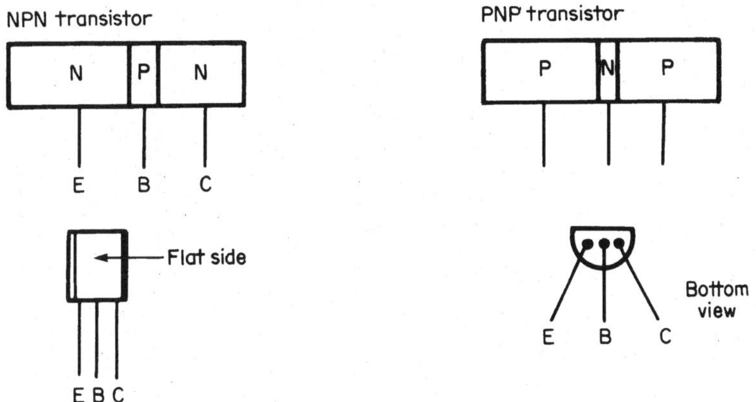

11-42 The base terminal (P), emitter, and collector (N) terminals of the NPN and PNP transistors.

and heat sink. Both transistor and metal heat sink should have silicon grease applied to help dissipate heat to the heat sink. The metal TV chassis might serve as a heat sink (Fig. 11-43). Always replace the metal heat sink on the replaced transistor to prevent damage to the semiconductor.

Testing

A NPN transistor will have a common measurement if the red probe (+) of the DMM is placed on the base terminal in the diode-transistor test of the DMM. When you receive a reading with the positive terminal common to the emitter and collector, you may assume you have located the base terminal, and the transistor is an NPN type (Figs. 11-44 and 11-45). Most transistors found in today's electronic products are NPN types. Remember, the collector terminal of an NPN transistor will have a positive reading, while the PNP transistor has a negative voltage reading on the collector terminal.

The base terminal is always common to the emitter and collector terminals. With the black probe at the base, and a comparable resistance reading on the emitter and collector, the transistor is a PNP type.

When the correct terminals of a suspected transistor are noted, start first with the NPN diode test. Place the red probe (+) to the base and the black probe (−) to the emitter. Switch the two test leads if there is no reading; the transistor may be a PNP type. Double-check the schematic to see if the transistor is an NPN or PNP transistor. If the transistor is an NPN type and the red probe is at the base terminal, you should get a resistance reading between the base and emitter terminals. No reading indicates that the transistor is open between these two elements (Fig. 11-46). Replace the open transistor.

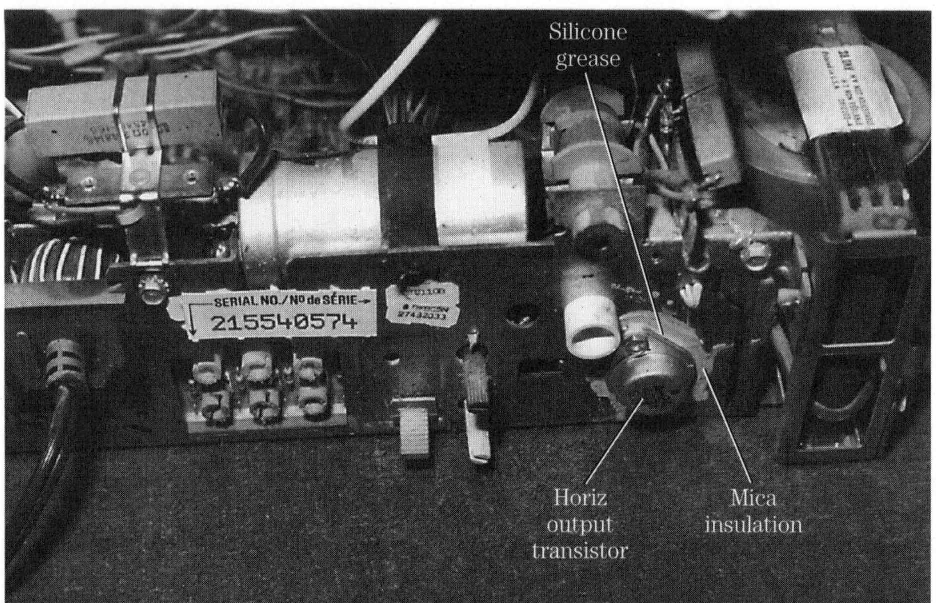

11-43 The horizontal output transistor is insulated from the chassis heat sink in an RCA TV.

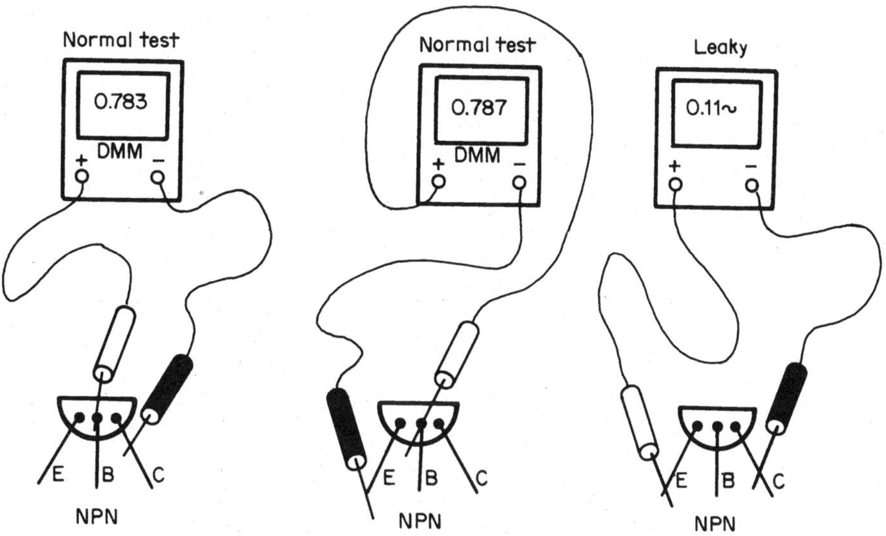

11-44 The normal and leakage test of an NPN transistor on the diode test of the DMM.

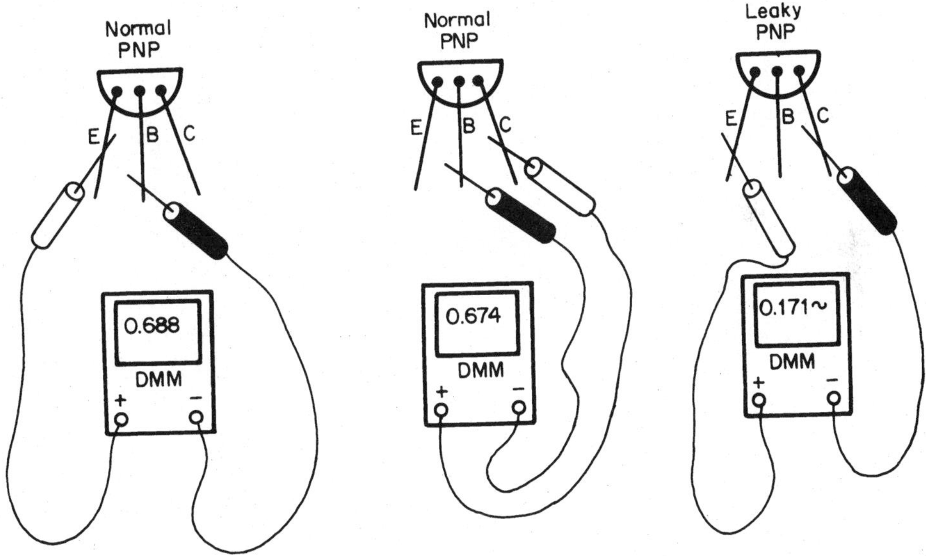

11-45 A normal and leakage test of a PNP transistor with a diode test of the DMM.

Now place the black probe (−) to the collector terminal, with the red probe (+) still at the base terminal. You should have a comparable resistance reading to the base and emitter measurement. If not, the transistor is open between the base and collector terminals. Double-check the same test procedures because a poor connection of either terminal might give a poor or no-resistance measurement.

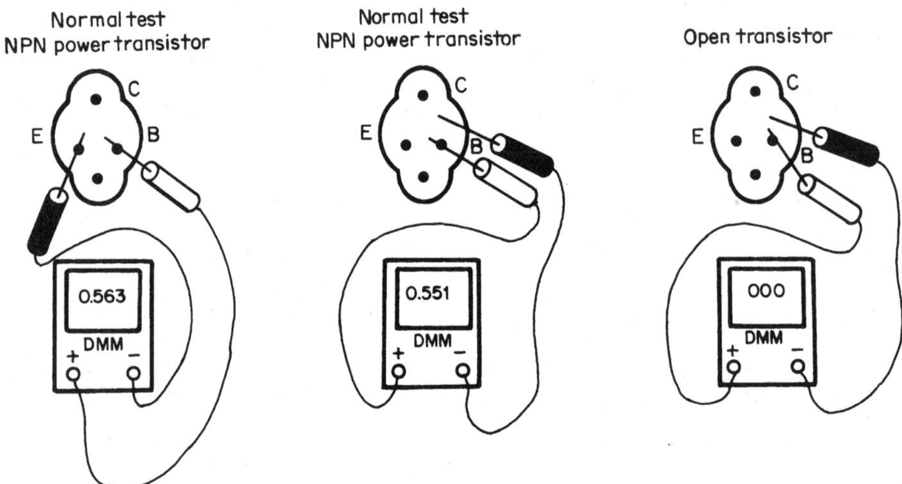

11-46 The normal and open test of a power transistor with diode-junction tests of the DMM.

When testing a transistor for normal measurement, you might find that a transistor has a very low reading, indicating leakage between two elements. In some cases, the transistor may have a normal transistor diode test, but when the test leads are reversed, the transistor shows leakage (Fig. 11-47). Heavy leakage between the emitter and collector terminal indicates a leaky transistor. A very low resistance indicates a shorted transistor. Check for leakage between any two elements. If a low reading is noted in the diode-transistor tests, switch the DMM to the 200 or 2K-ohm scale for actual leakage in resistance. If in doubt, go to the 20K-ohm scale and check for leakage in the reverse direction. Replace the leaky transistor if any reading is noted on the 20K-ohm scale in both directions.

Darlington transistors

What and where

Two bipolar transistors directly coupled in series are called a *Darlington pair*. Actually, two transistors are found in a single component (Fig.11-48). The two transistors are evenly matched and should be replaced with the original part number. Darlington amplifier transistors are found in the high-gain amplifier circuits of the receiver and in the high-powered amplifier circuits. The Darlington transistor might have bias resistors enclosed inside with the two transistors of a single component.

Testing

Test the transistor with the diode test of the DMM. Notice the low-resistance test between elements and with reverse leads that show infinite resistance. A leaky transistor will indicate a low measurement between any two elements. The base-to-emitter-diode

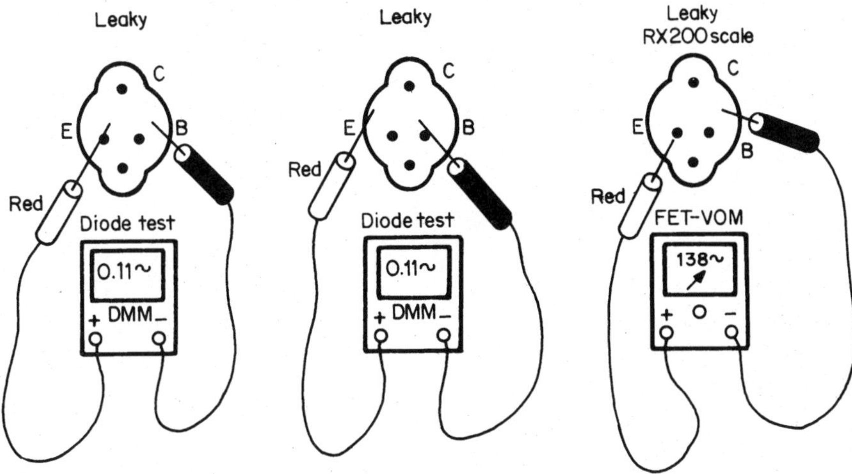

11-47 A leakage between emitter and collector, base and collector with a DMM and FET-VOM ohmmeter scale.

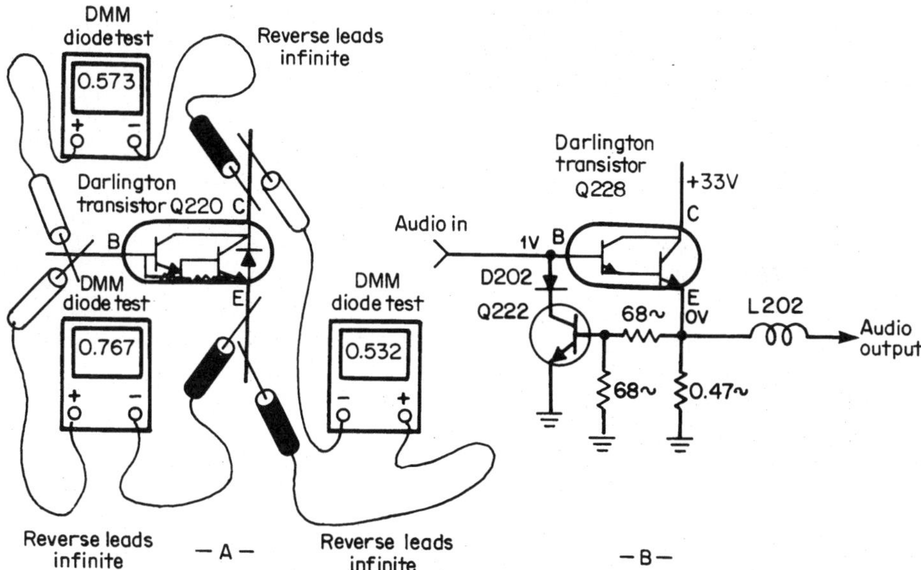

11-48 Transistor Q228 serves as an audio output amplifier in the early audio circuit. Notice that all reverse test probes have an infinite measurement.

test will indicate a high resistance, which is normal. A low resistance between the base and emitter might indicate a leaky transistor or an internal bias resistance. Critical in-circuit voltage measurements will indicate a defective Darlington transistor. If in doubt, sub a new one.

FET and JFET transistors

What and where

FET is the abbreviation for field-effect transistor. Field-effect transistors are found in high-frequency, FM-RF stages, mixers, IFs, TVs VHF tuners, shortwave receivers, VHF/UHF amplifiers, chopper switching, low-power switching, power output, and test equipment. The FET-VOM test instrument employs a field-effect transistor for sensitivity and high-impedance input, like the vacuum tube.

The FM-RF stage within the early FM receiver might have a bipolar or FET transistor as an RF amplifier. The input rod antenna might connect through the RF coil or filter network to the gate (G) terminal. The gate terminal has a very high impedance to the incoming signal. The drain (D) terminal is comparable to the transistor collector terminal with a positive voltage, while the source (S) is at ground potential (Fig. 11-49).

Testing

Check the FET transistor in the circuit with accurate voltage measurement. Test the FET in or out of the circuit with the diode test of the DMM. A leaky FET transistor might show leakage between the drain and source or gate and source (Fig. 11-50).

Check the resistance of the three terminals with the diode test of the DMM. The resistance between the gate and source and the gate and drain is quite similar. An infinite resistance is measured with reversed test leads. Likewise, the resistance is similar with the RX10 scale of an FET-VOM on the same two elements. The resistance between the drain and source is quite different between the drain and source terminals with a test of the FET-VOM. Around 85 ohms is measured with the red probe at the drain and with the black lead at the source terminal with reversed FET-VOM test leads.

MOSFET transistor

What and where

MOSFET is the abbreviation of a metal-oxide silicon field-effect transistor. The MOSFET is found in RF stages, IF stages, switching, RF mixers, small signal amplifiers, UHF amps, RF amps for UHF TV tuners, chopper-switches, and high-power switching. Today the MOSFET is found in high-powered audio output amplifiers in auto systems and high-speed switching. An N-channel enhancement-mode high-speed switch NTE2385 has a TO220 case style (Fig. 11-51). The power N-channel MOSFET is for switching power supply inverters, relay drivers, motor controls, and audio power amps.

The dual-gate MOSFET transistor is found in VHF amp-mixer, UHF/VHF preamp, and FM-RF stages. Gate 1 of the FM-RF amp connects to the input signal, while gate 2 contains a bias voltage. The source (S) terminal is at ground potential through

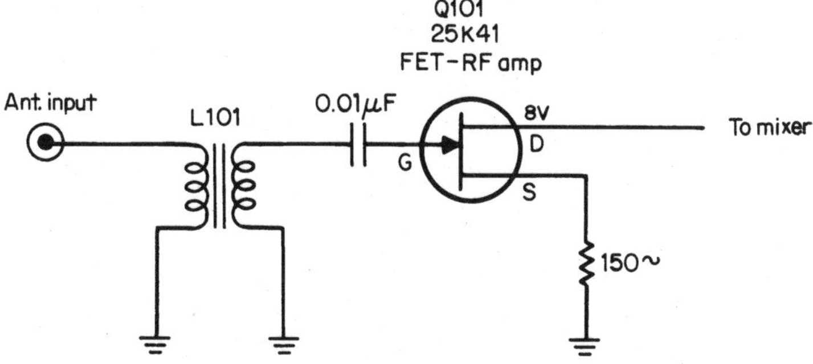

11-49 The FET transistor is found as an RF amplifier in the AM-FM-MPX receiver.

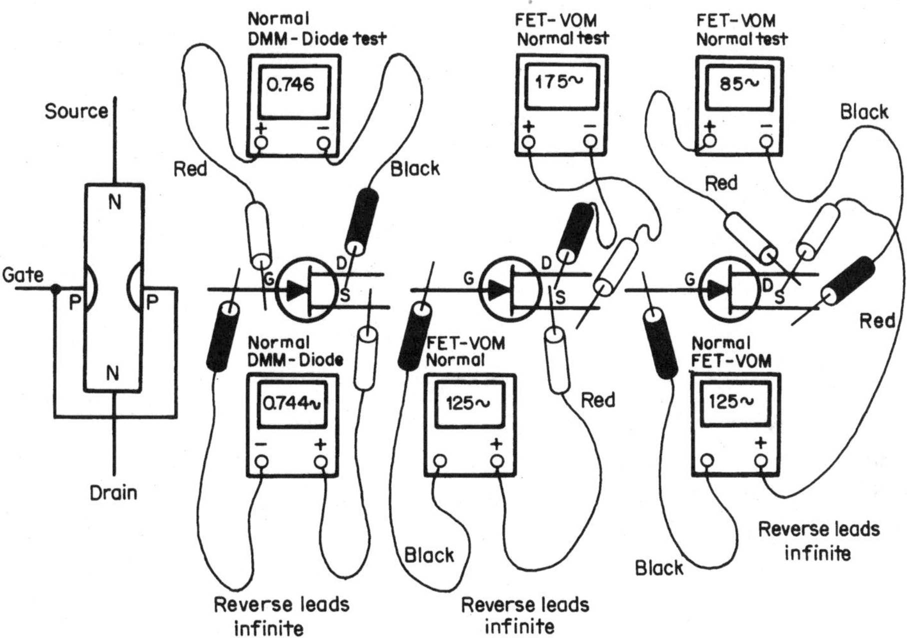

11-50 All reversed test leads on the suspected FET transistor are infinite, and a very low resistance measurement between drain (D) and source (S) indicates a leaky transistor in both directions.

a small 100-ohm resistor. The amplifier RF signal output is taken from the drain (D) terminal, which has the highest B+ voltage (Fig. 11-52). The source terminal might be attached to the metal case. Keep your body grounded when working around or testing MOSFETs.

Testing

Check the MOSFET transistor with accurate voltage measurements and diode-DMM tests. A resistance measurement is noted on any two terminals when

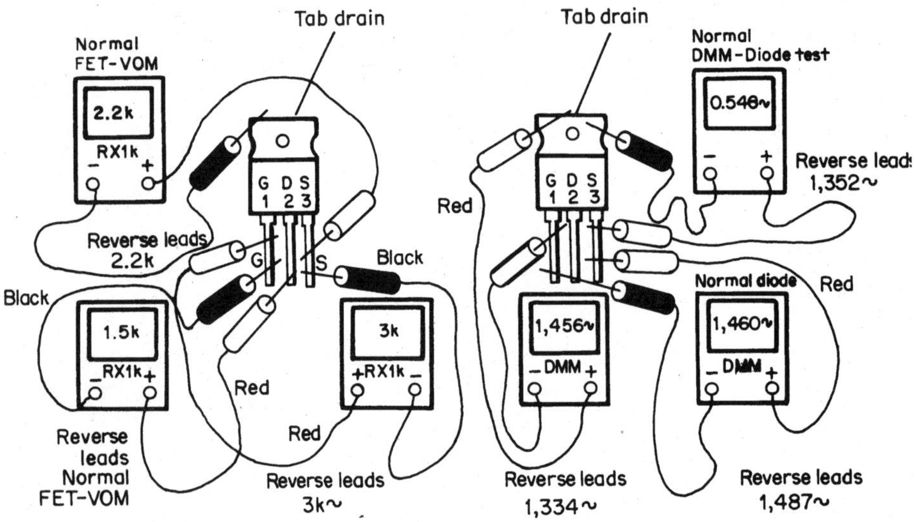

11-51 The normal and reverse test lead resistance measurements with the diode test of the DMM and the RX1K scale of the FET-VOM.

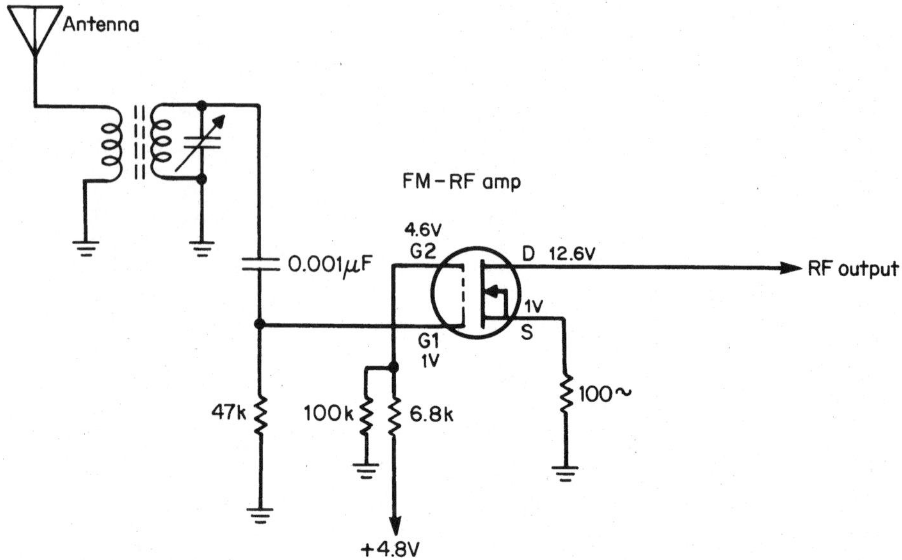

11-52 The dual-gate (G1 and G2) MOSFET transistor serves as an FM/RF amp in the deluxe AM-FM-MPX receiver.

testing the power MOSFET. Notice that the resistance measurements are above 1K ohms with both the DMM and FET-VOM terminal measurements. When the test leads are reversed, notice that the FET-VOM measurements are about the same. The reverse measurement of the DMM between drain and source is much higher (Fig. 11-53). Suspect a shorted or leaky power output MOSFET if there is a very low ohm measurement.

Unijunction transistor (UJT)

What and where

The unijunction transistor is a semiconductor component with a silicon bar and a single PN junction as the emitter terminal. Two base terminals at each end might be called a double-base diode (Fig. 11-54). Unijunction transistors are found in oscillator circuits.

Testing

Take critical voltage measurements while in the circuit. Check with ohmmeter tests like those found with the FET transistor. You should measure a normal constant value (100 to 10K ohms) with FET-VOM probes across the drain and source terminals. When test leads are reversed, you should read the same value. Replace the transistor if resistance is not the same.

Connect the probes to the gate and source terminals, and you should have a normal low-resistance measurement. Reverse the test probes and you should have

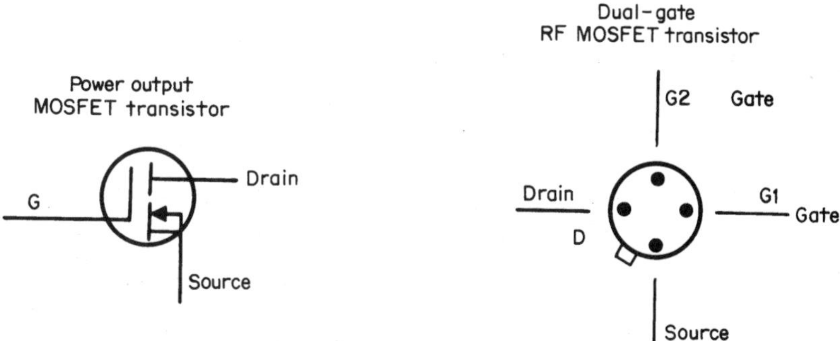

11-53 The gate, source, and drain terminals of a power MOSFET and dual-gate MOSFET transistor.

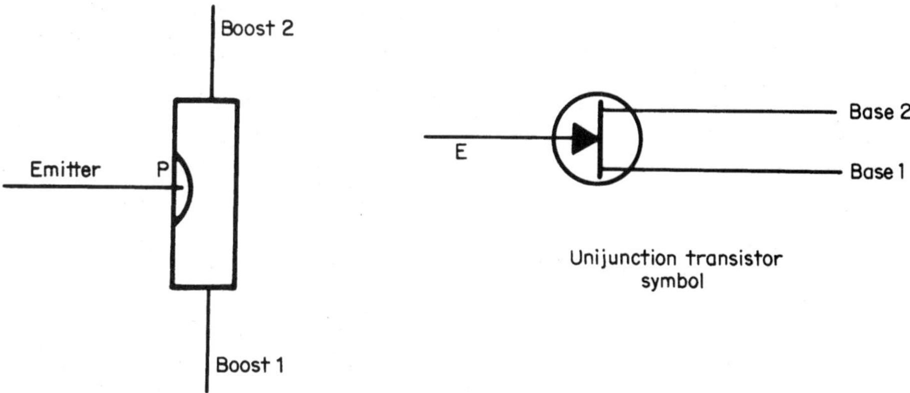

11-54 The unijunction transistor is used in the oscillator circuits of receivers and projects.

either a normal very high resistance or open measurement. Replace the transistor if a low measurement is found in both directions.

Special transistors

Signal transistors (RF power)

What and where

The RF power transistor might be used in the HF/SSB/CB amateur, low-band FM, VHF AM/FM, VHF marine/mobile, and UHF-FM circuits. These signal transistors might appear in RF power output and driver circuits. The frequency might range from 2–30 MHz, 27≈50 MHz, 136–175 MHz, 407–512 MHz, and 806–950 MHz. Most of the high-powered RF transistors have stud-bolt mountings.

Testing

Check the high voltage applied to the RF transistor. Test the RF transistor as with any NPN or PNP general-purpose transistor with the transistor tester or diode test of the DMM.

Power amplifier transistors

(See Signal RF power transistors)

RF amplifiers (radios)

What and where

The RF amplifier within the AM radio circuit might be a bipolar, FET, JFET, or a MOSFET transistor. RF amplifiers are found in both AM and FM sections of a deluxe receiver, while the AM converter transistor serves as both RF and oscillator transistor. The conventional AM clock radio might have a ferrite antenna coil that is transformer-coupled to the base terminal of Q6 (Fig. 11-55).

If only one local station can be tuned in, you may assume the problem relates to the front-end circuit. When touching the base or collector terminal of the RF transistor with voltage measurements of a dead radio and you can hear a weak local station, suspect a defective RF transistor. Inspect the antenna coil for breakage or broken connecting wires.

Testing

Measure the small dc voltages on the converter transistor. Ignore a possible weak IF stage if the FM stations are normal in the AM-FM receiver. Lower voltages will be measured with a leaky RF transistor. Test the transistor in circuit with a transistor tester or diode-junction test of the DMM.

Remove the suspected transistor and test out of the circuit. If the transistor appears intermittent, spray it with several coats of coolant. Then place the iron tip near the transistor and see if music is restored. Sometimes when a defective transistor is found with in-circuit tests and it is normal when removed, replace it. Just heating the transistor terminals during removal can restore an open internal junction. Cut the replacement transistor leads the same length as the original. Universal replacement transistors work well in radio circuits.

TV deflection transistor (horizontal output)

What and where

The horizontal output transistor receives the oscillator or countdown signal and drives the flyback or horizontal output transformer. A damper diode is found in the collector terminal to dampen oscillations to the power supply. Today, the damper diode is enclosed within the horizontal output transistor (Fig. 11-56). The horizontal output transistor is mounted on a separate heat sink or metal chassis.

The horizontal output transistor causes most of the problems within the TV chassis. A leaky or shorted output transistor might blow the fuse, shut down the chassis, open isolation resistors, and blow a secondary fuse. Sometimes the output transistor destroys the flyback, or vice versa. The output transistor can be damaged if the drive voltage or waveform is not found on the base terminal (Fig. 11-57).

Testing

Take a quick waveform with the scope probe alongside the flyback transformer (Fig. 11-58). If there is no waveform, check for a drive waveform at the base terminal. Suspect a defective driver transistor or horizontal deflection IC if there is no drive voltage or waveform. Check the voltage applied to the primary winding of the flyback. Low or improper voltage might be caused by a leaky output transistor or improper supply voltage. Check the transistor for leakage between collector and

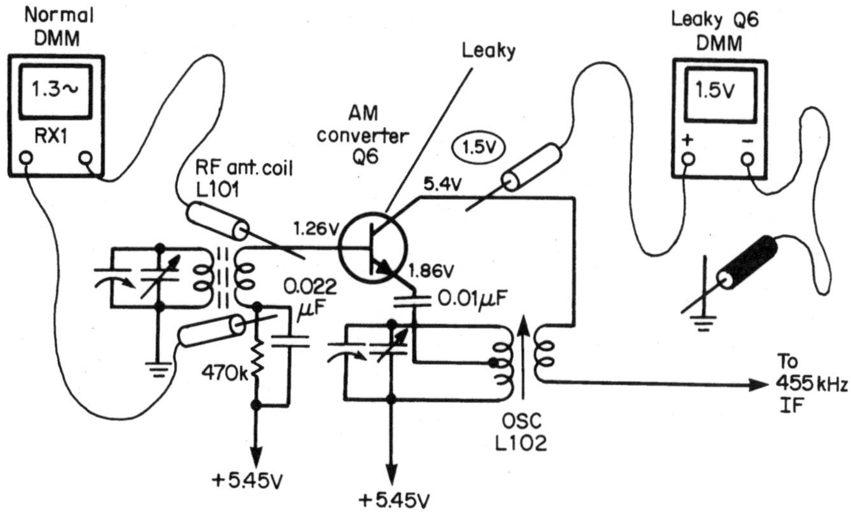

11-55 First check the RF transistor with voltage and resistance measurements. Then take in-circuit transistor tests.

11-56 The damper diode is found mounted inside the latest horizontal output transistor.

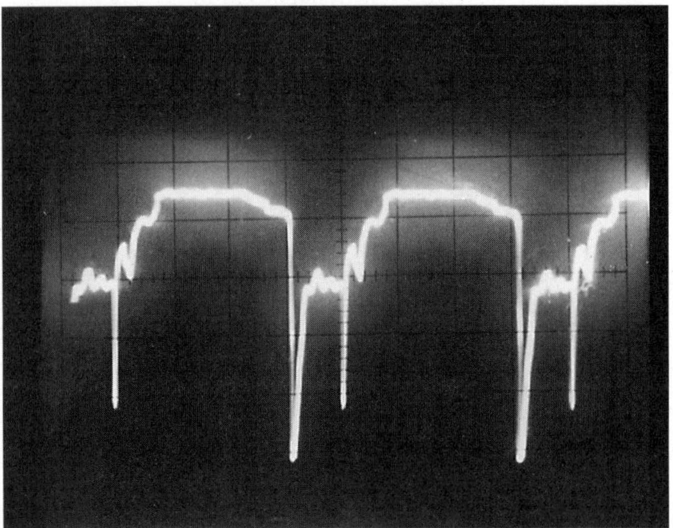

11-57 The normal base scope waveform of the horizontal output transistor in the TV chassis.

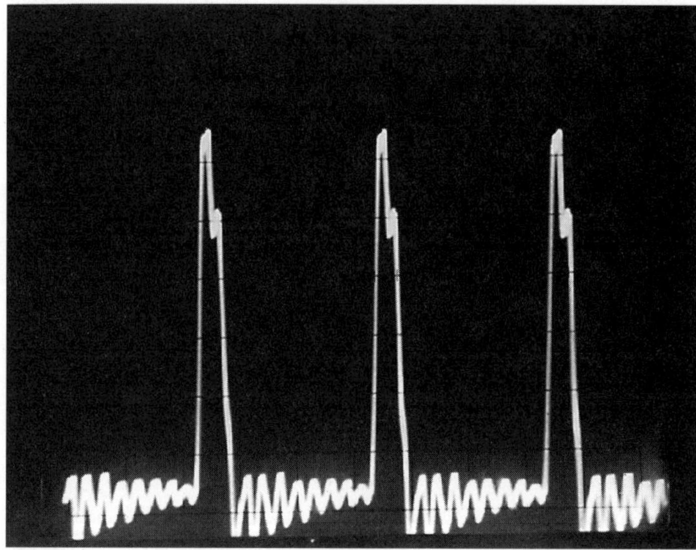

11-58 Place the scope probe alongside the flyback to determine if the horizontal output circuits are normal.

emitter terminals. If the output transistor is found leaky or shorted, make sure the correct drive waveform is present or the new replacement can be damaged.

Replace leaky output transistors with the exact part number. If not handy, replace with a universal replacement. Sometimes the original output transistor of a foreign TV is difficult to locate. Check the Howard Sams PhotoFacts for universal replacement. Also, check the universal semiconductor manual for a universal replacement. Often, the part number is stamped on the body of the output transistor.

TV power regulator (SCR)

What and where

The power regulator found in the TV chassis is usually an SCR type, except the power line regulator given in Chapter 9. The high-voltage regulator provides a regulated dc voltage to the flyback and horizontal output transistor. The anode terminal has a high dc voltage from the bridge circuit with the gate voltage controlling the +129-volt output (Fig. 11-59). The gate circuit is controlled through a small transformer from an error amp and sawtooth generator in the RCA CTC167 chassis.

When the SCR regulator becomes open or leaky, the output voltage might be zero. If the gate circuits malfunction, no or low voltage will be found at the cathode terminal of SCR. Improper voltage at the anode and defective gate circuits might remove the collector voltage found on the horizontal output transistor.

Testing

Take critical voltages on the three SCR terminals. If there is no or low dc voltage at the anode (A), suspect a defective low-voltage supply source. If there is no

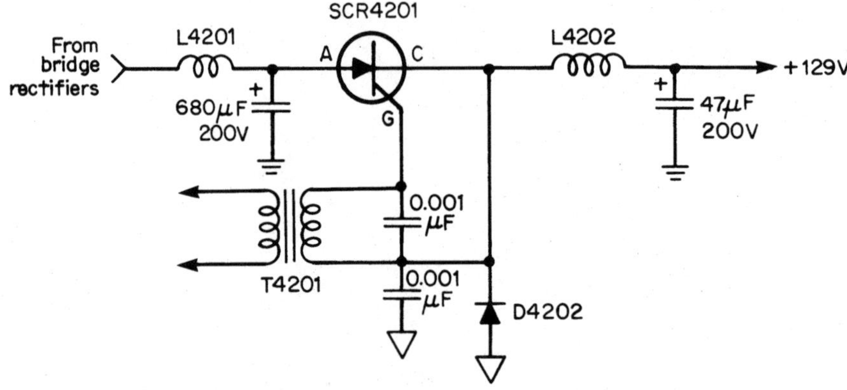

11-59 SCR 4201 is found in the +129 voltage source that supplies voltage to the horizontal output transistor in the TV chassis.

voltage at the cathode (C) terminal, suspect a defective SCR or gate circuit. Compare the waveforms found on the gate transistor circuit. Check the SCR with the SCR tester. Take critical resistance tests on the SCR after removing it from the circuit (Fig. 11-60).

TV vertical output transistors

What and where

Although the recent TV chassis might have an output IC in the vertical output circuits, you might still find two output transistors in some TV chassis. One transistor (Q501) affects the top sweep, while the bottom (Q502) provides the bottom sweep of the raster (Fig. 11- 61). The top transistor might have the highest collector voltage, while the bottom transistor emitter terminal is at ground potential. You can easily spot the two vertical output transistors mounted on separate heat sinks on the PCB.

Often, the output transistors are mounted directly on the heat sink without the insulator, since the heat sink is not at ground potential. You can identify each transistor by measuring the voltage on each heat sink. Transistors, diodes, and electrolytic capacitors cause the most problems within the vertical output circuits. When one transistor is found leaky or shorted, replace both vertical output transistors.

Testing

Take critical voltages on each output transistor. Test each transistor with in-circuit transistor tests (Fig. 11-62). A high dc voltage measurement at the bottom transistor (Q502) collector terminal might indicate an open Q502 transistor. Intermittent output transistors might indicate a normal transistor when chassis is shut down. A leaky output transistor (Q501) might show a high dc voltage on the collector terminal of Q502. When the original transistor part number is not available, replace with universal replacements.

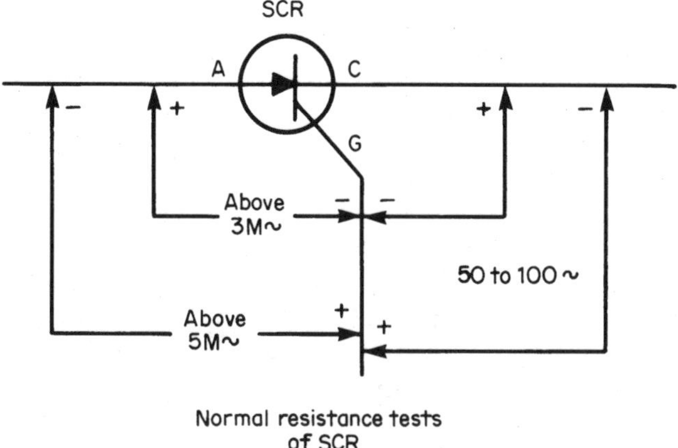

11-60 Test the SCR by taking resistance measurements on each terminal. A very low resistance between gate and anode indicates a leaky SCR.

11-61 The two vertical output transistors might be mounted on separate heat sink on the PCB.

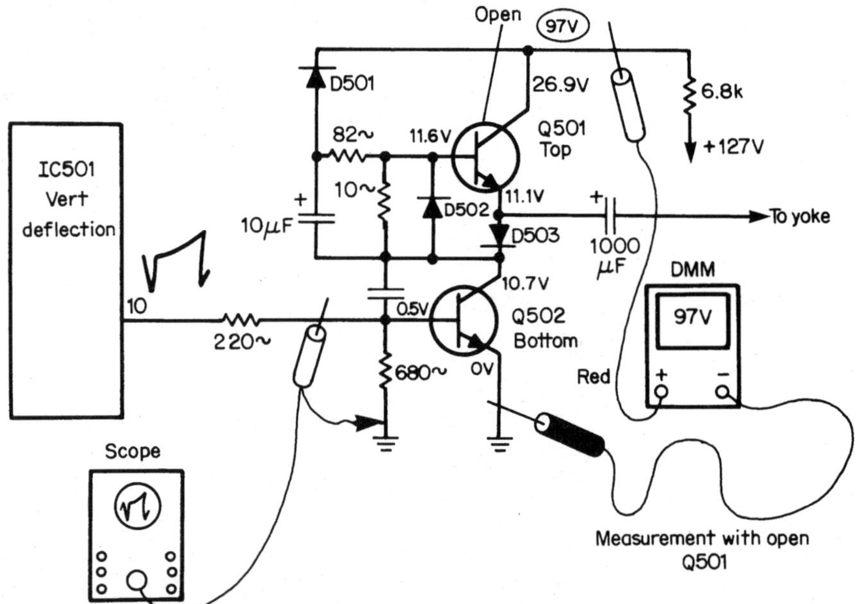

11-62 Take critical voltage measurements on vertical output transistors and scope waveforms to determine if transistors are defective.

TV video amp transistor

What and where

Today, the video amp transistor might be called a buffer amp located between the IF/DET and video IC circuits. You might find one or two video amps or buffer stages within the video circuits. Often the transistor might be a PNP type. The collector terminal of the video amp is at ground potential, while the emitter has the highest voltage within the PNP transistor video circuit (Fig. 11-63).

Testing

Check the suspected video transistor with in-circuit transistor tests using a transistor tester or diode-junction test of the DMM. If in doubt, scope the input video waveform at the base and at the output collector terminal. When a normal video signal is found at the input and there is no output, suspect an open or leaky video transistor (Fig. 11-64).

Tripler

(See Chapter 6)

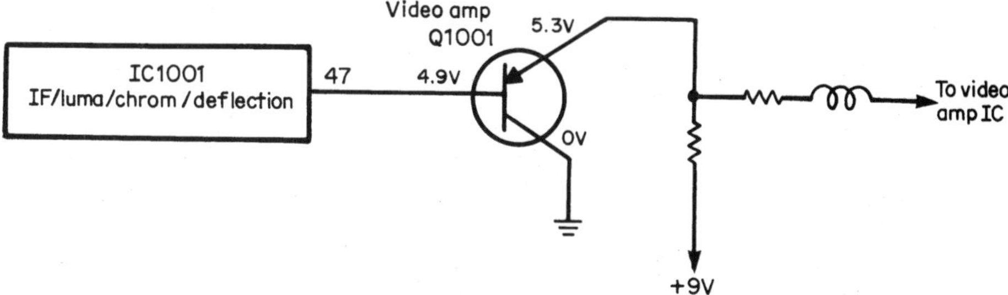

11-63 The video amp might be a PNP or NPN type located between the IF and video amp IC in the TV chassis.

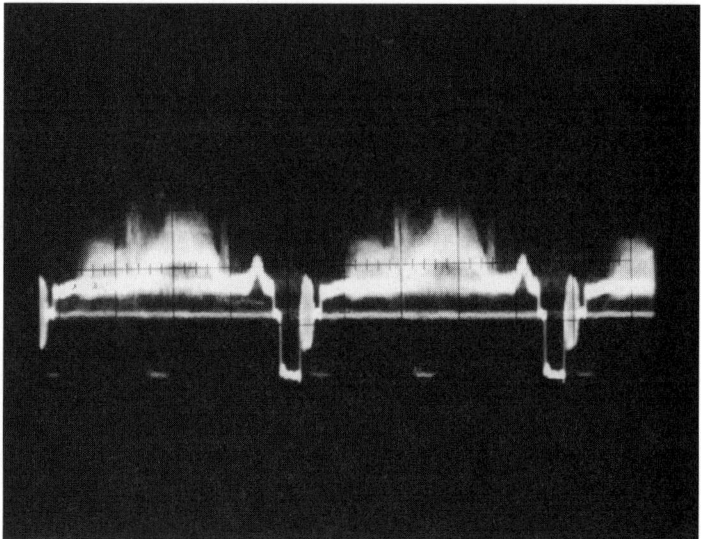

11-64 The on-air signal waveform at the base terminal of the video amp transistor.

12 CHAPTER

Testing electronic components S through T

(Signal devices through switches)

The signal device that produces a noise, signal, or warnings might be a buzzer, Sonalert(r), piezo, or electronic siren (Fig. 12-1). The soldering equipment might consist of a low-wattage soldering iron, battery iron, or a solder station that is heat regulated. Most electronics technicians have a regulated soldering station that is on all the time when the bench switch is flipped on. Speakers come in many sizes and shapes and might be tweeters, midranges, or woofers. There are many different switches used to turn electronic products on and off; there are pushbutton switch assemblies for different operations and many kinds of function switches. All of these electronic components are discussed in this chapter.

Signal devices

Sonalert(r)

What and where

The Sonalert(r) is a solid-state device that produces a loud alarm when voltage is applied. The Sonalert(r) is found in audible sound-alarm systems and projects. The audible signal device might send out a continuous, multipulsating, or intermittent tone. The tone frequency might operate from 1900 to 4500 Hz, and these signal devices operate from ac or dc voltages. The ac voltage ranges from 6 to 250 Vac, while the dc voltage varies from 6 to 30 Vdc. Sonalert(r) is a registered trademark of a signal-type

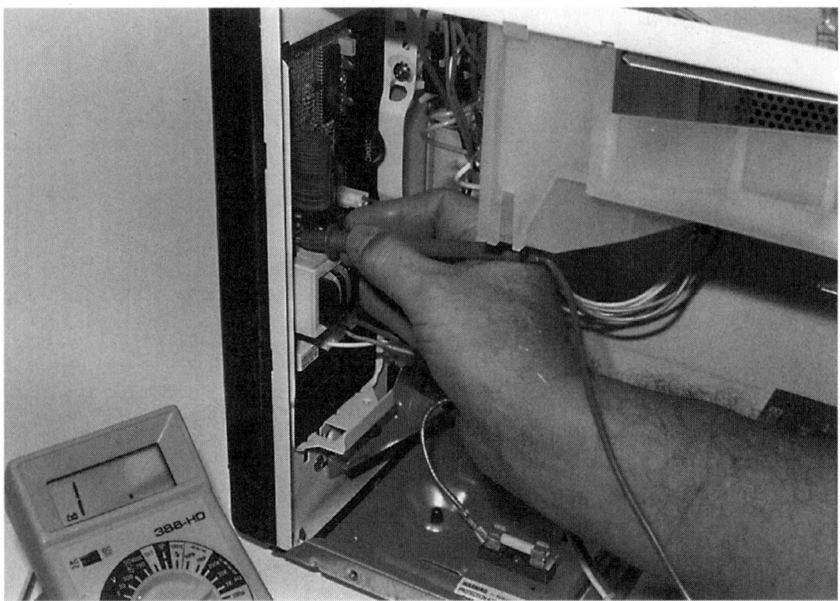

12-1 The electronics technician is testing the piezo warning circuits in the microwave oven.

buzzer. The standard Sonalert(r) operates from 3 to 20 Vdc with a maximum current of 3 to 16 milliamperes.

Testing

Check the ac or dc voltage applied to the signal device terminals. Suspect a defective power supply source or batteries when improper voltage is measured at the buzzer terminals. Replace the signal device if there is no sound and voltage source is normal

Solid-state buzzers

What and where

The solid-state buzzer might consist of piezoelectrics, piezo indicators, piezo transducers, Piezo-A-Lerts, piezoelectric loudspeakers, and audio transducers. These audio solid-state devices might be found in microwave ovens, telephone ringers, telephone answering machines, and smoke detectors as buzzers and alarm systems, to name a few. The solid-state piezo buzzer or alarm operates from 1.5 to 30 Vdc at a 1.5-kHz to 6-kHz frequency. They might have self-drive, external drive, or built-in type circuits. The piezo indicators and transducers pull very little current (1.5 to 20 mAs).

Testing

Measure the dc voltage applied to the dead solid-state buzzer (Fig. 12-2). Remove the indicator terminal leads and apply an external dc voltage to the solid-state device. Replace the piezo buzzer with correct voltage applied, and check the dc voltage source when improper voltage is applied.

Piezo-a-lert or audio transducers

What and where

The Piezo-A-Lert buzzer and audio transducer might have a continuous, intermittent, or dual continuous tone. The audio transducer is used in burglar alarms, paging systems, smoke detectors, gas detectors, cash registers, alarm clocks, fish finders, metal detectors, test instruments, cameras, automotive warning systems, and medical equipment. A continuous-tone internal circuit of an audio transducer might appear as in Figure 12-3.

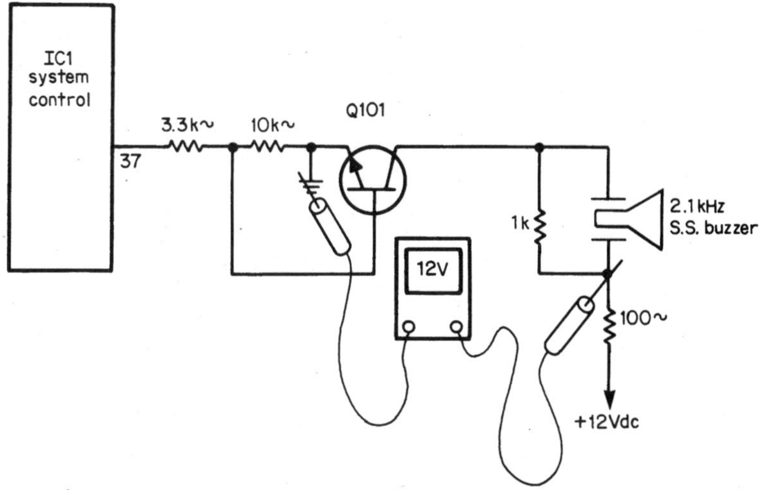

12-2 Taking a voltage test with the DMM within the microwave oven piezo buzzer circuits.

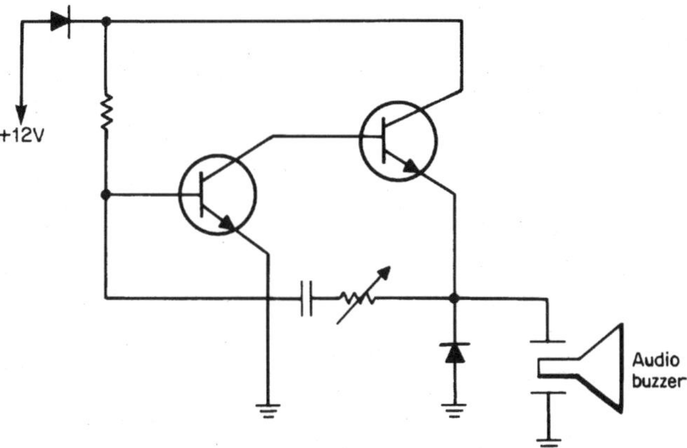

12-3 The internal circuit of a continuous tone of the audio transducer.

Testing

Measure the dc voltage applied to the transducer or piezo buzzer terminals, and recheck for correct polarity. Usually, a positive (+) sign is found on the positive terminal of the indicating device. Replace the defective unit that has no sound and correct applied voltage. {{author query: Previous sentence okay?}}

Piezoelectric ceramic buzzers

What and where

The piezoelectric buzzer is similar to the audio transducer unit that is found in smoke detectors, alarms, detectors, home appliances, radio beepers, electronic games, keyboard signals, telephone ringers, and cameras. The operating frequency might appear between 2.5 and 5 kHz. Most piezoelectric buzzers operate from batteries and a dc voltage source of 3 to 20 Vdc.

These piezoelectric buzzers have a low power consumption, long life, thin elements, external drive, and a high sound-pressure level (Fig. 12-4). Often, a transistor drives the piezoelectric buzzer.

Testing

Measure the voltage across the piezoelectric ceramic buzzer. Suspect a defective unit if correct voltage is found applied across the buzzer terminals. Check the drive transistor with in-circuit tests. Remove one end of the resistors and diodes for a correct resistance measurement, if accessible.

Spark gaps

What and where

Spark gaps are found in TV circuits when a high voltage might arcover and destroy components. You will find spark gaps on picture tube elements within the CRT socket. When the picture tube element arcs over, high voltage occurs at the gap areas and shorts out to ground, protecting the CRT and components tied to the CRT elements (Fig. 12-5). Sometimes a spark gap is found across a tripler unit in the early TV chassis.

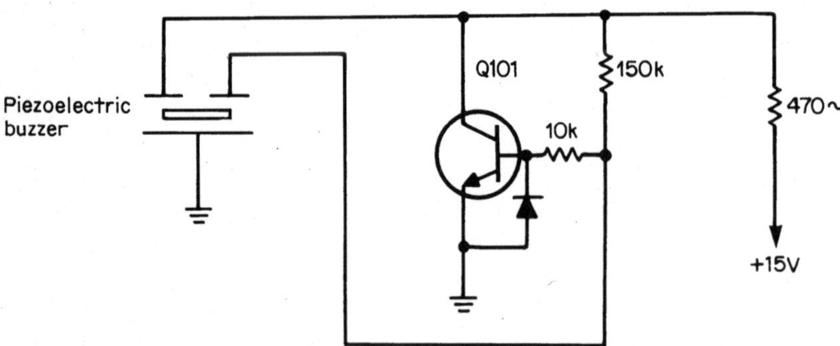

12-4 The piezoelectric driver circuit found within the ceramic buzzer.

12-5 Spark gaps are built right inside the CRT socket to protect the tube element circuits and components.

You will find a spark gap on each element of the picture tube to ground. These small spark gaps are constructed right inside the socket area. When the spark gap areas are loaded with dust, the TV chassis might start up, arcover, and then shut down. Usually, one of the color guns will arcover and show a predominate color before shutdown (Fig. 12-6). Sometimes, just blowing the dust out of the socket cures the problem. If not, replace the picture tube socket assembly and harness.

Testing

Check for correct voltage at each CRT pin for a possible shorted spark gap. Be very careful, as the focus and screen voltage should be measured with an HV meter probe. Shut the chassis down. Discharge the picture tube at the anode terminal to ground wire around the CRT. Remove the CRT socket. Take a low resistance measurement between each pin and chassis ground. Take a ¼"-resistor lead and stick it in each socket hole to measure the resistance to common ground. Replace the CRT socket and harness with the original part number when a spark gap is arced over to ground.

Solar cells

What and where

A photovoltaic cell converts light to electrical energy. The voltage developed by each cell is 0.45 volts or less (Fig. 12-7). Some efficient cells deliver 0.55 Vdc in full

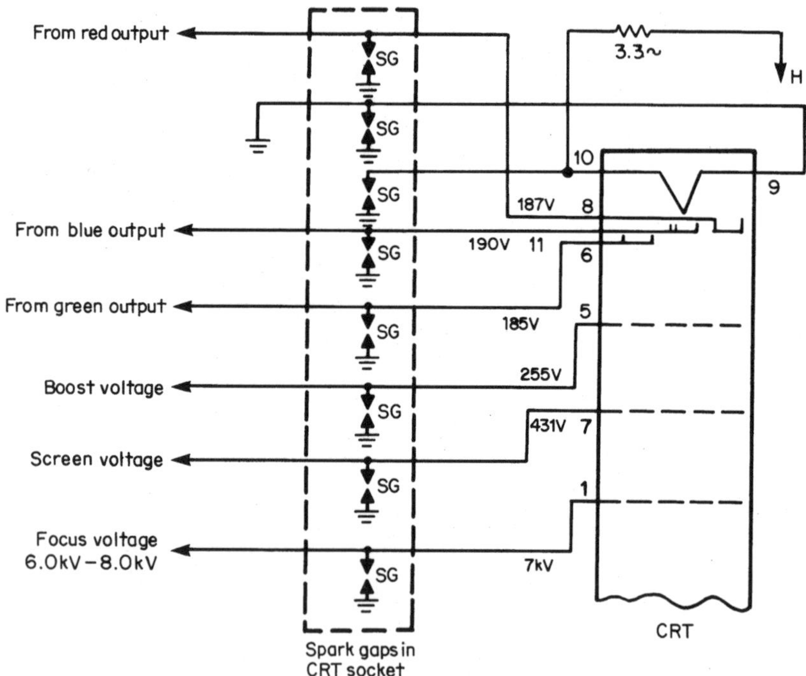

12-6 Each element of the picture tube has a spark gap so the excessive high voltage can arc over to ground.

12-7 The solar cells can be ¼, ½, completely round, or in broken pieces to provide voltage from outside light.

sunlight. When light strikes the surface of the silicon cell, current begins to flow from the negative pole to the positive. The top of the cell, with gridlike lines, is negative, and the bottom soldered side is positive. The current capacity of the cell is always equal to the physical size of the cell. The larger the size of the cell, the greater the current capacity.

Most separate solar cells consist of half-round or round 2-, 3-, or 4-inch solar cells. The cells are connected in series for a higher voltage. A solar panel consists of many separate solar cells. Connect solar cells or solar panels in parallel for greater current capacity. The solar panel might be rated at 21 Vdc with an open circuit and when loaded down, 14.5 Vdc is measured.

Today, solar cells are used to power radios, science fair experiments, lights, water pumps, railroad warning lights, road signals, commercial lighting, toys, navigational aids, electronic projects, communication systems, calculators, cabins, and home electric systems. A large solar panel might be used to charge batteries in a radio, cabin, or home electrical system.

Testing

Measure the required voltage of a solar cell (0.45V) or a solar panel under a strong light or while operating in the sun. Place the solar panel in the sun for maximum voltage output (Fig. 12-8). A poorly soldered connection can quickly reduce the low voltage between cells. With the diode test of the DMM, double-check the isolation and polarity diode if found in series with the solar cells. With a low- or no-voltage measurement, suspect a cracked or broken cell and poor connections.

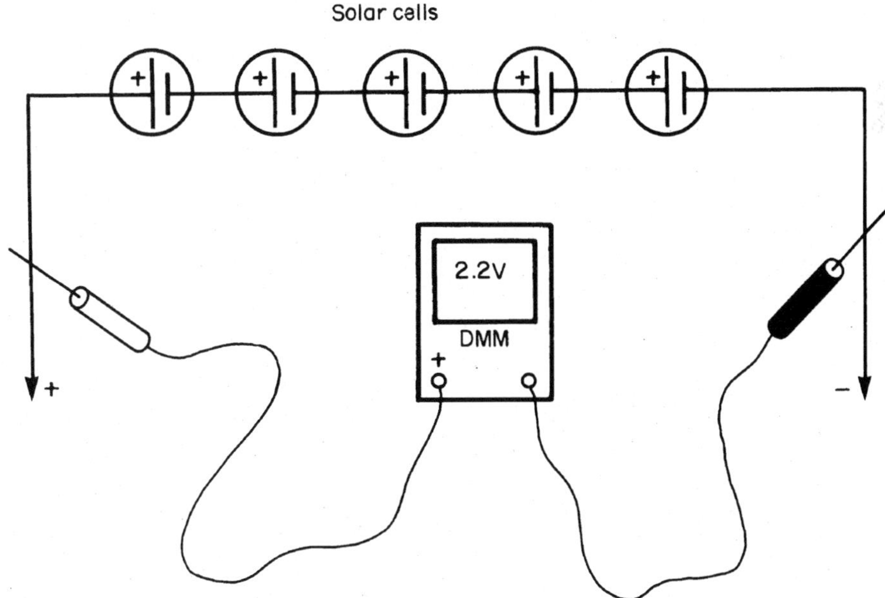

12-8 Solar cells can be connected in series to add voltage and in parallel to increase the current rating.

Soldering equipment

Low-wattage iron

What and where

Operating from a power line or battery, the low-voltage soldering iron applies heat to melt solder and connect electronic components within a circuit. A 25–30 watt soldering iron is ideal for precision work or soldering up solid- state components. The dual-wattage soldering pencil iron operates at 15 watts, and for more power, just flip the switch to 30 watts. A 15-watt pencil iron is ideal for IC work (Fig. 12-9). Today, a gas-powered portable soldering iron operates from standard butane lighter fuel. The portable battery-operated soldering iron is ideal for SMD connections.

Testing

Most problems with the low-wattage iron occur in replacing the solder iron tips. Check for a broken power cord or dirty ac plug when the iron has low or intermittent heat. With a low-ohm scale of the DMM, check continuity at the ac plug to determine if the cord is broken or the iron element is open. The 25-watt ac-operated soldering iron might have a resistance under 500 ohms. Make sure the iron is grounded while soldering up ICs and microprocessors.

The battery iron

What and where

The cordless soldering iron has an isolated tip, is ground free, eliminates electrical leakage, and has low-voltage design, making it ideal in soldering up SMD components,

12-9 The 15–30 watt iron is safe for making connections on ICs or microprocessors.

semiconductors, and ICs. The 25-watt iron sits in a charging stand and can be charged in 60 minutes. The nicad battery and soldering tip are replaceable (Fig. 12-10). The cordless battery iron is ideal in replacing SMD components.

The rechargeable cordless soldering gun has a trigger lock that prevents it from accidentally turning on. The iron is powered by Ni-Cd batteries, so you can use it anyplace. This iron has a trigger-activated light that illuminates the work on the chassis. It is ideal also for hard-to-reach places. The 20-watt gun can be replaced with a 20- or 40-watt blade tip.

Testing

Replacement of the cordless soldering iron and gun tip causes the most problems. Choose a fine-pointed soldering iron tip when soldering SMD components. When the Ni-Cd battery will not hold a charge, replace it. Replace the gun trigger switch with a new part of exactly the same number.

Solder gun

What and where

The soldering or desoldering gun is used for heavy work in the electronic chassis. This soldering gun might be a 100 or 250-watt iron with trigger action. A dual soldering trigger switches in a different coil for 150- or 250-watt heat. The soldering gun operates from the 120-volt ac power line (Fig. 12-11).

12-10 The battery iron is ideal for soldering SMD components on the PCB.

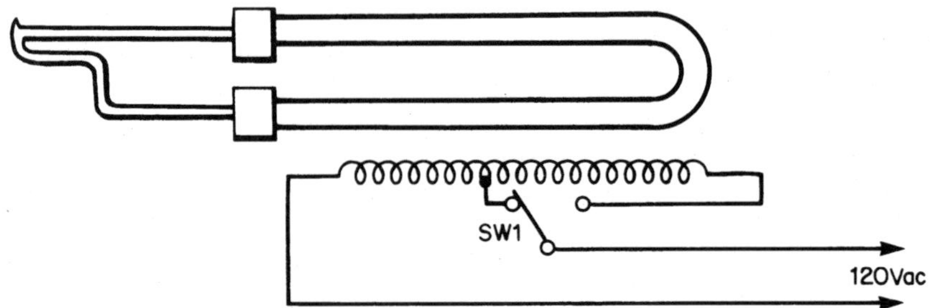

12-11 The soldering gun circuit might have a 150- or 250-watt dual-switching switch of heat.

Testing

Replace the soldering iron tip when it breaks after many hours of soldering. Replace the trigger action switch when it will not make contact or operates intermittently; use the original part number. These can be purchased at regular electronic parts stores. Check the cord for breakage at the ac plug and where it enters the iron. Retighten the soldering iron tip nut when it takes awhile for the iron to heat up. Measure the soldering iron winding for low ohm continuity with the DMM. Check for coil wire breaks at the switch or cord input terminals.

Solder station

What and where

The electronically controlled soldering iron offers the ultimate in temperature control at all times. The low-priced soldering station might have adjustable temperatures from 350 to 850 degrees Fahrenheit. Usually, these irons are 20- , 40-, and 45-watt units. A soldering iron with replaceable tip plugs into the temperature-controlled iron stand (Fig. 12-12). Often you will find a soldering station on the electronics technician's service bench, operating all the time.

The deluxe variable soldering station might have digital display with dual-soldering and a water reservoir that keeps the sponge moist. These units are electrostatic discharge safe (ESD safe), with a grounded iron tip. An electronically controlled desoldering station has a desoldering unit with an internal vacuum pump. Expensive modular soldering stations might have many SMT replacement tips for removing ICs, microprocessors, and gull-wing type SMD components.

Testing

Check the iron resistance after removing it from the station stand. Some soldering stations have replaceable irons and tips. It is best to return the soldering station to the manufacturer when no or improper voltage is found at the iron plug-in socket. Check the ac cord for breaks at the ac plug or where it enters the control station. Keep the sponge moist and replace it when torn or full of holes.

Solenoids

What and where

The solenoid has a multilayer wound coil used in electromagnetic relays. Often the solenoid is wound over an iron core. Relays are found in microwave ovens, TV chassis, audio equipment, automotive electronics, security systems, telecommunications, and power switching (Fig. 12-13). Most relays have a solenoid winding that closes the switch contacts. Locate the relay by a clicking or energizing sound.

Testing

A defective solenoid might have a charred winding, broken connection, poor switching contacts, and open winding. Check the resistance of the solenoid with the FET-VOM or DMM. Compare these measurements with those found on the schematic.

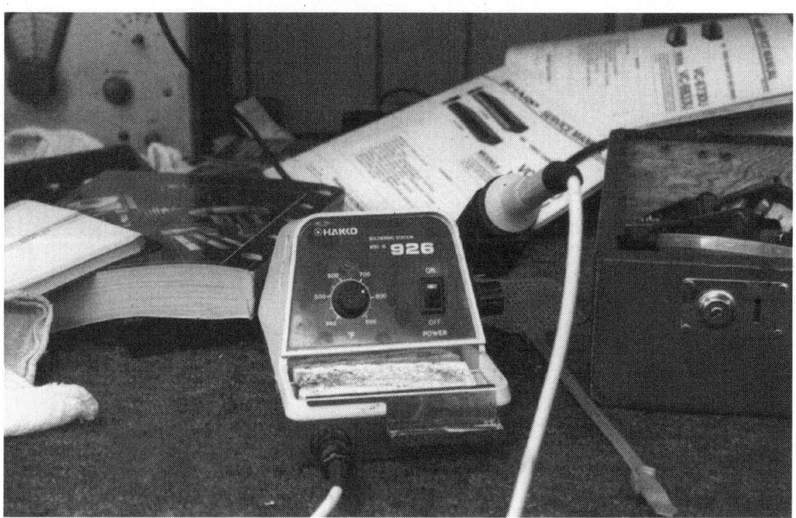

12-12 The temperature control iron is found on the electronic technician's service bench.

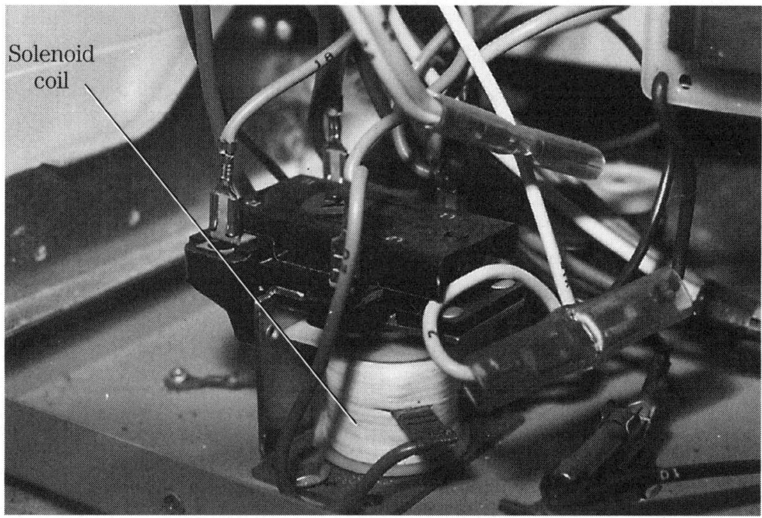

12-13 The solenoid winding is part of a power relay in the microwave oven.

Suspect a defective winding with a very low- or high-resistance measurement. Check the diameter of the wire coil. The larger wire will have a low resistance, while a solenoid wound with fine wire might have a greater resistance measurement. Check for broken solenoid wires at the solenoid terminals.

A screwdriver blade held near the solenoid will indicate if the coil is energized. The relay contacts can be energized manually by removing the cover and holding down the relay assembly with an insulated tool. If the relay points are now making contact, the circuit is completed, indicating a defective relay coil or improper voltage applied to the solenoid.

Speakers

A speaker converts electrical energy back into sound waves. The permanent magnet (pm) speaker has a voice coil, magnet, cone, and framework. Common speakers are listed as tweeters, midrange, and woofers.

Auto speakers

What and where

Early auto radio speakers had a field coil instead of a pm magnet and a voice coil. Today, all auto speakers are permanent magnet (PM) types (Fig. 12-14). The early car radio had only one speaker mounted within the front panel. Now with high-powered amplifier systems, you might find 14 or more speakers. The deluxe system might have four tweeters, four midranges, and six woofers. Besides being mounted in the front panel, speakers are located in the front doors and trunk area. A three-way speaker might consist of a rectangular woofer with a midrange and tweeter mounted in the center of the large speaker.

12-14 The auto radio speaker might have a tweeter speaker in the center of the cone area.

When

Since outside moisture in the auto creates a problem with speakers, the voice coil might become warped, or water might leak into the speaker area and damages the cone. The voice coil drops against the center magnet piece and results in a frozen cone or speaker. Too much volume might destroy the voice coil and cone area. Small articles stuck into the speaker baffle area might punch holes in the cone area and cause vibration. The standard auto speaker installation might have two speakers in the front door panels, and two as rear-deck speakers.

Testing

When one speaker seems to be intermittent, sub another pm speaker to determine if the speaker or radio is intermittent. Check the continuity of the suspected speaker. Remove the speaker leads for accurate measurement. For instance, an 8-ohm impedance voice coil will measure around 7.5 ohms on the DMM. The auto speaker impedance might be 4, 8, 10, and 16 ohms.

Tweeter speaker

What and where

The dual-coned speaker with a small cone in the center can reproduce high frequencies, but not as well as separate tweeter speakers. A tweeter is a small-diameter speaker or driver that reproduces high-frequency sound. Sometimes a dome tweeter is mounted separately or bracketed inside a larger speaker. The tweeter in a portable cassette-CD player might be mounted in a slotted area. The tweeter speakers are found in cassette and CD players, auto receivers, and high-fidelity speaker systems. A 1-inch tweeter in the auto might have damped-silk dome, neodymium magnets, and ferrofluid cooling. The tweeter speaker can handle from 2 to 15 watts of power.

The small tweeter reproduces the high frequencies, while the woofer reproduces the low frequencies. A crossover network separates the signals to the tweeter, midrange and woofer speakers. A high-pass filter bypasses or blocks the low and midrange frequencies from the tweeter. The tweeter cannot handle the high sound volume like the woofer speaker. A tweeter speaker can be mounted flush or at an angle. The cone tweeters are cheaper, and dome types are smoother. The tweeter size runs from 1 inch to 3 inches in diameter.

Testing

Quickly test the voice coil terminals for open or intermittent winding with the DMM. Intermittent sound might result from a poor voice coil connection. Sometimes the flexible wire from the speaker terminal to the voice coil is broken by vibration. Push evenly up and down on the speaker cone area and notice if the ohmmeter resistance changes or goes open. Sub another speaker to determine if the amplifier or speaker is defective.

Midrange speaker

What and where

The midrange speaker produces sound in ranges between the tweeter and woofer speakers. The loudspeaker produces frequencies in the middle of the audio frequencies. The midrange speaker might be a 3-½ to 8-inch-diameter speaker. A speaker system might contain a tweeter, midrange, and large woofer speaker (Fig. 12-15).

The midrange and midbass speakers might have a maximum peak power between 50 to 200 watts. The provided frequency midrange averages 200 to 3150 Hz with a recommended crossover of 100/5 kHz. The midrange speaker is found in most high-fidelity speaker systems. You might find four separate midrange speakers in a large stereo high-powered auto system.

Testing

Sometimes the midrange speaker is found damaged, with too much power applied to the midrange and woofer speakers. Check for a damaged voice coil or dragging cone. Measure the resistance of the midrange speaker of 4- or 8-ohm impedance. The resistance of the voice coil will always be less than the impedance rating on a DMM. Replace the speaker that has an open voice coil or damaged cone.

Woofer speakers

What and where

The woofer speaker reproduces the bass or lower midrange frequencies. A woofer is the largest speaker in all speaker systems (Fig. 12-16). The woofer cone is

12-15 The top speaker is the tweeter, the middle a midrange, and the bottom speaker in the column is a large woofer type.

12-16 A woofer speaker is the largest speaker in any audio system.

made of polypropylene or poly mixed with other materials for excellent sound. The surround-flexible ring around the edge of the woofer might be made up of foam or rubber. This surround ring lets the woofer cone travel freely in and out.

An excellent subwoofer speaker might have a unique U-beam rigid cone that dramatically reduces distortion. The oversized voice coil of 2 inches might be wound on a heat-resistant form for better transients and higher power-output handling. A separate shock ring absorbs excess energy and cushions the sub from abuse. The butyl rubber surround and copolymer cone can withstand environmental extremes.

The average subwoofer speakers have an 8-, 10-, 12-, and 15-inch diameter at 4-ohm impedance. The 15-inch woofer speaker might have a frequency range of 20 to 2000 Hz and up to 750 watts of peak power. An 8-inch subwoofer might have a peak power of 150 to 400 watts at 20 to 6000 Hz.

When

A defective woofer speaker might have an open or damaged voice coil. Mushy or tinny music might be caused by a dropped cone or by the voice coil riding on the center pole piece. A frozen cone might result from too much applied power or dc voltage from the amplifier circuits. When the speaker column is dropped, the speaker magnet can break away, causing a frozen cone. A blown voice coil is the result of too much power applied from the amplifier. The erratic or intermittent speaker might have a poor connection at the voice coil or speaker terminals.

Vibrations might be caused by holes torn or poked into the cone area. Loose packing or damping material can also cause vibrations. Loose sections around the rim area of the cone might come loose and vibrate. Small parts or excessive dirt found on the speaker cone can produce vibrating noises. A broken or unglued speaker or shock ring might vibrate and produce blatting sounds. Small holes poked in the cone can be repaired with speaker cement without any decrease in speaker performance, but a speaker with large holes or a cracked cone must be replaced.

Testing

Inspect the speaker for a damaged cone area. Check the voice coil for open winding at the speaker terminals, and remove one lead for correct resistance test. Replace the damaged speaker with one of the same physical size, speaker impedance, and same weight magnet. Properly phase the new speaker with the other so they go in and out all at the same time. A positive signal is with the cone going outward, and a negative signal is with the cone moving inward. The damaged cone of an expensive woofer speaker can have a new cone installed for about half the price of a new speaker. Check for a speaker repair depot in your area.

Special speakers

What and where

A special speaker might be classified as one with a standard 4-×-6, 5-×-7, 6-×-8, 6-×-9, or 4-×-10-inch size found mounted in the automobile. A two-hole mounting-pin speaker might be found in the Toyota, Mitsubishi, and Chevy Corsica dashes. Some special speakers have a higher-impedance voice coil.

Testing

Inspect the speaker for a torn or loose cone, and check the voice coil continuity with the low range of a DMM. Replace the special speaker with exact size, magnet, and impedance or a new part with the original part number. Two-hole or special-type speakers should be replaced with exact replacements. Often, the special auto speaker can be obtained at the car dealership.

Stylus phono needles

What and where

A dirty stylus or phono needle might cause distortion or fuzzy music. Excessive dirt between the stylus and cartridge might produce tinny and weak sounds. The needle is found in all phono products. First clean the stylus with a small brush, then look at the stylus under a magnifying glass to determine if the point is worn or chipped. Always replace the stylus with one having a diamond needle. A chipped diamond stylus might cause scratchy music and gouge out the recording.

To remove the defective stylus, drop the cartridge down, straighten the stylus clip upward, and then pull outward (Fig. 12-17). Some needles are removed by pulling straight out. The stylus might slip under a metal keeper to hold it in position. When removing the stylus, be careful that the small saddle attached to the crystal or moving vane is not broken. Replace the new stylus in reverse order.

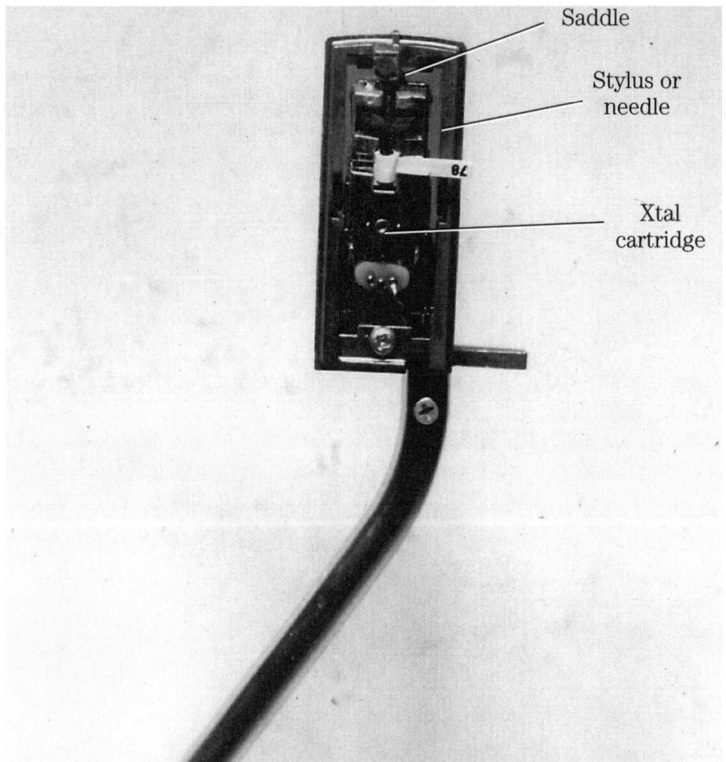

12-17 The stylus or needle rides in a groove and is connected to the piezoelectric crystal.

Surface-mounted parts

(See Transistors, Capacitors, and Resistors)

Surge suppressors

(See Chapter 6)

Switches

A switch might turn the power off and on in a TV set, cassette player, compact disc player, auto receiver, or any other consumer electronic product. A switch closes or opens a circuit, changing the state of a circuit to off or on position. Different mechanical, electrical, and electronic switching is found throughout electronic circuits. A multiposition switch might be a SPST, SPDT, DPST, DPDT, and DP4T position. The telephone might have a rotary switch (where you place your finger in a hole and rotate to a certain number) or a pushbutton switch (where you simply touch a button that switches in each number or letter). A defective switch can be checked with resistance and voltage measurements.

DIP switch

What and where

The DIP switch is a group of miniature switches mounted in a dual-in-line package or soldered directly to the PCB. These switches are found in digital and low-current circuits. The DIP switch might mount as a low profile, extended activator, rocker activator, piano, or slide activator style switch. The DIP switch must have a rotary action and SMT mounting. These DIP switches might switch from 2 to 10 positions. Most DIP switches have an SPST contact, while a preprogrammed DIP switch might have up to 16 switching positions.

Testing

Check each switch position with the low-ohm scale of the DMM, and replace the DIP switch assembly that has one or more defective positions. Treat each contact with care, as the terminals can break off like an IC component.

Function switch

What and where

A function switch might be a multifunction switch that switches in several different functions. The cassette player function switch might switch in the AM radio, FM radio, FM MPX, and cassette tape heads. The digital multimeter function switch rotates to the voltage, resistance, current, capacitance, diode, inductance, and transistor tests. A function switch is found in just about every electronic product.

The function switch might have several different positions and poles of switching in various stations (2 to 13) of the TV tuner. The function switch works in many different circuits of an electronic product. Often, the multiposition switch is a rotary or slide function type. A slide function switch is found in the small portable cassette player and large cassette AM-FM-MPX receiver (Fig. 12-18).

Testing

When the TV station cannot be tuned in or is intermittent, suspect a dirty wafer switch. If the FM radio station fades out after switching from an AM station, check the worn or dirty switch contacts. Spray cleaning or tuner lube fluid right down inside each switch contact. Rotate or flip the switch back and forth to clean up the switch contacts. Replace the switch for excessively worn or broken switch contacts.

Key switch

What and where

An external key is required to unlock or switch on and off an electrical instrument, burglar alrm, security system, computer, or protective device. The inserted

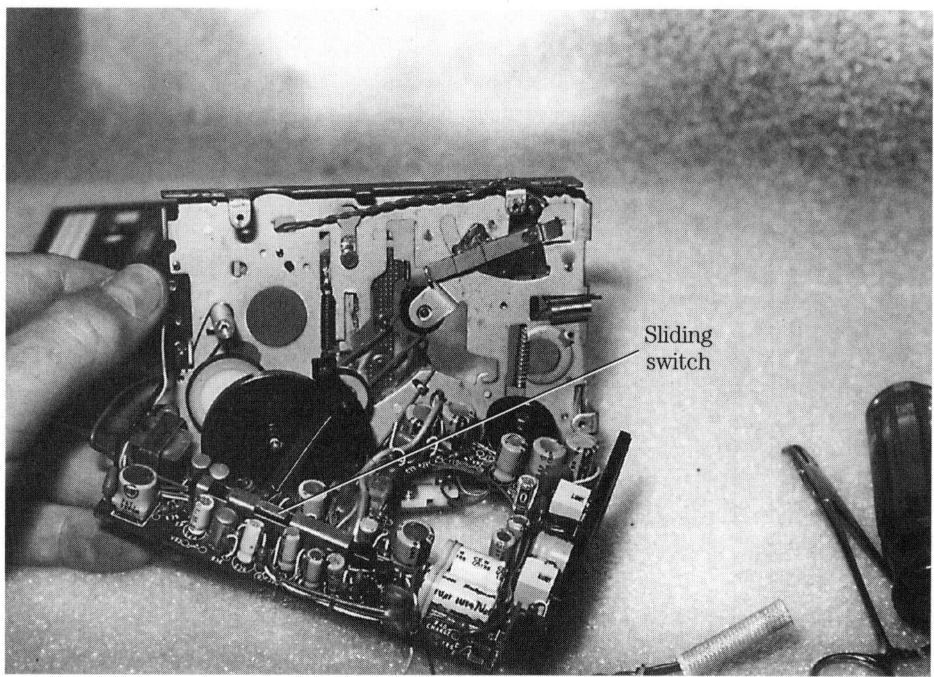

12-18 The portable cassette player has a sliding bar-type function switch.

key might switch in the electrical or electronic circuit. A cash register key switch might have several color-coded wires coming from the switch area. Most key switches are on/off types, with some as SPDT, and at the center in off position.

Testing

Check for broken cables or wires at the switch terminal when the switch will not close a circuit. Take a low-ohm continuity measurement with the DMM. Clean up the switch contacts with cleaning spray, and place the spray tip right down inside the switch contact areas. Replace the defective switch when the switch will not close or has worn contacts.

Magnetic switch

What and where

The magnetic or reed switch will close contacts when a magnet is placed nearby. The reed switch might have a coil or magnet that pulls the movable contacts together. A reed switch might be found in a cassette player as a counter or to shut off the unit. The magnetic alarm switch might be used on windows and doors as detectors for alarm systems. They are also found in auto, marine, and moisture environment applications. The magnetic switch might be recessed, press-fit, and surface-mounted. A magnetic alarm switch might contain both normally open (N/O) or normally closed (N/C) switch contacts.

Testing

Check the reed switch contacts with the low-ohm scale of the DMM. Bring an external magnet against the glass envelope and notice if the contacts are closed. Replace the reed or magnetic switch if it has dirty or burned switch contacts. These contacts cannot be cleaned up like a standard off/on switch, since they are enclosed within the glass tube.

Mercury switch

What and where

The mercury tilt-switch has two switching electrodes and a drop of mercury inside a glass tube. When the tube is tilted toward the switch contacts, the mercury closes the circuit between the two elements, providing a conducting path. Mercury switches are found in tiltable components such as furnace bonnet controls and early electronic products. Often, the mercury switch is only an SPST switch.

Testing

Tilt the mercury switch so the liquid is on the element contacts. Check for low continuity with the DMM. If the mercury switch glass envelope is black with intermittent contacts, replace it.

Micro switch

What and where

The microswitch might have momentary, normally open (N/O), or closed (C/O) contacts. These switches make a quick action contact with very little outside pressure. The microswitch is found in microwave ovens to activate circuits when the door is opened or closed. A microswitch is also found in alarms, security systems, and appliances. These switches are normally open or closed with SPST or SPDT action. The microswitch will switch circuits from 50 milliamperes to 3 amps.

Testing

Activate the lever or small button on the microswitch and clip an ohmmeter to each terminal. An open microswitch will have no reading. Sometimes the microswitch spring breaks inside the switch, leaving a loose lever. Replace the intermittent or open microswitch.

Pushbutton switch

What and where

The pushbutton switch is found in audio receivers, instrumentation, telecommunications, and test equipment used as multicontact or function-type switches (Fig. 12-19). The push-on push-off button switch is used in power supplies, receivers, radios, cassette players, VCRs, computers, and camcorders. A power pushbutton switch might have an SPST, SPDT, DPDT, or 4PDT action. The momentary pushbutton switch is on only when pressed. When the button is released, no contact is made. A defective or dirty switch contact can cause a loud popping sound in the audio of the radio or amplifier. Replace the worn switch that finally on a few seconds after the button has been released.

12-19 A pushbutton switch with many contacts might be in a gang of switches within the AM-FM-MPX, phono, and cassette receiver.

Testing

Clip the ohmmeter leads across the suspected switch terminals. Set the meter to a 200-ohm range, and push the switch off and on to make contact. Sometimes an intermittent switch will not make contact when rapidly pushed. Notice if the meter goes to zero and stays there, or if there is intermittent action, indicating poor contacts. Clean up the switch by placing the cleaning fluid tube down into the switch area. Push the switch back and forth to clean up the contacts, and replace the switch when contacts are excessively worn or open. You can check the current of the piece of equipment by clipping across the switch contacts with the unit off. When the switch is pushed on, no current can be measured. Then push the switch to off position and measure the operating current.

Reed switch

(See Magnetic)

Relay switching

(See Relays)

Rocker switch

What and where

The rocker switch is somewhat like a toggle switch with a shaped bar which is rocked back and forth to turn the switch off and on. This type of switch might have on-on, on-off, and on-off-on positions. A lever switch might be connected to a rocker-type switch. Rocker switches appear in SPST, SPDT, DPDT, and 4PDT circuits. These switches are used to turn on and off receivers, transmitters, test equipment, communications equipment, electrical devices, and computers. They operate in 3- to 15-amp circuits.

Testing

Replace the rocker switch when it intermittently turns on an electronic product. With poor switching contacts, sometimes a loud popping sound can be heard. Check the low-resistance measurement across the switch contacts. This resistance should be less than 0.15 ohms on the digital multimeter (DMM). Often, clean-up of a high-amperage on/off switch with cleaning fluid is fruitless. Replace it.

Rotary switch

What and where

The wiping blade of a rotary switch moves in a circle to make contact with the various circuits. A nonshorting rotary switch disconnects the circuit before the next contact is made. The shorting contact rotary switch makes the next contact before it breaks away. A rotary switch might have several poles with many different positions. The rotary switch is found in radios, large receivers, test equipment, measurements, telephone base stations, hospital controls, rotary DIP switching, and PC mounting. The early TV tuner had a rotary action with many gangs and different positions to switch different circuits in to receive a television station. The on/off switch found on the back of a volume control might be considered a rotary switch.

Testing

First try to clean up the switch contacts with cleaning fluid or tuner lube. Place the plastic tube down on each switch control and spray into the switching contacts (Fig. 12-20). Rotate the switch back and forth to clean up the controls. Replace the switch or tuner that has broken or worn contacts on a wafer section. Check suspected contacts with the ohmmeter.

Thermal switches

What and where

A thermal switch (such as a thermostat) is activated by a change in temperature. You might find a thermal switch in a transformer lead, on a microwave magnetron, or in a microwave thermal oven sensor. The thermal switch contacts or sensor found on the magnetron within the microwave oven opens up when the magnetron is excessively too hot and shuts down the cooking process (Fig. 12-21). When the oven cavity becomes too warm, the oven sensor shuts down the oven operation.

Testing

Thermal switches can be tested with the ohmmeter or pigtail light-bulb test within the microwave oven. The intermittent or erratic thermal cutout switch can be detected with a light bulb. Simply clip a 120-V 100-watt pigtail light across the thermal

12-20 The early TV rotary tuner has many switching contacts and wafers.

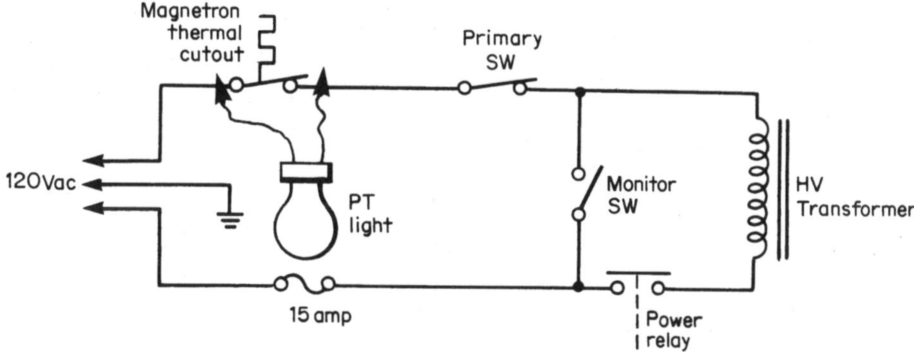

12-21 The thermal switch mounted on the magnetron tube protects the tube from damage and overheating.

cutout terminals. When the switch opens up, the light will come on. Replace the thermal switch if more than one ohm is measured across the switch terminals.

Toggle switch

What and where

When the lever of a switch is flipped one way or another, the switch terminals are turned off and on. Toggle switches might have momentary-off, center-off, on-none-on, on-off-on, on-on, SPST, SPDT, DPST, DPDT, 4PDT, and a lighted toggle switch. The toggle switch is held into position with a large nut around the toggle area. Toggle switches are usually rated in 3 to 10 amps. They might be surface-mounted or mounted at a right angle or vertical right angle. The toggle switch might be a heavy-duty, miniature, submini, or microtoggle.

Testing

The defective toggle switch might appear open, intermittent, and lock in one position. Spray cleaning fluid down inside the switch area to clean up an intermittent or erratic toggle switch. Check the switching action by clipping a low-ohm meter across the switch contacts. For open or erratic conditions, replace the switch. A 0.1-uF 250-volt bypass capacitor can be placed across the switch contacts to help eliminate a popping or cracking sound when the receiver or amplifier is turned on and off.

13 CHAPTER

Testing electronic components T through Z

(Tapes through zener diodes)

Magnetic tape is used for recording and playback functions within the cassette player, camcorder, and VCR. Test tapes are used to help repair the defective cassette player and video tape recorder. A tape eraser might erase the whole cassette or cartridge from a previous recording.

The electronics technician might have a DMM volt-ohm meter, oscilloscope, color-pattern generator, capacitor tester, signal tracer, sweep generator, and frequency counter on the service bench (Fig. 13-1). For special electronic service, you might find a wow-and-flutter meter, laser-power meter, Magnameter, video-head tester, CRT restorer and analyzer, and laboratory power supply.

Besides knowing how to test tapes and equipment, the electronics technician must know how to locate, test, and replace various transformers. Rounding out this chapter, you will also find how to test varistors and voltage dividers.

Tape

Magnetic tape might appear in a cassette or cartridge. The cassette tape records sounds while tape in the VCR and camcorder records video and sound. Magnetic tape is a plastic-coated tape with a film of magnetic particles that can be magnetized to record sound, computer information, and video signals. The magnetic recording head receives current impulses and converts them to magnetic impulses on the tape. The magnetic particles are picked up and played back by a recording head. The

13-1 Besides a schematic diagram, basic test equipment is needed to repair the many different electronic products.

record/playback head in the cassette player is used both for recording and playback functions. A stereo head has two separate windings.

Audio tape

What and where

The cassette player might record and play back audio tape or just play the tape. The cassette holds a reel of magnetic tape that can be rotated forward or in reverse mode. These cassettes are inserted into a holder for playback or recording modes. The width of the tape in the cassette is very small compared to the VHS or VHS-C tape in the camcorder or VCR.

In an audiocassette player, the stereo cassette might record in one direction and, when reversed, record in the reverse direction. Two different audio tracks and two different tape-head circuits are found within the stereo cassette player. You might find an audio stereo tape head with four different connecting terminals. The tape head should be cleaned at least six times per year if the cassette player is played constantly.

When

The iron oxide particles from the magnetic tape will build up on the components that come in direct contact with the tape. This excessive oxide can produce garbled or muffled sound during playback. Intermittent or erratic sound can be recorded with a dirty tape head. One channel gap can be closed with oxide, which results in no sound or recording from that channel. Oxide deposits can cause an improper erase function and prevent automatic stop operations. Clean up the tape and erase heads with a cleaning stick and alcohol.

Testing

Sub another cassette to determine if the tape is at fault or if the trouble is in the cassette player. Of course, a broken tape will not rotate and no sound is present. The wrinkled or bent tape might cause intermittent or erratic recording. Background noise might be caused with a magnetized tape head or tape. Always degauss the tape head after making repairs.

Improper head azimuth adjustment can cause distortion and loss of high frequencies. A worn tape head can cause a loss of high frequencies. You can adjust the azimuth screw located on one side of the tape head by playing a recorded cassette of violins or high-pitched music to maximum into the speakers. Accurate azimuth adjustment can be made with a 3-kHz, 6-kHz, or 10-kHz test cassette with an 8-ohm load, instead of the speakers (Fig. 13-2). Use the scope or frequency counter as an indicator.

Test tapes

What and where

One or two test tapes are handy when checking audio stages, tape speeds, and head alignment in the cassette player. Today, some of these test cassette tapes are difficult to find. You can record your own test cassettes with an audio-signal generator. Check Table 13-1 for the different cassette tests and test cassettes.

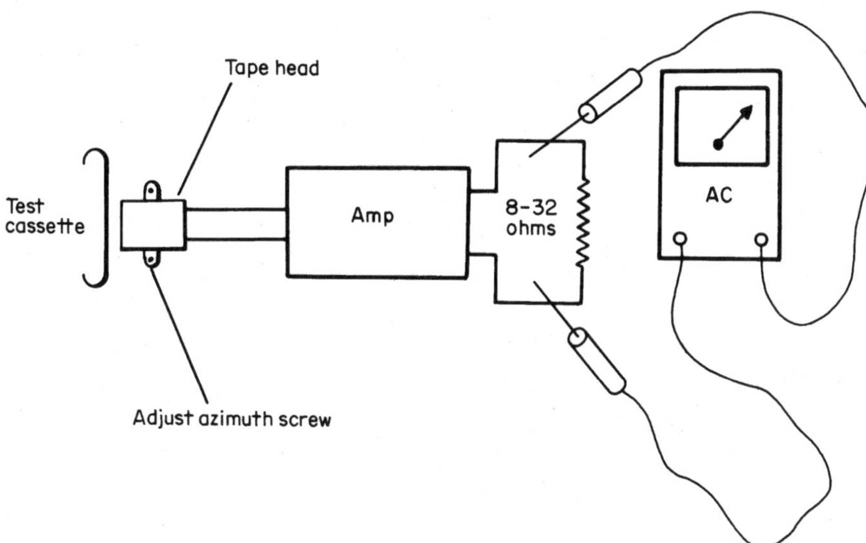

13-2 Insert a 3-kHz test cassette to make azimuth adjustments of the tape head with an ac meter as indicator.

Table 13-1. Test tape frequencies and functions

Test tape	Frequency	Function
VTT-658	10 kHz	R/P head function
MTT-114	"	"
MTT-216	"	"
Standard	6.3 kHz	Head azimuth and sensitivity
MTT-111	3 kHz	Tape speed adjustment
MTT-118	1 kHz	Tape speed adjustment
MTT-150	400 Hz	Playback level

The speed of the cassette player can be checked with a test tape and a frequency counter. Insert a 1-kHz or a 3-kHz test cassette and place it in the play mode. Connect an 8- or 10-ohm resistor at the earphone jack. Measure the frequency with the frequency meter or counter (Fig. 13-3). If the reading is at 1 kHz or 3 kHz, the speed is correct. A higher reading indicates a faster speed and a lower measurement indicates a slower speed. Some larger cassette players have regulated speed adjustments, or you can find a speed adjustment in the end bell of the dc motor. Slow speeds can be caused by a dirty, oily, or loose belt. Oil deposits on the capstan/flywheel can produce slow speeds.

VCR and camcorder test tapes

What and where

The VHS hi-fi alignment tape or cartridge is a high-precision tape for hi-fi RF adjustments, such as RF amplitude, hi-fi muting level, and critical tape-path alignment. Always check the hi-fi head performance before and after replacing the upper head drum or cylinder. Also, check the multiburst, modulation ramp, IRE gray-field-test signals, and SMPTE color bars (Fig. 13-4).

The VCR playback test tape presents test signals for adjusting the tape path, audio/control head alignment, luminance/chroma circuit, and tracking. The playback tape or cartridge includes a gray raster, 30-Hz audio tone, convergence, dots and grids, and NTSC color bars.

A professional alignment tape includes tracking for all manufacturers in the SP mode. The tests include an NTSC color-bar test signal and 1-kHz linear-stereo audio signal. The professional tape is used for VCR tracking, tape path, A/C head height, and azimuth adjustments. The professional alignment tape is available in extended play (EP) and standard play (SP).

The 8mm video alignment tape contains standard NTSC video patterns of color bars, convergence, dots and lines, and a red raster. The audio signal is a linear format and contains a wide variety of test tones that include 20 Hz, 100 Hz, 250 Hz, 400 Hz, 1 kHz, 2.5 kHz, 4 kHz, 8 kHz, 10 kHz, 15 kHz, and 20 kHz. A full 20–25-kHz sweep tone is also included.

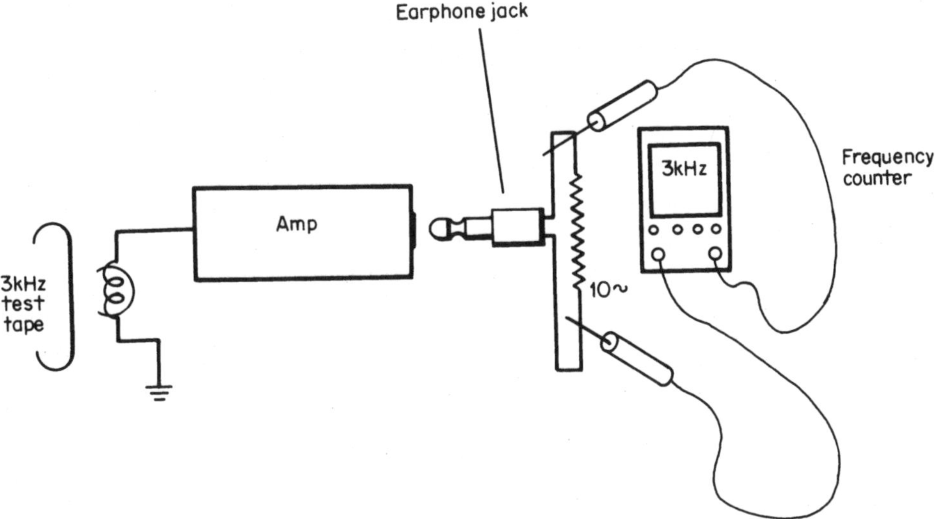

13-3 Connect the frequency counter to the amplifier output to determine the correct speed of the cassette motor.

13-4 Use the different test cartridges to align and make adjustments on the camcorder and VCR.

Testing

Although test tapes cause very few problems, an excessively used test tape might contain washed-out features and worn areas. Sometimes the start of the tape might be wrinkled with erratic tests. Cut out portions of defective tape and repair with a splice repair kit. Discard a tape that has intermittent or erratic tones.

Tape eraser

What and where

The audiocassette head demagnetizer demagnetizes the record/play tape head when the cassette is inserted. Demagnetizing the tape head can improve playback and recording greatly. The noisy reception might be blamed for a magnetized tape head. The latest cassette demagnetizer uses a DC-DC converter to treat the head and the process takes only about one second.

The head demagnetizer operates from the 120-Vac power line with a curved demagnetizer tip to reach down into those difficult to access record/playback heads. A plastic cover is placed over the tip to protect the head surface. The hand-held demagnetizer is ideal for the cassette player or a reel-to-reel recorder.

A bulk-type eraser operates directly from the power line and erases audio/video tapes and computer floppy discs. This bulk-type unit erases the entire tape at once.

Testing

Place a screwdriver blade near the bottom of the demagnetizer to determine if the unit is operating. The blade should vibrate. Check the ac cord and plug for no power to the bulk eraser. Double-check for a broken cord where it enters the body of the unit or at the ac plug. If the plug is defective, cut off two inches of the ac cord and install a screw-type ac plug.

Telephone

A simple telephone has a microphone, switching and ringing devices, earphone, wire line or cable, and power supply. A cordless telephone can be taken anywhere in the house or yard, transmitting the messages to a receiver at the telephone rest station. A cellular telephone system is operated from the auto or outdoors and is transmitted over a high tower via the airwaves.

Telephone answering machine

What and where

The telephone answering machine automatically answers the telephone after a number of rings and records the message from the caller on a small tape cassette (Fig. 13-5). The answering machine is plugged into the telephone and is usually mounted nearby. Some answering machines have dual cassettes, multifunction remote systems, and voice-operated recordings.

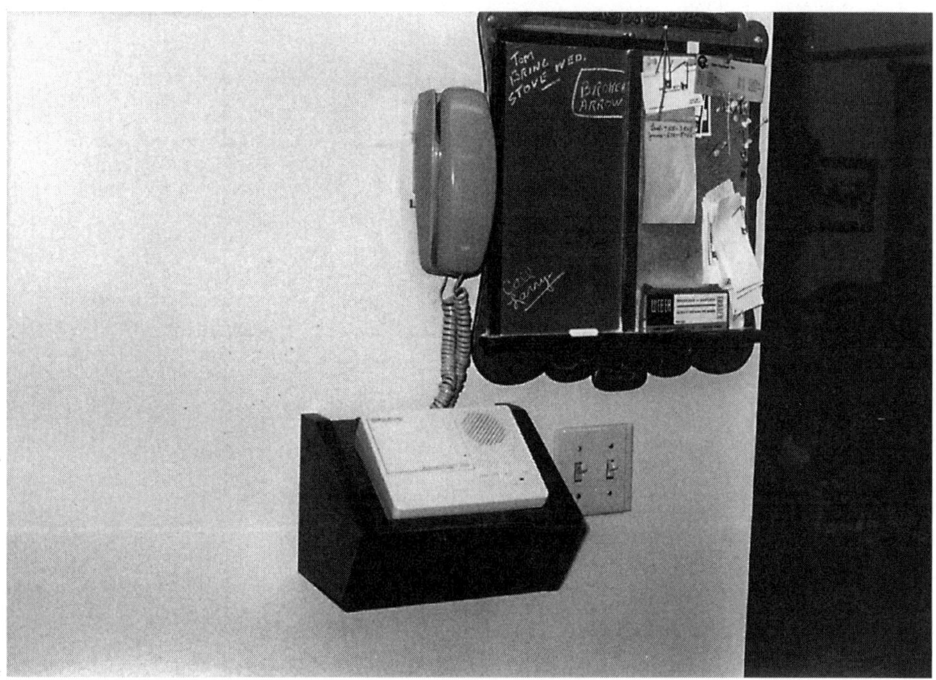

13-5 The telephone answering machine records messages.

The typical telephone answering machine has a ring detector, switching transistor, electronic governor, ac power supply, cassette motor, record amp, outgoing amp, preamp, main amp, beep-tone oscillator, controller, microphone, and timing circuit. A recorded message is left on the cassette tape and it plays back this message after several rings. The caller then leaves a recorded message on the recording cassette. A blinking light indicates a message has been left on the answering machine. When the owner returns home, the message is received by playing back the recording (Fig. 13-6).

Testing

Clean up the tape head with alcohol and cloth, head-cleaning sprays, and solvents. Remove the built-up oxide on the tape head to prevent weak, noisy, or distorted messages, loss of high-frequency response, and a dirty head might slow down or jam the tape. A weak sound or hiss can be cured with a good tape-head cleanup.

Likewise, clean up all components that are touched by the tape. Clean the grease or oxide dust off the flywheel assembly and drive belt. Check the idler and end pulleys for slick spots. Clean off the motor pulleys and both forward and rewind reels. Visually inspect the different rubber parts for cracks or worn areas while cleaning the various moving components.

Check the end foil on the tape and note if it is worn, cracked, or torn. Replace the cassette tape if it doesn't stop or if it pulls. Keep all magnetic material away from the cassette and tape heads. Demagnetize the tape head if the recording is noisy. Check for a worn or loose belt with slow-speed symptoms.

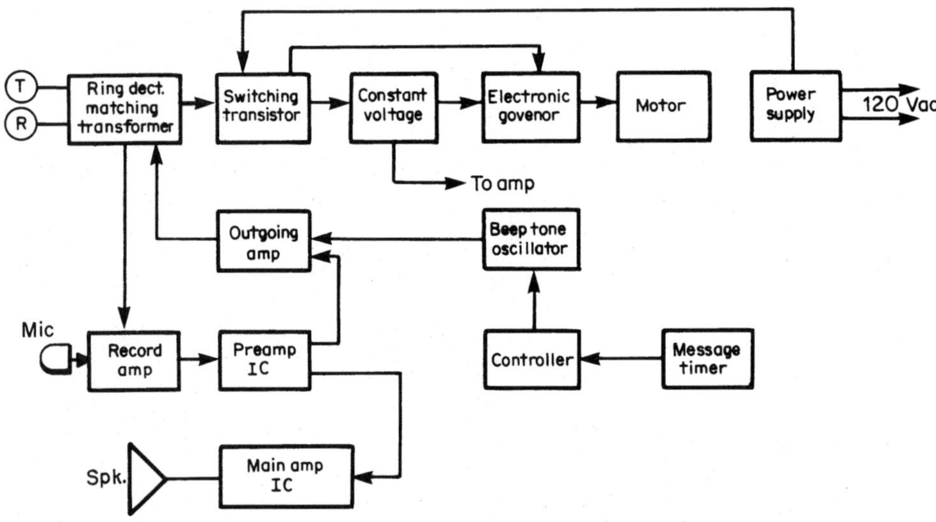

13-6 The block diagram of a typical answering machine.

Signal trace the audio circuit with a signal generator or injector to locate a defective amplifier stage. The early answering machines contained only transistors while the latest have mostly IC components. Suspect a defective transistor or IC if no audio output exists. Check for dried-up coupling capacitors for weak or no reception. Measure the dc voltage on the electret microphone if there is no recording. Sub a new cassette for one with noisy or scrambled recordings.

Cordless telephones

What and where

The cordless telephone can be operated anywhere in the house when the phone rings. This cordless telephone can make or receive calls from the unit to a base station. You can place it beside your chair and when it rings, pick it up, turn it on, and start talking (Fig. 13-7). The portable telephone operates as a transmitter to the base station and is connected to the regular telephone system in the home. Some cordless telephones can operate over 50 feet from the base station.

Testing

Suspect a weak or dead battery when the phone will not light up or receive a signal. These Ni-cd batteries will not operate if left out in the cold or in a cold room. Remember, the batteries are being charged as the cordless telephone is resting on the master unit.

Take a dc voltage measurement at the two contact charging points. This voltage might measure around 16.5 volts without a load or if the cordless phone is out of the cradle. Replace the battery pack when the battery voltage is low and cannot be charged up. Usually, the cordless telephone battery packs have three Ni-cd batteries in series with a total output voltage of 3.6 V.

13-7 The cordless telephone can be operated in any room of the house or office without connecting wires.

Clean up all switches by spraying cleaning fluid down into the switch area. For intermittent reception, check all connections for poor or broken wires. Inspect for loose components on the board if the phone has been dropped. If the phone is intermittent while transmitting (talking), suspect a defective microphone or loose connection. Take a continuity measurement of the mike unit. The resistance should be around 5 to 10 ohms. Likewise, check the speaker in the same manner. Signal-trace the audio section with a 1-kHz signal from the generator. All transmitter repairs should be made by a first-class electronics technician.

Telephone wireless extension

What and where

The telephone wireless outlets provide telephone conversation without any wires attached between the telephone extension and the telephone. These extensions work quite well where telephone lines cannot be run in the house due to solid walls or impossible room additions. This requires installation tools, drilling of holes, or stringing of wires to control the wireless jack system. Simply plug one unit into the power line and telephone, and the other into an outlet at the difficult location. Telephone conversation occurs between the two units like a wireless intercom unit. Both RCA and GE manufacture a wireless phone jack system.

Testing

The telephone wireless extension must be set to correspond to the telephone transmitting unit and extension. The extension must be on the same power line wiring in the house as the telephone transmitter. The telephone system might not operate if it is tied to another circuit with the extension unit in the outside garage.

Make sure ac power is applied to both units. Check the telephone line entering the transmitter unit. Determine if the telephone line is entering the transmitter unit and if the telephone wires are fed into the wireless telephone outlet. Double-check all telephone cords and plugs. Sub another telephone to determine if the wireless outlet is operating. Since these units are low in cost, it's best to replace instead of repair.

Test CD

What and where

There are many CD test discs on the market for correct alignment and making adjustments. Most manufacturers have their own test discs, or they might recommend one for alignment procedures (Table 13-2). The test disc is used to make EFM RF signals, grating, track offset, focus gain, and tracking gain adjustments. Some manufacturers use a regular recorded disc to make the required test adjustments.

Testing

Insert the CD test disc and play the disc (Fig. 13-8). The EFM test disc signal can be checked with the scope at the output terminal of the RF amplifier. A very complex encoding scheme is used to transform the digital data to a form that can be placed on a disc. This information is modulated by the EFM. This EFM signal must be present or most CD players will automatically shut down if focus and tracking are missing (Fig. 13-9).

Table 13-2. the various CD players and recommended test disks

Make	Model	Test disks
Dennon	DCH-500	CA-1094
Goldstar	GCD-616	YEDS-7 (Sony)
Onkyo	DX-200	YEDS-18 (Sony)
Pioneer	PD-710	YEDS-7 (Sony)
Realistic	14-529	YEDS-43 (Sony)
Sanyo	CP-500	800104
Sony	CDX-5	YEDS-1, YEDS-7, YEDS-18 & YEDS-43

13-8 Insert the CD test disc when making alignments on the CD player.

Test equipment

Servicing electronic products can be most difficult without good test equipment. A combination of electronics know-how, experience, and correct test equipment will solve most electronic problems found in consumer electronics. The electronics technician might have only three or four pet test instruments that are used daily (Fig. 13-10).

Test equipment is quite expensive. At first, only a few test instruments are required to do the job. You can add a new test instrument to the bench each year to suit your electronics budget. Read the test instrument manual over and over. Know how the equipment works and know what tests are made. Remember, the test instrument

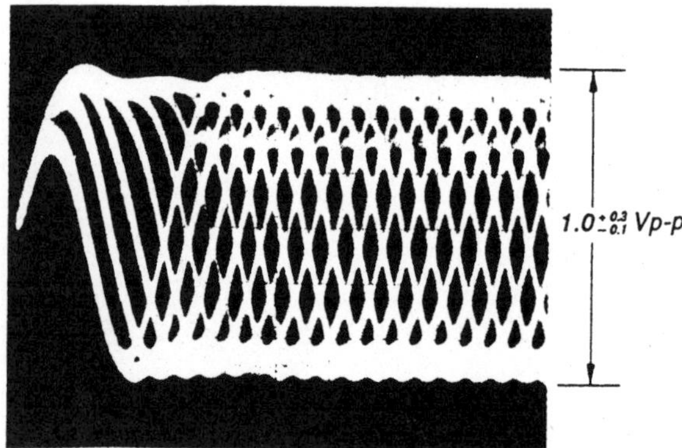

13-9 The EFM waveform indicating that the optical assembly and RF amplifier are normal.

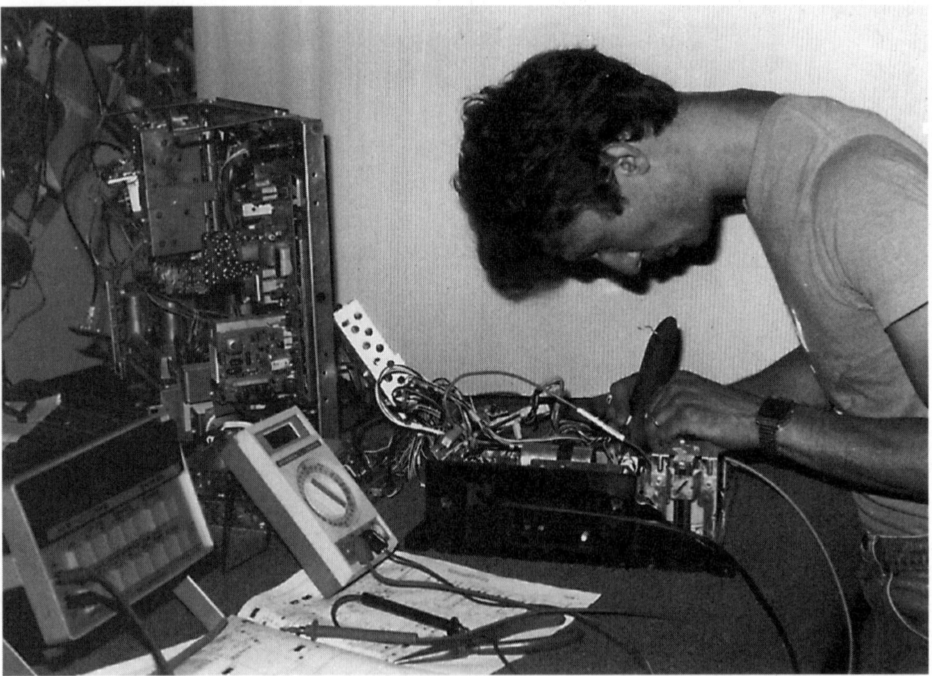

13-10 The electronics technician might have three or four pet test instruments that are used daily to service the many electronic products.

performs no better or worse than the electronics technician operating it. The more you use it, the more products you can service in shorter periods of time. The descriptions of the test equipment listed in the following pages does not tell you how to operate the test equipment, but instead tells what, where, and how tests are made.

VOM

What and where

You can purchase a VOM for under $20.00 (Fig. 13-11). Some VOMs have a reset pushbutton release when overload occurs and will not reset until the overload is eliminated. Choose a more expensive VOM ($175 to $250) with overload protection. The Simpson model 260 VOM series has provided rugged electronic servicing for many years.

The VOM is ideal when connected as an output meter in sound adjustment procedures. The analog meter can track the volume and audio signal of an audio oscillator for gain, level, and head alignment. The VOM outshines the DMM in making these adjustments, because the numbers that move up and down the scale are difficult to follow on the DMM display.

The VOM can be easily damaged with excessive high voltage, wrong polarity, or improper function setting. If the VOM is left in resistance measurements and accidentally touches a voltage circuit, the meter hand might wrap around the peg

13-11 The low-priced VOM can make resistance, voltage, and current tests of electronic circuits.

and result in a damaged meter. Do not place a VOM or DMM test instrument in the high-voltage circuits of a microwave oven or TV chassis. Most VOMs load down the circuit to be tested, providing inaccurate measurements. The test probe must be placed on the right polarity when taking voltage measurements. The meter hand goes in the wrong direction with the wrong polarity of the test probes. The VOM is not as accurate in low resistance and voltage measurements as the digital multimeter (DMM).

Digital multimeter (DMM)

What and where

The digital multimeter (DMM) is the most versatile test instrument found on the service bench (Fig. 13-12). You can check ac or dc voltage, resistance, continuity, ac and dc current, diodes-transistors, transistors, capacitance, inductance, logic probes, and check the frequency in the deluxe DMM. The accuracy of a dc voltage test can be from 0.5% to 0.1%.

Some DMMs have an auto-ranging method to determine the range of the meter. Resistance measurements are accurate under 1 ohm. Beeper sounds are heard on the 200-ohm range of some DMMs. The DMM is ideal for measuring critical voltages in semiconductors and solid-state circuits. The forward bias of a silicon transistor might be 0.6 volts, while a germanium transistor bias is 0.3 volts.

The DMM can quickly make in-circuit tests of diodes and transistors. Besides the diode test, the DMM can check the junctions of a transistor. The diode-transistor test of the DMM can eliminate a defective transistor within minutes. Some larger DMMs have a separate transistor socket and HFE measurement. You can determine if the transistor is a PNP or NPN type with diode-transistor tests (Fig. 13-13). Often, the small pocket DMM with many different functions can only test capacitors up to 40 microfarads (40 uF). The digital multimeter is a very accurate and valuable test instrument for the electronics technician.

FET-VOM

What and where

Before the FET-VOM meter appeared on the service bench the VTVM (vacuum tube volt meter) was king. Today, some service benches might have a VTVM, although they are no longer manufactured. Because the FET transistor input has high impedance, like the VTVM, the circuits are not loaded down when connected into the circuit to be measured. The electronic voltmeter, VOM, and VTVM are analog meters (Fig. 13-14).

The FET-VOM is quite accurate when taking low-voltage and resistance measurements. Very low forward transistor voltage can be taken with this meter. The FET meter is ideal as an indicator for alignment procedures since a large analog meter is used. The FET-VOM does not load down the circuit being tested since it has a high-impedance input (10 megohms). Some FET-VOMs have a reversing polarity switch instead of reversing test probes.

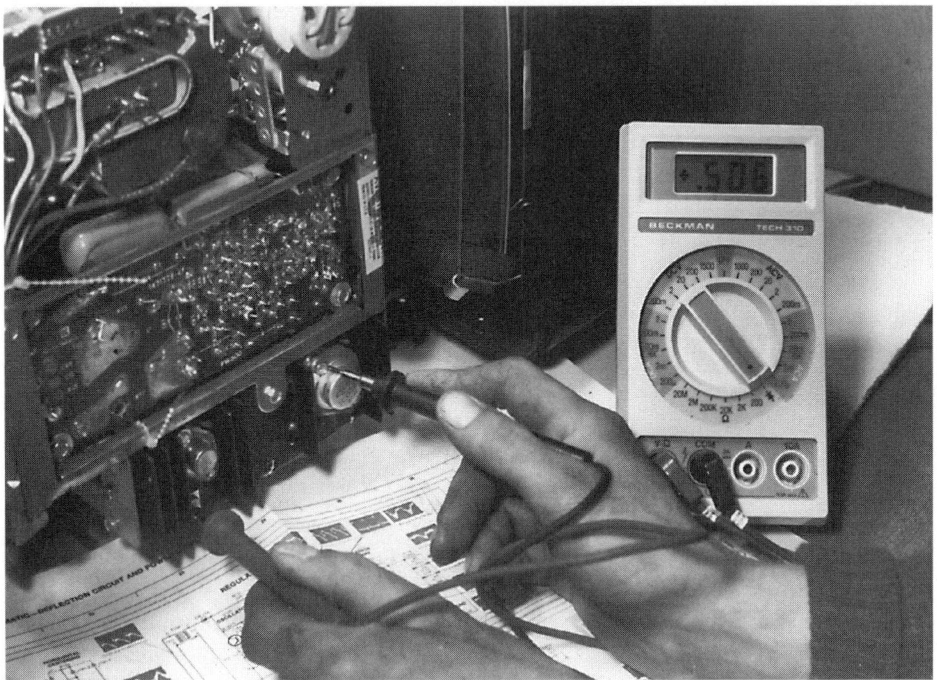

13-12 The digital multimeter (DMM) is the most used test instrument found on the service bench.

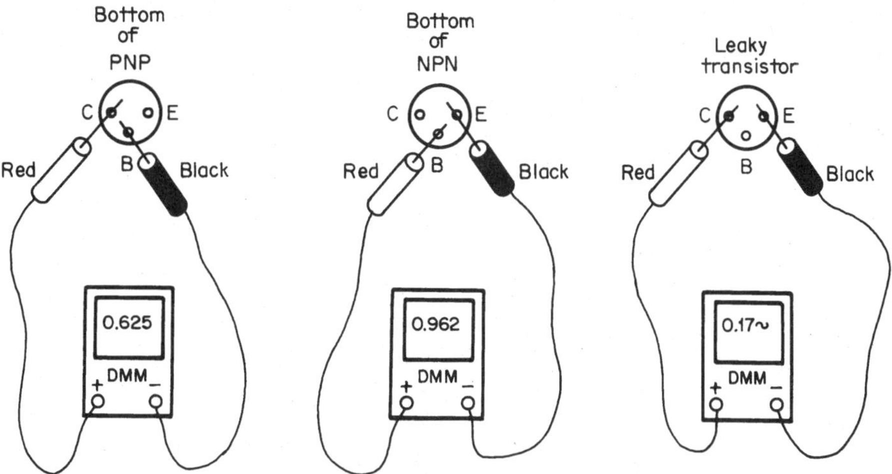

13-13 To identify an unknown transistor, place the positive probe to the base of the NPN transistor and the negative probe to the base of a PNP transistor.

13-14 The analog FET-VOM can provide accurate voltage measurements in the TV chassis.

The oscilloscope

What and where

The cathode-ray tube (CRT) or oscilloscope provides visual pictures of waveforms, time, frequency, phase, and voltage applied to the input terminal. The scope is most valuable in taking waveforms in the TV, CD player, and VCR. Use the scope as an indicator when signal-tracing video and audio circuits (Fig. 13-15).

The scope can determine if a signal is present; the wave slope of square, sine, triangle, and pulse waveforms; base voltage and peak-to-peak measurements; and frequency at the same time with dual-trace operation. A loss of gain in one audio channel can be compared with the normal stereo channel. The amplitude of the waveform might be controlled by the vertical gain control.

The most critical waveforms in the TV chassis are the horizontal, vertical, video, and color (Fig. 13-16). Critical waveforms taken in the CD player are EFM, tracking error (TE), tracking coil, focus error (FE), focus coil, PLL-VCO, and motor waveforms. Along with the DMM, the scope is the electronics technician's best friend.

Semiconductor tester

What and where

Besides the diode-transistor test on the DMM, there are many other components tested by a semiconductor tester. A good semiconductor tester should test FETs, SCRs, transistors, diodes, and triac components. The low-priced semiconductor

13-15 The oscilloscope can take critical waveforms when signal tracing video and audio circuits.

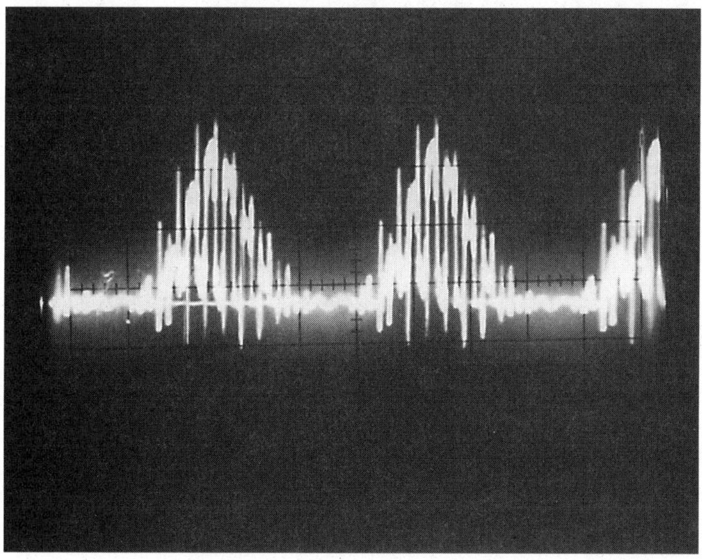

13-16 The color output waveform indicating that the color signal is fed to the red output transistor

tester might have LED indicators, while more expensive testers have a large analog meter (Fig. 13-17). Some of these testers can make in-circuit tests and the most accurate test is with the semiconductor removed from the circuit.

The semiconductor test can determine if a transistor is open, leaky, shorted, or intermittent. This tester can check the junction of a diode for shorted, leaky, high-resistance, and open conditions. A good semiconductor tester can test the SCR and triac for leaky, shorted, or open conditions. The semiconductor tester can solve 85 % of the problems found in the electronic chassis.

Audio signal generator

What and where

The audio signal generator provides a sine- or square-wave signal to troubleshoot the audio amplifier circuits. The audio signal generator can be used for signal tracing, signal injection, and audio alignment in audio amplifiers, stereo audio circuits, CD players, VCRs, camcorders, car stereos, and cassette players. A sine-waveform might be used for signal tracing and signal injection in any audio circuit. Locating distortion in the audio circuit can be located with a square-waveform signal and oscilloscope.

Inject a 1-kHz audio sine-wave signal into the input of an audio amplifier or cassette player and check each stage with the scope for loss of signal, dead, or weak

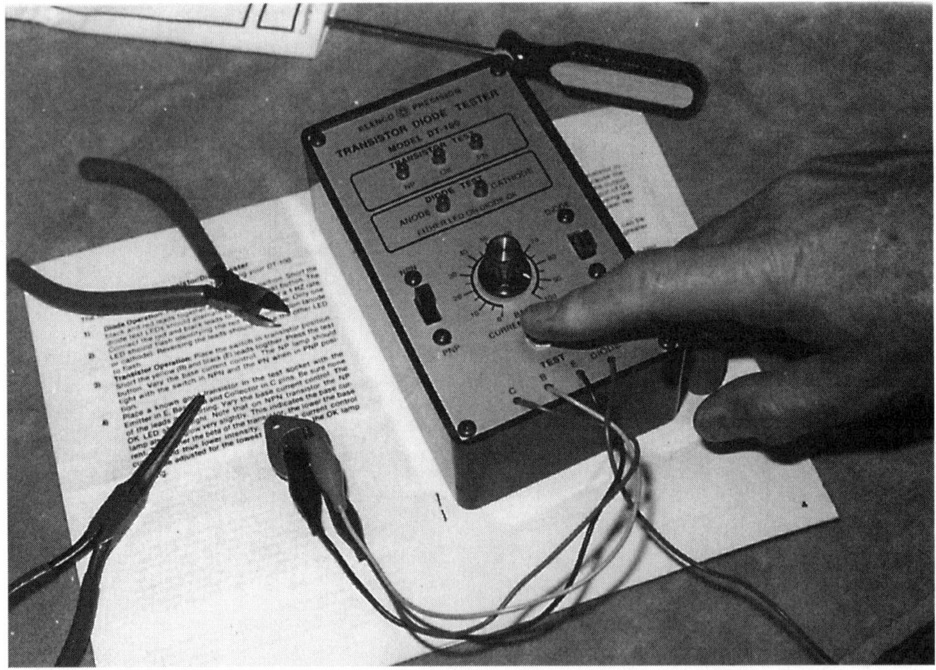

13-17 The semiconductor tester can test FETs, SCRs, transistors, diodes, and triac components.

reception. Connect the 1-kHz sine-wave signal to the audio input of a cassette player and check the output at the speaker with a frequency counter to determine the amplifier frequency response and level adjustments.

By injecting a square-wave signal into the input or output stages of a stereo amplifier, you can determine what stage has distortion with clipped or rounded corners of the square wave on the oscilloscope. When a square-wave audio signal is injected into both the right and left channels of the stereo amplifier, both channels can be seen with a dual-trace oscilloscope. Each stage can be checked in the stereo stages by connecting both scope inputs into the audio circuits at the same point in each audio channel to show gain, weak, or distorted square-wave forms.

The frequency counter

What and where

The basic frequency counter-test instrument counts the range of frequency in most consumer electronic products. A normal counting range might result from 10 Hz to 100 MHz. Often, the digital frequency counter has a digital readout display.

The frequency counter can be used to check or align the frequency in RF or audio circuits. Besides checking frequencies, the counter can be used to check the speeds of the tapes, VCRs, camcorders, and cassette player motors. The frequency counter can be used to determine the correct output frequency of an audio or RF signal generator. The frequency counter accuracy should be checked every two or three years by sending the unit to the wholesaler or factory for correct calibration.

Capacitance meter

What and where

Most capacity meters found in the digital multimeter (DMM) might have a capacity range of 1 pF to 40 uF. Of course, this meter will not measure large filter capacitors found in today's electronic consumer products (Fig. 13-18). Choose a capacity meter or one with extended range to accurately measure capacitors (0.1 pF to 20,000 uF).

Before measuring capacitors in or out of the circuit, make sure the voltage is discharged across the capacitor terminals. Keep the ac product turned off with correct capacitor measurements. Select the correct range of the capacitor to be tested, found on the body of the capacitor. Connect the test leads across the capacitor terminals and read the capacity value on the digital display.

An open or dried-up electrolytic capacitor will have a very low reading compared to the capacitor markings. The shorted or leaky capacitor might not indicate properly. You might assume that the electrolytic capacitor is normal if it is within a few microfarads of the original capacitor markings. Remove one end of the bypass or the coupling capacitor for accurate measurement. Capacitors removed from the pc board provide more accurate tests than those in the circuits.

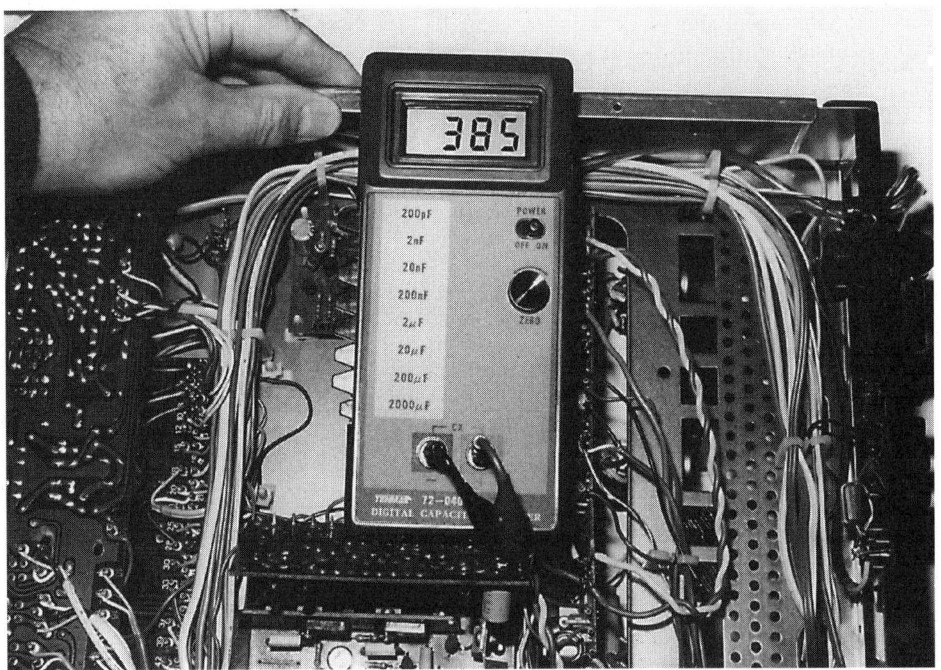

13-18 The small capacitor tester can indicate the correct capacity up to 2000 uF.

RF signal generator

What and where

The RF signal generator provides stable RF and IF signals in servicing AM/FM/MPX receivers, signal injection, and alignment. A wide range of frequencies from 100 kHz to 450 MHz covers most RF circuits. Five or six individual bands might be switched in to cover the wide range of frequencies. The RF signal generator can be purchased separately or in a combination frequency counter and generator.

The RF signal generator can be used to provide complete RF and IF alignment for the AM and FM bands. The IF signal can be injected at the mixer or at the RF stages for IF alignment or troubleshooting. The RF signal generator might use a radio speaker, scope, frequency counter, and output meter as indictors. They are connected to the speaker connections. Most RF and IF alignments are adjusted for maximum output of sound in the speaker or output meter. You might find a combined frequency counter and generator in one test instrument.

Color pattern generator

What and where

A color pattern or NTSC color generator provides a variety of test signals and patterns for TV monitors, camcorders, video disc players, and VCRs. The generator might provide a dot, crosshatch, color bar, full-color raster, blank raster, purity, wide white bar, or a half-screen white bar. The dot, crosshatch, and color bars were used to set up the TV color picture tube. The crosshatch or lines are used to level the picture and make correct vertical and horizontal linearity adjustments.

The blank raster, purity, wide white bar, and half-screen bars are used for purity, static, dynamic convergence, and screen adjustments. Correct adjustment of the yoke and position on the neck of the picture tube are used for purity adjustments. The color bar indicates what color is missing on the TV screen. Chroma adjustments can be made from red to green or from only a color raster.

The color pattern generator can be used to service the various IF, video, and color circuits in the TV chassis. With the color generator connected to the TV antenna terminals, scope waveforms taken in these circuits will quickly locate the defective stage. The color pattern generator waveforms are found in Howard Sams Photofacts. The color pattern signal generator has a horizontal sync output of 15,750 Hz with a chroma subcarrier of 3.563795 MHz (color burst frequency).

Additional test equipment

Sweep generators

What and where

The sweep generator can be used as a standard function or sweep generator. This generator produces square, sine, triangle, ramp, and pulse waveforms. The sweep generator can be used for audio, stereo, TV, VCR, and camcorder repairs. Besides injecting the sine, square, or triangle waveform into audio electronic products for signal-tracing and alignment, the ramp and pulse-sweep waveforms can be used in the TV and VCR circuits. The sweep-marker generator can be used as an FM-stereo generator for troubleshooting FM circuits in the AM/FM/MPX chassis.

Isolation transformer

What and where

The isolation transformer provides power-line insulation while servicing the "hot" chassis. This isolation transformer should be used while servicing any electronic product that operates from the power line (Fig. 13-19). The TV chassis is plugged into the isolation transformer to prevent shock, and damaged test equipment and TV circuits.

A variable isolation transformer is ideal to service the chassis that might shut down from excessive HV, defective shutdown circuits, and intermittent symptoms. Some isolation transformers have a variable ac voltage with a switchable transformer winding. Others might use a separate isolation transformer with a variable power transformer plugged into the isolation transformer to provide isolation and also a variable-line voltage. A built-in current and voltage meter might be found in several variable-ac-supply test instruments.

CRT tester and restorer

What and When

The combination picture tube tester and analyzer that tests the condition of the CRT might also restore the picture tube. Most CRT testers will check picture tube

13-19 Plug the electronic product into the isolation transformer to prevent shock and test-equipment damage.

emission, leakage, and shorted gun assembly. Intermittent tests can be made with the tester connected, and especially when tapping lightly on the end of the picture tube. The new CRT restorer-analyzers will test most picture tubes with several socket adapters (Fig. 13-20).

A restoring function consists of removing shorts in the gun assembly cleaning and balancing the gun, and rejuvenating the cathode. One color might have low emission or weak color and can be restored by removing electrons that pile up on the cathode element. By stripping the cathode element, the tube might be restored for several months or years of service. You can see the firing inside the gun assembly while it is in the restoring process.

Variable power supply

What and where

The variable power supply should include at least two different variable power sources of at least 3 amps. Both power sources should be capable of variable voltage of 0 to 30V. Use the variable power supply to provide voltage for different power sources, motors, camcorders, and low-voltage-powered auto radio repairs. Two different voltage sources might be needed in servicing vertical and horizontal circuits in the TV chassis. Providing a dc source to the vertical or horizontal sweep IC might determine if the sweep IC, circuit, or stage is defective when the regular voltage source is taken from a derived or secondary winding of the flyback. Variable dc voltage injection can help you to isolate and determine if a certain circuit is defective or normal.

Signal/injector/tracer

What and where

The signal injector tracer can troubleshoot audio circuits from input to the speaker. Some injection testers have a switchable detector for troubleshooting AM circuits. Usually, a 1-kHz tone provides injected signal with a VU meter and speaker as indicator.

Start at the volume control and proceed towards the speaker until a sound is heard. The tone should be loud at the volume control if the audio stages are normal. When the signal is heard at a point in the audio circuit, take voltage, resistance, and semiconductor measurements to locate the defective component. Remember, transistors, ICs, and coupling capacitors cause the most audio problems.

Laser power meter

What and where

The laser power meter measures the output of the laser and infrared sources used in servicing CD players, video disc players, VCRs, and remote control units (Fig. 13-21). A test of the laser head might determine if the compact disc player

13-20 A CRT tester and restorer test instrument can weed out and restore the picture tube.

13-21 The laser power meter determines if the optical assembly must be replaced.

is worth repairing. Most CD laser optical assemblies are so costly that many owners will dump the CD player instead of having it serviced.

Place the laser meter probe over the CD player optical light beam. First determine if an interlock must be shorted out, so the laser beam can be measured. Move the laser probe back and forth until the highest reading is obtained. Check the meter reading against the manufacturer's laser measurement. A weak or dead infrared signal might not indicate an EFM waveform. The laser power meter can also test the infrared transmitter remote control for the TV, VCR, CD player, and receiver.

Magnameter

What and where

The Magnameter is a specialized test instrument used to quickly and safely test microwave ovens (Fig. 13-22). This instrument will test the oven high voltage (1.8 to 4.5 KV) and plate current with one setup. Once the test leads are connected, both current and high-voltage tests can be made. The Magnameter is a safe instrument for taking high-voltage measurements.

The Magnameter can locate a leaky, low-emission, open-filament, and overheated magnetron tube. Besides the magnetron, a leaky diode and HV capacitor, shorted or open HV transformer, and fuse can be located with the Magnameter. The biggest advantage of this meter is the safety factor in testing the high voltage applied to the magnetron. If the meter hand appears in the red area, suspect an open magnetron, open filament, bad filament transformer, poor filament lead contacts, and open HV wire. For a low- or no-current measurement, suspect no high voltage or a defective magnetron.

LCR meter

What and where

The LCR meter is a convenient way to accurately measure inductance, capacitance, and resistance. Inductance can be measured from 10 to 200 H. Measure capacitance from 10 to 200 uF in seven or eight different ranges. Resistance measurements might be checked from 20 ohms to 20 megohms. The dual-display LCR meter provides simultaneous measurement of inductance and Q, or capacitance and dissipation factor.

Wow-and-flutter meter

What and where

The wow-and-flutter meter is ideal when servicing any record/playback equipment such as VCRs, VTRs, reel to reels, and home or auto cassette players. The wow-and-flutter meter will indicate improper speed in motor-controlled audio and video equipment. This test instrument is quite expensive and is found on the specialized bench of servicing precision products.

The wow sound might be a slow flutter or an uneven, wavering audio. Flutter and wow are changes in pitch of an audio recording or reproduction system. The

13-22 Check the magnetron with the Magnameter to determine if high voltage is present or the tube is leaky.

wow fluctuation in sound occurs in a longer duration, while the flutter sounds are shorter in duration. These sounds are very annoying and can be detected with a wow-and-flutter meter.

Video head tester

What and where

The VCR and camcorder video head tester for VHS video heads provides a bridge-measuring circuit to determine if the head is worn or defective. The analog meter might have a red-and-green scale to indicate if the head is good or bad. The measuring frequency might be 1MHz with three different ranges: A = 0.2 to 3.5 uH, B = 8 to 3.0 uH, and C = 0.5 to 15 uH.

High-voltage probe

What and where

The high-voltage probe can measure the HV found at the anode terminal of the picture tube in the TV chassis. The latest high-voltage probe measures up to 42 kV with a factory-calibrated marking at 25 kV. A metal probe must contact the HV button or socket on the CRT anode terminal (Fig. 13-23). Push the metal probe under the rubber cover for HV measurement.

Always keep the ground wire of the test probe clipped to the metal TV chassis or to the ground strap on the picture tube. If not, you can receive a terrible shock by holding on to the probe tester when measuring high voltage. Slip the probe tip under the HV rubber anode connection, and keep the ground wire attached at all times while taking HV measurements.

Specialized test equipment

What and where

Most specialized professional test equipment is quite expensive. These test instruments are designed for servicing special products and components in the consumer electronics field. For video work, choose a Sencore (CVA94) audio tracker, VG91 universal video generator, SG60 AM/FM stereo analyzer, and a Sencore (TVA92) TV video analyzer for tough TV work. For camcorder and video servicing, select a Sencore VA62A universal video analyzer. Sencore VC93 makes VCR troubleshooting and repair easy.

Testing

Most simple things that occur can be repaired by the electronics technician. Test probes, broken wires, and cable leads can easily be replaced or repaired. Soldering iron tips and points can be replaced. Testing of semiconductors and rectifiers in the test equipment can be done by the electronics technician. Power supplies and simple test equipment circuits are easily repaired.

Specialized test equipment that needs special parts should be sent to the factory for repair. Test equipment that needs frequency adjustments must be sent in for ser-

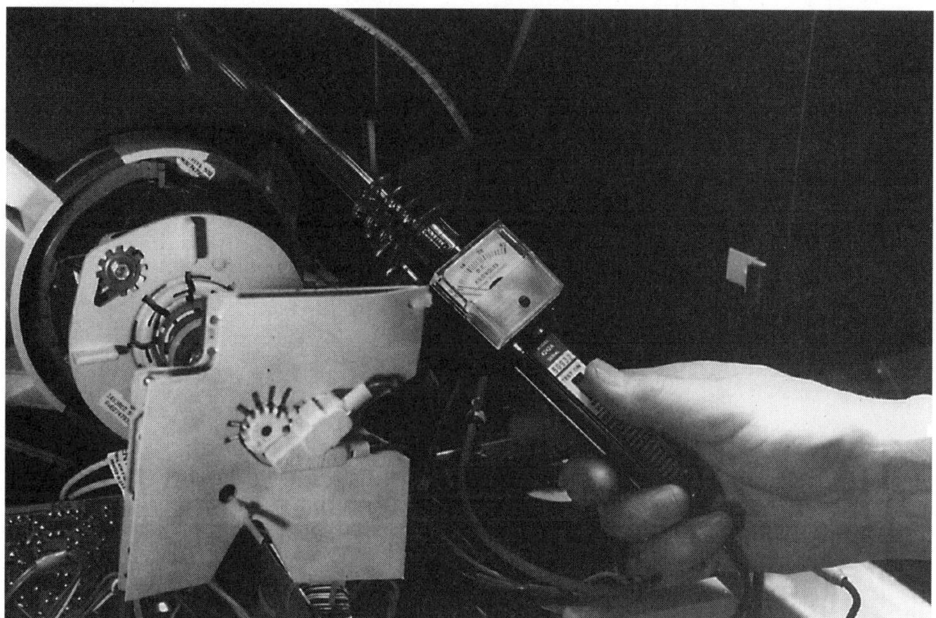

13-23 Slip the high-voltage probe under the rubber covering of the anode terminal to determine if correct HV is present.

vice. Send the microwave oven leakage tester, frequency counter, and oscilloscope to the factory for frequency tests and repairs.

Thermistor

What and where

A *thermistor* is a temperature-sensitive resistor. The resistor changes value in response to a temperature change. The thermistor might be made up of processed oxides of nickel, uranium, cobalt, magnesium, and manganese. A thermistor might be found in electronic thermometers, thermostats, fire alarms, TV degaussing circuits, and computer monitors.

A *posistor* is a thermistor device found on transformers and output transistors that may run too hot. The posistor protects them from overheating. The 120-ohm thermistor was found in the degaussing circuits of the early TV chassis. In-rush current limiters and thermistors are found in the ac lines of ac/dc switching power supplies. Thermistors might be found in telephone line foul protectors, overcurrent protection for transformers, locked rotor protection for FHP motors, wind gauges, liquid level gauges, and communication equipment, to name a few.

Testing

Check the resistance of the thermistor when cold. Precision thermistors might have a resistance from 3K to 10K ohms. The thermistor resistance in the TV degaussing circuit might be from 5 to 120 ohms (Fig. 13-24). Replace the thermistor if it has higher resistance than normal, broken leads, or burned terminals and cracked areas.

Thyristor

What and where

A thyristor is a bi-stable semiconductor device that has a cathode, anode, and gate terminal. It is somewhat like the triac and SCR devices. The thyristor is used in high-current circuits. Replacement thyristors are found in 25V-25A to 600V-25A circuits.

Testing

Check the thyristor as a triac or SCR, and replace with the exact voltage and a high-current rating.

Toroids

What and where

A toroid or toroidal is a doughnut-shaped coil. The coil wire is wound in and out of the round-shaped core. Toroid coils are small in size, have a high Q, and can operate at extremely high frequencies (Fig. 13-25). Toroid coils are found in shortwave receivers, high-powered auto amplifiers, chokes, noise suppressors, wide-band transformers, hash coils, power supplies, RFIs, and computers.

Switching power supplies require the use of high-permeability ferrite-type cores, rather than high-permeability iron-powder cores. A ferrite zinc core has a high Q factor and works up to the 100-MHz frequency range. The carbonyl powdered-iron core has an excellent Q factor of up to more than 200 MHz. Toroidal inductors condition the output signal by leveling out the current waveform, providing a more stable current supply.

Testing

Check the inductance of a toroid coil with an LCR meter. Take a continuity test on a suspected coil, and notice if the core is cracked or broken.

Transformers

Auto transformer

What and where

The auto transformer has a single winding, with the primary winding as a step-up and the secondary as the rest of the winding. The variac is a form of the auto transformer. The variable transformer might have primary and secondary taps, providing a different ac voltage range. The auto transformer or variable transformer is

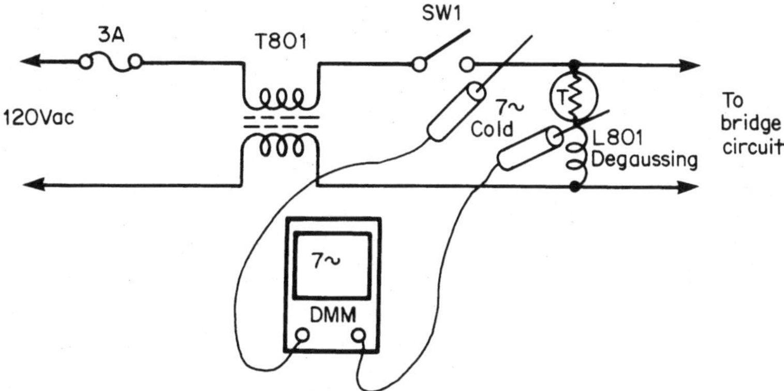

13-24 Check the resistance of a cold thermistor when the degaussing circuits are erratic or dead.

13-25 The coil wire is wound in and out of the ferrite or iron toroid coil form.

used to change the ac input voltage while servicing electronic components or products (Fig. 13-26). The auto transformer might be used in conjunction with an isolation winding to eliminate shock hazards when servicing a hot chassis.

Testing

Check the ac output voltage with the auto transformer plugged into the ac receptacle. Take a continuity resistance measurement with the auto transformer

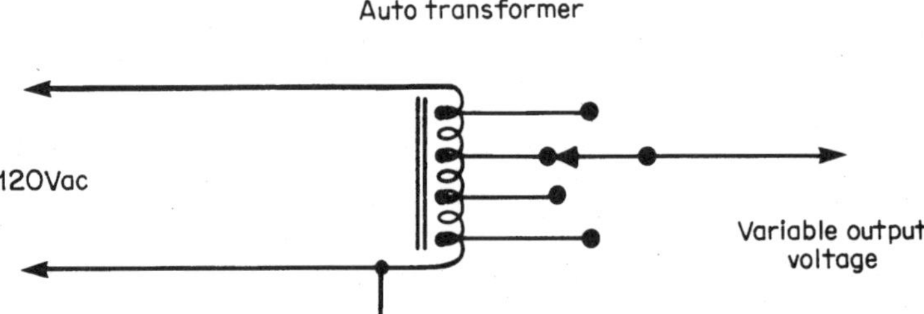

13-26 The autotransformer has variable taps switched in to vary the ac power line voltage.

unplugged from the ac power line. Insert the ac plug of an electronic product into the auto transformer outlet and see if it performs. The defective transformer might have burned windings with a hot-tar odor.

Coupling transformer

What and where

The coupling transformer transfers ac energy electromagnetically into another circuit. In the early radio chassis, a coupling transformer was used to couple the audio signal to another tube stage. The output transformer couples the speaker to the last audio output circuit with matched impedance. The telephone coupling transformers are used for interconnecting data/voice to telephone lines. A typical data/voice coupling transformer might have an impedance of 600 to 600 and 600 CT to 600 CT. The IF transformer couples the intermediate frequency in the receiver to another IF stage. The modem transformer is an impedance coupling device. The line transformer with audio and telephone applications might be called a coupling transformer.

The audio interstage transformer couples the audio from one stage to another in the transistor or tube circuits. Remember, the impedance resistance is different than the actual measured resistance. Check Table 13-3 for the primary and secondary impedance and resistance measurements of audio coupling transfomers.

Testing

Check the continuity resistance of each winding for openness or an increase in resistance. Sometimes the transformer lead will break inside where the terminal wire is connected. Scrape off the enameled wire, reattach, and solder. Replace miniature audio and coupling transformers that have open windings.

IF transformer

What and where

The IF transformer is called the intermediate-frequency transformer. The intermediate frequency of a superhet broadcast receiver is 455 kHz, while in the early auto receivers it was 262 kHz (Fig. 13-27). The IF frequency signal is coupled to another IF stage. The IF stages are located between the mixer and detector circuits. You might find two IF transformers in the low-priced radio or three or more in the deluxe shortwave receiver. Today, ceramic filters are used instead of IF transformers in some radios. You can find a combination of ceramic filters and IF transformers in the deluxe AM-FM-MPX and shortwave receivers.

Testing

Check each winding continuity with the DMM. A high resistance measurement might indicate a poor coil internal connection or right at the terminal connection. Signal-trace each IF stage with the signal generator or injector. Connect the signal generator to the antenna input terminal or mixer stage, at 455 kHz, and notice if a tone is heard in the speaker. Inject a 455-kHz signal at each stage to determine if the transformer is defective.

Table 13-3. Typical primary and secondary impedence with actual resistance measurements of audio coupling transformers.

Impedence		Actual ohm measurements	
Primary	**Secondary**	**Primary (ohms)**	**Secondary (ohms)**
200K CT	1K CT	5300	100
50K	1K CT	3800	75
10K CT	1. 2K CT	1050	200
10K CT	2K CT	1050	350
7.5K CT	12	796	2.9
500 CT	600	67	98
1.2K CT	600	168	92
600	600	72	92

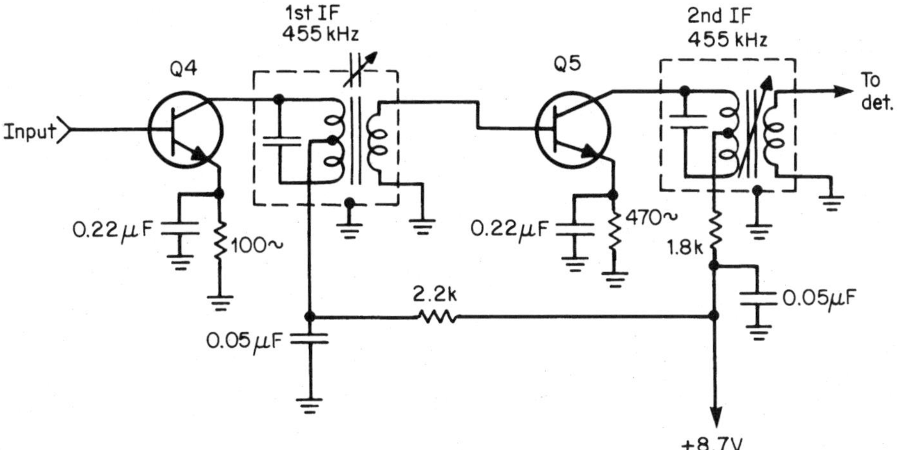

13-27 The intermediate 455-kHz transformer is found in the early radio circuits.

Isolation transformer

What and where

The isolation transformer usually has a 1:1 turn ratio for isolating the electronic product from direct connection to the power line. You might find an isolation transformer winding inside a variable transformer. Today, many of the electronic products have a "hot" chassis. If not isolated with the isolation transformer, the TV set can be damaged, along with the connected test equipment. Always plug the electronic product into an isolation transformer for safe servicing. The isolation transformer might have a 115 to 115V and 115/230 to 115V primary and secondary winding.

Testing

Check the ac output voltage with an ac meter. Take a resistance measurement of each winding. Inspect and test the ac cord and plug for breakage. Insert an electronic product or test instrument and see if it operates from the isolation transformer outlet. Notice if the ac voltage changes as the transformer is adjusted.

Matching transformer

What and where

The matching transformer matches the correct impedance of the product to be connected to a line or another electronic product. A telephone coupling transformer, antenna transformer, or antenna-to-transmission line can be called a matching impedance transformer. The matching transformer might be AF or RF to match one impedance to another. For example, the TV antenna terminals should match a 300- to 72-ohm impedance coupling transformer, before it is connected to the 72-ohm shielded cable. Likewise, a 72-ohm to 300-ohm transformer must be attached to the shielded cable to match the 300-ohm TV antenna connection for good TV reception (Fig. 13-28).

Testing

Check both windings with the low-ohm range of the VOM or DMM. Inspect the transformer for broken or corroded connections, and replace the transformer instead of trying to repair it.

Power transformer (step-up and step-down)

What and where

The power transformer is designed to step up or down the ac voltage for the power supply of an electronic product. The step-up power transformers were located in early tube radio receivers and amplifiers. Today, most electronic products operate from a lower dc voltage and have ac step-down transformers. The typical step-down transformer from the 120-volt power line is 6.3 V, 12.6 V, and 25.6 V. A step-down transformer has an output voltage lower than the input voltage, while a step-up transformer has a higher output voltage than the power-line voltage.

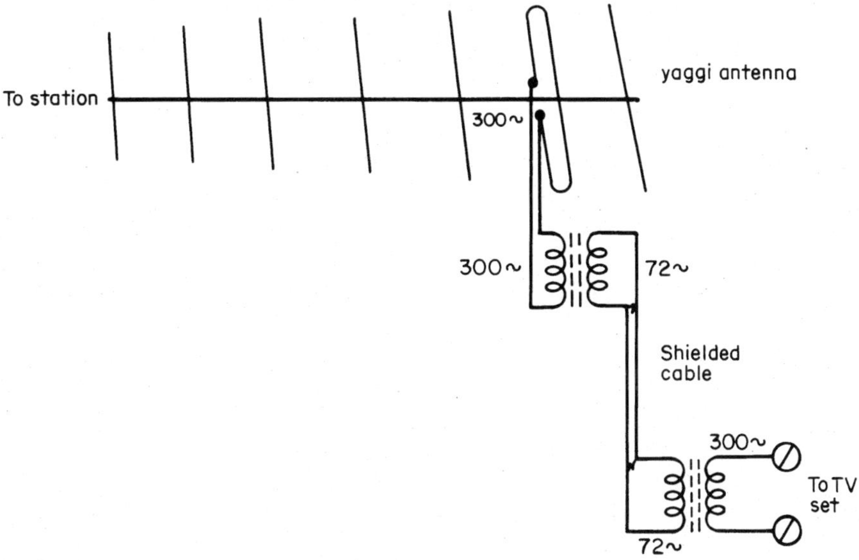

13-28 The matching 72- to 300-ohm transformers are found at each end of the TV shielded lead-in.

In the early step-up transformers, the power supply contained one or more secondary windings. A couple of windings might be used for high voltage at two different power sources and a low-voltage winding might be used for several pilot lights. The step-down transformer might have a single, dual, or center-tapped secondary winding. A multiple secondary winding might be found in a tube tester or CRT tester for different filament or heater voltages.

When

The defective power transformer might be shorted, open, burned, or have intermittent connections. Often, in small appliances and electronic products, the power transformer has no fuse protection. The transformer will become hot with burned windings and load down the power line if a fuse does not blow open. The step-down power transformer without a fuse might open the primary winding when one or more silicon rectifiers becomes leaky or shorted. The defective transformer can have black substance oozing out of the transformer windings. To determine if an overheated power transformer has shorted internal windings, remove all leads of the transformer from the circuit and notice if the transformer runs warm without any load. Now check the ac voltage of each secondary winding.

Testing

Quickly take a continuity low-ohm measurement of each winding. Remove the power plug and check the resistance across the ac prongs. Take ac voltage measurements across corresponding windings to determine if there is no or low output ac voltage. Remove the leaky diode or shorted electrolytic capacitor from the circuit when the fuse keeps blowing after replacement. Often, the small defective ac step-down transformer, found in small electronic products, has an open primary winding. Replace the transformer that has an open winding.

Sound output transformer

What and where

The sound output transformer couples the signal or power from the output tube, transistor, or IC component to the speaker. The output impedance should match the correct impedance of the speaker. A universal output transformer has several secondary taps to match the impedance of the various speakers. The output transformer is found in sound circuits of receivers, TVs, and early tube products (Fig. 13-29).

The primary winding of audio output transformer might have an impedance of 500 ohms CT (center tapped) and an 8-ohm secondary winding. The actual resistance of the primary winding might be 44 ohms, with the secondary resistance of less than 1 ohm. A 1-K CT primary winding might have an actual resistance measurement of 55 ohms.

Testing

Check the primary and secondary resistance with the low-ohmmeter range of the DMM. You may assume the winding is normal if it is within a few ohms of the manufacturer's measurements. Sometimes a high-resistance measurement might be caused by a poor terminal-soldered internal connection. Remove the outside covering and check each connection. Clean off the enameled wiring and resolder each terminal wire. Replace the transformer if the winding is burned or has a change in resistance.

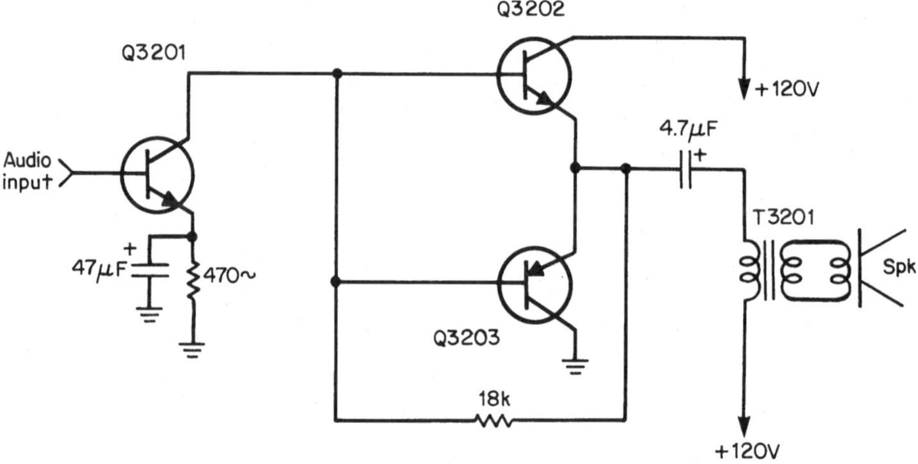

13-29 The audio output transformer is still found in some of the latest TVs.

Telephone transformer

What and where

The telephone coupling or line-matching transformer can have a center-tapped primary and secondary winding. The telephone transformer might have an impedance of 600 CT–600 CT, 600–600 CT, 600–600/600, 600 CT–900 CT, 600–300/300, and 600–600 primary and secondary ohm impedance (Fig. 13-30).

Testing

Check the primary and secondary windings with the low-ohm range of the DMM. Sub another telephone to see if the transformer or telephone is defective. Replace the transformer that has an open winding.

TV driver transformer

What and where

The TV driver transformer couples the drive signal from the horizontal driver transistor to the base terminal of the output transistor in the TV chassis. The driver transformer has a higher voltage of the primary winding and less than 1 volt on the secondary winding. Likewise, the actual primary resistance is high compared to the secondary winding (Fig. 13-31).

Testing

Check the primary and secondary drive waveforms on the transformer to determine if the transformer or transistors are defective. An overheated driver transformer indicates no drive signal or a leaky driver transistor. Poorly soldered connections on the driver transformer might cause poor start-up or chassis shutdown. Improper voltage on the driver collector terminal might indicate a high resistance or poor terminal connection of the primary winding. Check both windings' resistance with the DMM. If in doubt, replace the driver transformer.

TV flyback transformer

(See Chapter 6, IHVT)

Variable transformer (auto)

(See Isolation transformer in this chapter)

Transducer

What and where

The ultrasonic, speaker, microphone, and piezo transducers are used as pickup and signal devices. The ultrasonic transducer can be used as a receiver or transmitter. The transducer might pick up the remote transmitter signal to the TV, receiver, VCR, and CD player. A piezo transducer might be used as a signal device in the microwave oven or fire alarm. The transducer can also be a microphone device that picks up sound and transfers it into electronic energy.

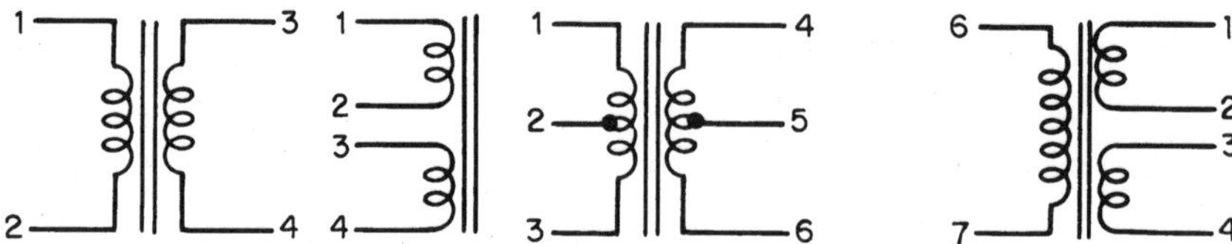

13-30 A typical telephone matching transformer circuits.

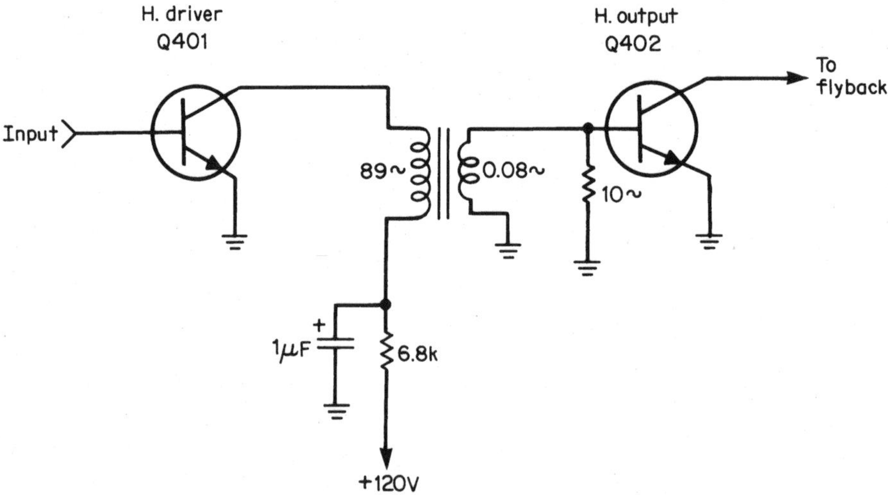

13-31 Check the primary resistance of the driver transformer for a burned winding.

The transducer might be found in a cassette player, amplifier, microwave oven, camcorder, telephone, and TV set. The transducer microphone operates with a low dc voltage (2 to 10 volts) (Fig. 13-32). The electret microphone might be considered a transducer. The piezo transducer might have an external or self-drive circuit operating between 2500- to 6000-Hz resonant frequency.

Testing

Measure the dc voltage applied to the transducer element, and troubleshoot the transducer with an external amplifier. Sub another electret or piezo device to determine if the amplifier or transducer is defective. Check the signal in and out of the transistor amplifier.

Transistors

(See semiconductors)

Triacs

(See semiconductors)

Tripler

(See Chapter 6)

Trim pots

(See Potentiometers)

Tubes-audio

(See Chapter 6)

Turntables (phono)

What and where

The phono turntable might have a crystal or magnetic cartridge. The turntable might be a fully automatic, a single play, or a record changer. Some turntables have gear-and-belt-driven mechanisms. Three different record speeds were found on the early turntables.

When

With a dirty or worn belt and idler wheel, the motor speed might change. Inspect the belt and motor pulley for oil spots. A sluggish motor might result from a dry or gummed-up motor bearing. Apply light grease to the cams and sliding areas (Fig. 13-33). Erratic shutoff might be caused by a dry cam assembly. Improper adjustment of the height screw might prevent the arm from setting down properly on the record. Adjustment of the motor pulley is needed when the speed is too fast or slow. To set the speed or pitch control, check the motor circuits for a speed-control VR adjustment.

Testing

Place a screwdriver blade next to the motor assembly to determine if power is applied to the motor windings. The screwdriver should vibrate if ac power is applied

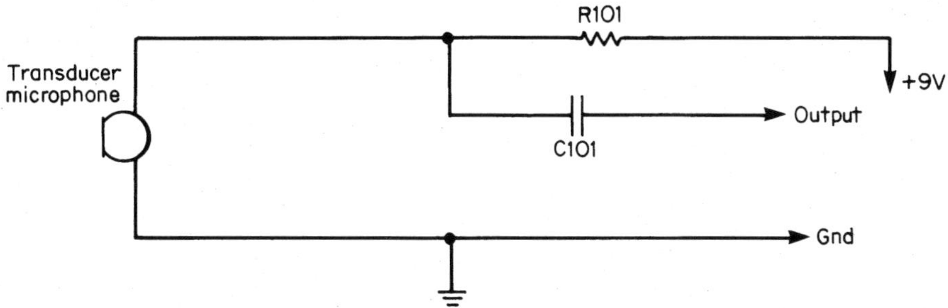

13-32 A dc voltage is found on the terminal of a transducer found within the cassette player.

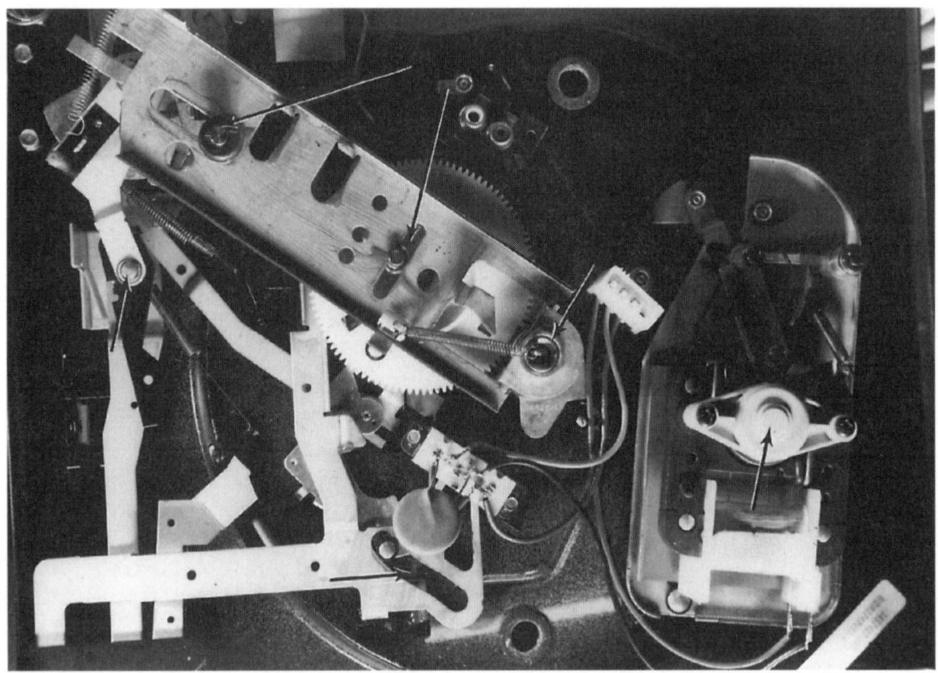

13-33 Clean up and lubricate sliding areas with light oil or grease on the turntable.

to the motor winding. Check for power-line voltage across the motor windings, and check for gummed-up motor bearings when ac voltage is applied to the motor terminals. Remove the motor assembly, disassemble the motor, and wash out the bearings with cleaning fluid. Lightly oil the motor bearings and replace the motor assembly. Check the speed of the turntable with a test disk placed on the platter and with a fluorescent light overhead.

Tweeter

(See Speakers)

Varistors

What and where

The *varistor* could be called a voltage-dependent resistor. A varistor is a resistor whose value varies with the voltage drop across it. The varistor can be found in ac or dc circuits, to provide protection from positive or negative transient surges. The varistor protects the circuit from power-line surges in the control circuits of microwave ovens, TVs, VCRs, computers, and receivers.

Testing

Often, the defective varistor is blown apart or has melted lead terminals. Remove one varistor lead and test with the high-megohm range of a VOM or DMM.

Replace the varistor that has a damaged body, is leaky, or shows any measurement of resistance.

Voltage regulators

(See Semiconductors)

Zener diodes

(See diodes)

Voltage dividers

What and where

The voltage divider network might appear in any circuit of an electronic product. This voltage divider might have several resistors and bypass or decoupling capacitors to form a voltage divider circuit. Within a voltage dividing network of a TV power supply, you might find a transistor and zener-diode regulator, R1001, R1002, R1003, C1001, C1002, C1003, C1004, and C1005 (Fig. 13-34).

Testing

Check the dc voltage at the emitter terminal of Q1001. If the dc voltage is low at terminals 2 through 7, suspect a burned R1001 (100 ohms) and possibly leaky C1002. Take critical voltage measurements on each power source, and measure the resistance of each resistor. Remember, an open or dried-up filter or decoupling electrolytic capacitor can cause a lower voltage source.

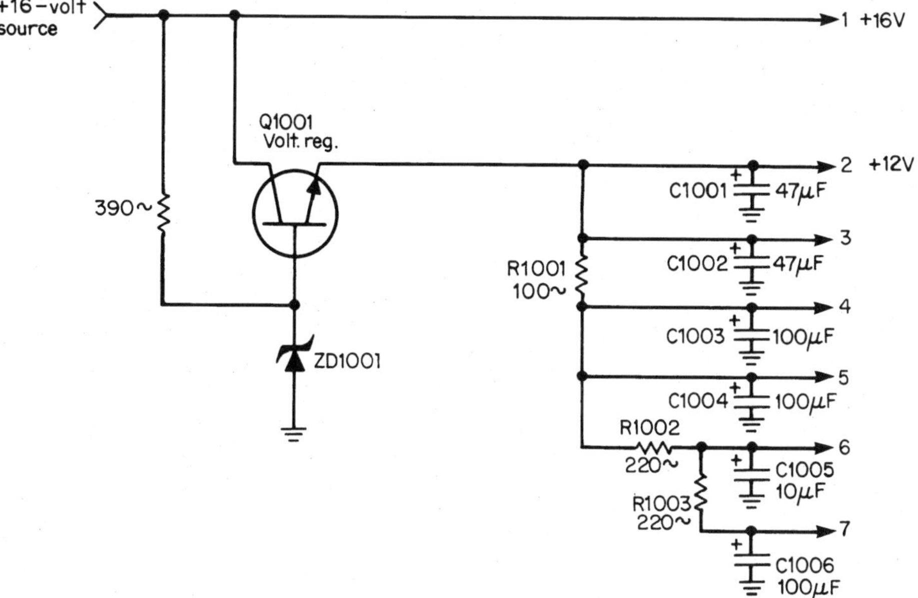

13-34 A typical dc voltage divider network source found in the TV chassis.

Index

D

E

F

H

Q

R

U

V

W

Z

About the Author

Homer L. Davidson is one of the best-selling electronics authors of all time, with over 35 technical books and more than 1,000 magazine articles to his credit. Among his books are *TV Repair for Beginners*, now in its fifth edition, and *Troubleshooting and Repairing Microwave Ovens*, which has sold more than 40,000 copies. Both are published by McGraw-Hill. Mr. Davidson lives in Fort Dodge, Iowa..